Noise Theory of Linear
and Nonlinear Circuits

Noise Theory of Linear and Nonlinear Circuits

J Engberg and **T Larsen**
The University of Aalborg
Denmark

JOHN WILEY & SONS
Chichester · New York · Brisbane ·Toronto · Singapore

Copyright © 1995 by John Wiley & Sons Ltd.
Baffins Lane, Chichester
West Sussex PO19 1UD, England

Telephone *National* Chichester (01243) 779777
International (+44) 1243 779777

Other Wiley Editorial Offices

John Wiley & Sons, Inc., 605 Third Avenue,
New York, NY 10158-0012, USA

Jacaranda Wiley Ltd, 33 Park Road, Milton,
Queensland 4064, Australia

John Wiley & Sons (Canada) Ltd, 22 Worcester Road,
Rexdale, Ontario M9W 1L1, Canada

John Wiley & Sons (SEA) Pte Ltd, 37 Jalan Pemimpin #05-04,
Block B, Union Industrial Building, Singapore 2057

Library of Congress Cataloging-in-Publication Data

Engberg, J. (Jakob}, 1936-
    Noise theory of linear and nonlinear circuits / J. Engberg, T.
Larsen.
        p.   cm.
    Includes bibliographical references and index.
    ISBN 0 471 94825 X   : $72.00
1. Electronic noise.   2. Electronic circuits.   I. Larsen, T.
(Torben)   II. Title.
TK7867.5.E54   1995
621.382'24—dc20

94-39501
CIP

British Library Cataloguing in Publication Data

A catalogue record for this book is available from the British Library

ISBN 0 471 94825 X

Produced from camera-ready copy supplied by the author using LaTeX software.
Printed and bound in Great Britain by Bookcraft (Bath) Ltd.

Contents

Preface

This book consists of a linear part written by J. Engberg and a nonlinear part written by T. Larsen. Noise is anything unwanted by the hearer. Electrical noise in this work is the result of quantification of processes and electronic movements thus leaving out hum, intermodulation phenomena and regular harmonic oscillations, distortions etc. The book begins with a short historical review on the main developments in noise theory.

Chapter 2 concerns noise in one-ports. From Nyquist's expression for thermal noise, where the frequency dependency is mentioned, the expressions for noise resistance, noise conductance and noise temperature are developed and extended to active as well as passive one-ports. The noise expressions are defined for thermal as well as for other noise sources. The last section of the chapter is devoted to calculation rules for noise quantities of networks consisting of two-ports in various combinations.

Chapter 3 is devoted to the definition of noise quantities for multiports. Here all types of noise quantities – the noise factor, the effective noise temperature and the operating noise temperature – are extended to active as well as passive networks and terminations. The definitions are for passive networks and terminations in accordance with the IRE definitions except for the operating noise temperature, where a minor correction is included. The noise quantities are defined in two ways: spot frequency and average values. Also the equivalent noise bandwidth is discussed. In a discussion of the noise quantities an operating load noise temperature is introduced.

In Chapter 4 the noise parameters of a two-port are developed from the equivalent noise two-port of Rothe and Dahlke. Similarly noise parameters are developed on the basis of noise power waves. The chapter finally presents transformation formulae between the many types of noise parameters.

In Chapter 5 the extended noise measure is introduced. The chapter is devoted to graphic representations of the noise measure. In order to do this the exchangeable power gain, which takes part in the definition of the noise measure, is illustrated. The graphic representation is shown in both the source admittance plane and in the source reflection plane. Both representations lead to a third-degree equation from which the extreme values of the noise measure can be calculated.

Chapter 6 concerns embedded circuits. In the first section a three-pole embedded in lumped one-ports is discussed. This makes it possible to calculate the noise quantities of a transistor with bias and feedback elements included. Then a two-port with a lossy transmission line in front is considered, and from that the noise parameters of a transmission line are deduced. Then interconnection of two-ports is considered in various ways, and a method by Pucel et al. making it possible to compute the noise parameters at one high frequency from data at a lower frequency is presented. This method makes extensive use of embedding theory as well as matrix formulation. Two special circuits are considered next. One is a transistor with mixed transformer feedback and the other is a circuit with input at both base and emitter. Finally formulae for noise parameter transformations from common emitter to common base and common collector are presented.

Chapters 7 and 8 deal with the basic theory of non-linear noisy networks and systems. Chapter 7 presents the derivation of the theory, and chapter 8 presents examples and conclusion. Chapter 7 presents a unified method of analysis of low-level noise in non-linear networks and systems. Low-level noise refers to that the noise is a small pertubation of the deterministic signal regime. This book is the first to present a method based on Volterra series. The basic representation of noise sources is investigated. Both unmodulated (fundamental, independent) and modulated (dependent) noise sources are treated. The noise sources are represented as the noise response from a non-linear system with inputs given by a fundamental (unmodulated) noise source and one or more controlling variables. The controlling variables may be any system variables in the non-linear network. In this way it is possible to represent a wide variety of noise sources. Based on this representation a method is derived to analyse noise in general non-linear networks and systems. Expressions for the noise and deterministic response from the network are derived. To be able to determine average noise powers, expressions for the ensemble cross-correlation between Fourier series coefficients of the noise response at two arbitrary ports and at arbitrary frequencies are derived. The noise response may be determined as the dot product of a non-linear conversion vector and a noise vector at precalculated frequencies. The non-linear conversion vector is described by multi-port Volterra transfer functions determined from an equivalent circuit description of the network. Eexamples to illustrate the method are included in chapter 8. The practical applications of the method of non-linear noise analysis are expected to be in the analysis and optimization of noise in (near-sinusoidal) oscillators, in mixers with moderate local oscillator levels, and in frequency multipliers. Many oscillators are relatively weakly non-linear, but the non-linear $1/f$ noise upconversion is still very important. Because of this a non-linear noise analysis is required. More and more mixers are being used in low-power applications, e.g. in portable communications equipment, which means that the Volterra series based noise analysis may be almost ideal for this purpose. Even though this type of mixers are not being switched on/off by the local oscillator signals, the sideband noise is still of importance and the loading of the mixer ports may have a significant impact on the mixer performance. Also the method of non-linear noise analysis is expected to be useful in the determination of noise models of non-linear devices.

Chapter 9 deals with the determination of multi-port Volterra transfer functions. In the existing literature only one-port Volterra transfer functions containing one-port non-linear elements are allowed. The method developed in the present book allows the determination of multi-port Volterra transfer functions containing multi-port non-linear elements (subsystems). This is a fundamental requirement for the noise theory, since there are generally more than one input port in the noise description of the networks. Moreover, in the analysis of (noise free) non-linear circuits it has been pointed out that some models of MESFET's should contain multi-port elements, but the theoretical methods required to determine the Volterra transfer functions did not exist. The present method has been implemented in a symbolic programming language which allows determination of the Volterra transfer functions in algebraic form. Using this it is possible to determine Volterra transfer functions up an order of about 8–10. Several examples are presented to illustrate the method, and comparisons with the existing literature have been made in some special cases. The work on multi-port Volterra series may be of high interest in the development of accurate non-linear models of devices with multi-dimensional non-linear elements and in the analysis of systems with multi-port excitations.

J. Engberg &   T. Larsen

Part I

Linear systems

1

Some milestones in the development of noise theory

The first person to show the connection between "Spontaneous Fluctuations", as noise was called then, and thermodynamics experimentally was the Dutch scientist Geertruida Luberta de Haas-Lorentz [1,2]. Using very sensitive mirror galvanometers she showed that the electrons carrying the current behaved like molecules with temperature and she proved the thermal origin of noise on the basis of the thermodynamic theory which independently was developed by Albert Einstein [3,4] and Marian von Smoluchowski [5,6,7] and shown experimentally by Jean Baptiste Perrin [8]. In a third paper Albert Einstein [9] theoretically calculated a noise voltage on a capacitor. An interesting description of the development from the experiments of Robert Brown in 1827 and up to about 1906 is given by Haas-Lorentz in [2].

1.1   Johnson Noise

Quickly passing W. Schottky's 1918 paper [10] on the theory of shot noise, the next major development was Nyquist's and Johnson's papers [11,12,13] in 1927–28. Here Johnson showed experimentally and Nyquist theoretically the thermal noise from a one-port. An outline of Nyquist's proof is given here.

In Figure 1.1 everything is assumed ideal. The (long) coaxial transmission line is lossless and not radiating, and the switches are lossless and open at the beginning of the experiment. The two resistances have the same surrounding temperature and as they remain in thermodynamic equilibrium the noise power N_{21} transferred from 2 to 1 must equal the noise power N_{12} transferred the other way. Then the two switches are closed simultaneously. The two noise powers are perfectly reflected at the ends of the line and the energy thus trapped in the transmission line will form oscillations at the fundamental mode and its harmonics with a voltage node at each

3

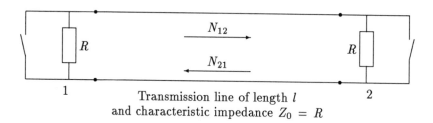

Transmission line of length l
and characteristic impedance $Z_0 = R$

Figure 1.1: Nyquist's theoretical model.

end. The frequencies of oscillations are

$$f_1 = \frac{c}{2l}, \quad f_2 = 2 f_1, \quad f_3 = 3 f_1, \quad \dots$$

where c is the speed of light and l the length of the transmission line. This length
is imagined to be large ($l \to \infty$).

At every frequency there are two degrees of freedom ($\sim$ electrical and magnetic
energy), and from the (classic) theory of thermodynamics it is known that each de-
gree of freedom has the energy of $\frac{1}{2} k T$ where $k = 1.38 \times 10^{-23}$ J K^{-1} is Boltzmann's
constant and T the absolute temperature in kelvins.

$\Delta f = f_n - f_m = n \frac{c}{2l} - m \frac{c}{2l}$ determines the number of frequencies between
f_m and f_n to be $n - m = (f_n - f_m) \frac{2l}{c}$. The energy in the frequency band from
f_m to f_n is then

$$[E_n]_{f_m}^{f_n} = 2 \cdot \frac{1}{2} k T (n - m) = k T (f_n - f_m) \frac{2l}{c} \quad \text{[J]}$$

This trapped energy must be equal to the noise energy delivered from the two
resistances in the time τ it takes for the power to be transferred from one end to
the other. Therefore

$$[E_n]_{f_m}^{f_n} = 2 [N]_{f_m}^{f_n} \tau \quad \text{[J]}$$

where $[N]_{f_m}^{f_n}$ is the thermal noise power from one resistor and $\tau = \frac{l}{c}$. This determines

$$[N]_{f_m}^{f_n} = k T (f_n - f_m) = k T \Delta f \quad \text{[W]}$$

In the quantum mechanical theory Nyquist suggested that the energy $k T$ is
replaced by $h f / (\exp \frac{hf}{kT} - 1)$ where $h = 6.626 \times 10^{-34}$ J s is Planck's constant and
thus the expression for thermal noise in the frequency band Δf is

$$N_{\Delta f} = \int_{\Delta f} \frac{h f}{\exp \frac{hf}{kT} - 1} df \quad \text{[W]}$$

Today most authors agree that the zero-point energy term ($\frac{1}{2} k T$) should be included
in the quantum mechanical expression of the energy: see [14,15,16].

1.2   Receiver noise

In the thirties and early forties noise in receivers was the great subject of interest. It took some time to separate the noise from the source (the antenna) from the receiver noise itself. One early attempt was made by Burgess [17], who introduced a K factor which was dependent only on the source resistance, the resistance of the input network and the equivalent noise resistance of the first valve.

A figure of merit for receiver noise – the noise factor – was introduced by D. O. North [18] and independently by K. Fränz [19]. Two years later H. T. Friis [20] wrote a paper on the noise figure (which today is called noise factor) and a lot of articles emerged discussing the definitions of North and Friis – a rivalry on which Okwit [21] has written an interesting article.

The definitions were expressed a little differently, but they were all on the familiar noise factor. Fränz did not call his definition anything, but he clearly used the concept of available power. North also introduced the "operating noise factor" which multiplied by the noise standard temperature, T_0 equals the modern operating noise temperature, T_{op}. Friis's definition was very stringent. He used available power and available gain and he also derived a formula for the noise factor of networks in cascade.

One more thing that was discussed was the value of the standard noise temperature. Values from 288.39 to 300 K had been proposed – see [21] and [22, pp. 54-55] – until 290 K was chosen as the standard noise temperature by IRE in 1962 [23].

1.3   Linear two- and multi-ports

In 1955 Rothe and Dahlke [24,25] enlarged the well-known (voltage and/or current based) small-signal parameters (four complex numbers) to include noise by adding four more numbers (two real and one complex). They also facilitated noise computations by replacing two partly correlated noise sources with two uncorrelated noise sources and a correlation immittance. The four noise quantities are called noise parameters and they exist in many forms. Later, noise power wave based noise parameters were developed by Penfield, Meys and others as explained in Chapter 4.

This theory was further developed to a linear noisy network theory in 1959 by Haus and Adler, collected in [26]. They introduced the noise measure and showed that the minimum noise measure was invariant by embedding in noiseless components. Also they introduced the extended noise factor for negative sources. In 1967 Bosma [27] introduced the characteristic noise temperature which can be related to the noise measure. It is, however, seldom used – perhaps because the characteristic noise temperature is negative for ordinary amplifiers.

A very good review of the development of noise research up to the year 1980 was written by A. van der Ziel [16].

1.4   References

[1] Haas-Lorentz, G. L. de: "Over de theorie van de Brown'sche bewegung en daarmede verwante verschijnselen", E. Ijdo, Leiden, 1912.

[2] Haas-Lorentz, G. L. de: "Die Brownsche Bewegung und einige verwandte Erscheinungen", F. Vieweg & Sohn, Braunschweig, 1913.

[3] Einstein, A.: "Über die von der molekularkinetischen Theorie der Wärme geforderte Bewegung von in ruhenden Flüssigkeiten suspendierten Teilchen", *Annalen der Physik*, vol. 17, pp. 549 – 560, 1905.

[4] Einstein, A.: "Zur Theorie der Brownchen-Bewegung", *Annalen der Physik*, vol. 19, pp. 371 – 381, 1906.

[5] Von Smoluchowski, M.: "Sur le chemin moyen parcouru par les molécules d'un gaz et sur son rapport avec la théorie de la diffusion", *Bulletin International de l'Académie des Sciences de Cracovie*, Année 1906, pp. 202 – 213, 1907.

[6] Von Smoluchowski, M.: "Essai d'une théorie cinétique du mouvement Brownien et des milieux troubles", *Bulletin International de l'Académie des Sciences de Cracovie*, Année 1906, pp. 577 – 602, 1907.

[7] Von Smoluchowski, M.: "Zur kinetischen Theorie der Brownschen Molekularbewegung und der Suspensionen", *Annalen der Physik*, vol. 21, pp. 756 – 780, 1906.

[8] Perrin, J.: "L'agitation moléculaire et le mouvement brownien", *Comptes rendues hebdomadaires des Séances de l'Académie des Sciences*, Paris, vol. 146, pp. 967 – 970, 1908.

[9] Einstein, A.: "Über die Gültigkeitsgrenze des Satzes vom thermodynamischen Gleichgewicht und über die Möglichkeit einer neuen Bestimmung der Elementarquanta", *Annalen der Physik*, vol. 22, pp. 569 – 572, 1907.

[10] Schottky, W.: "Über spontane Stromschwankungen in verschiedenen Elektrizitätsleitern", *Annalen der Physik*, vol. 57, pp. 542 – 567, December 1918.

[11] Johnson, J. B.: "Thermal agitation of electricity", *Bell Lab. Rec.*, vol. 3, no. 2, pp. 185 – 187, February 1927.

[12] Johnson, J. B.: "Thermal agitation of electricity in conductors", *Physical Review*, vol. 32, pp. 97 – 109, July 1928.

[13] Nyquist, H.: "Thermal agitation of electric charge in conductors", *Physical Review*, vol. 32, pp. 110 – 113, July 1928.

[14] Callen, H. B. & Welton, T. A.: "Irreversibility and generalized noise", *Physical Review*, vol. 83, pp. 34 – 40, July 1951.

[15] Buckingham, M. J.: "Noise in electronic devices and systems", Ellis Horwood – John Wiley, 1983.

[16] Van der Ziel, A.: "History of noise research", *Advances in Electronics and Electron Physics*, vol. 50, pp. 351 – 409, 1979.

[17] Burgess, R. E.: "Noise in receiving aerial systems", *Proc. Physical Society*, vol. 53, pp. 293 – 304, 1941.

[18] North, D. O.: "The absolute sensitivity of radio receivers", *RCA Review*, vol. 6, pp. 332 – 344, January 1942.

[19] Fränz, K.: "Messung der Empfängerempfindlichkeit bei kurzen elektrischen Wellen", *Hochfrequenztechnik und Elektroakustik*, vol. 59, pp. 105 – 112, April 1942.

[20] Friis, H. T.: "Noise figures of radio receivers", *Proc. IRE*, vol. 32, pp. 419 – 422, July 1944.

[21] Okwit, S.: "An historical view of the evolution of low-noise concepts and techniques", *IEEE Trans. on Microwave Theory and Techniques*, vol. MTT-32, pp. 1068 – 1082, September 1984.

[22] Mumford, W. W. & Scheibe, E. H.: "Noise performance factors in communication systems", Horizon House, 1968.

[23] IRE Standards Committee: "IRE Standards on Electron Tubes: Definitions of Terms, 1962", *Proc. IEEE*, vol. 51, pp. 434 – 435, March 1963.

[24] Rothe, H. & Dahlke, W.: "Theorie rauschender Vierpole", *Archiv der electrischen Übertragung*, vol. 9, pp. 117 – 121, March 1955.

[25] Rothe, H. & Dahlke, W.: "Theory of noisy fourpoles", *Proc. IRE*, vol. 44, pp. 811 – 818, June 1956.

[26] Haus, H. A. & Adler, R. B.: "Circuit theory of linear noisy networks", John Wiley, 1959.

[27] Bosma, H.: "On the theory of linear noisy systems", Philips Research Reports Supplements, 1967.

2

Noise in one-ports

With a background of Nyquist's theory of thermal noise this chapter gives the definitions of some noise quantities for one-ports. These are extended to allow negative immittances[1] and calculation rules for noisy one-ports are developed.

2.1   Thermal noise

H. Nyquist [1] has shown that the thermal available noise power from a conductor at the physical temperature T [K] is

$$N = N_{\Delta f} = \frac{h f}{\exp[\frac{h f}{k T}] - 1} \Delta f \quad [\text{W}] \tag{2.1}$$

where $h = 6.626 \times 10^{-34}$ J s is Planck's constant,[2] $k = 1.3807 \times 10^{-23}$ J K^{-1} is Boltzmann's constant,[3] f is the frequency and Δf is the bandwidth (of the measuring system), both in Hz. It is assumed that the noise power density is constant in the frequency range Δf around f_0 or $0 < f_0 - \frac{\Delta f}{2} < f < f_0 + \frac{\Delta f}{2}$. For $h f \ll k T$, a condition which at room temperature is fulfilled for $f < 600$ GHz, Equation (2.1) is reduced to

$$N = k T \Delta f \quad [\text{W}] \tag{2.2}$$

In the following it is – if not specifically stated – assumed that the frequency is low enough for using Equation (2.2). If $T = 290$ K (17 °C) the available noise power density is

$$N' = k T = 1.38 \times 10^{-23} \times 290 = 4.00 \times 10^{-21} \quad \text{W Hz}^{-1}$$

---

[1]The term immittance is used when it is not necessary to distinguish between impedance and admittance.

[2]$h = 6.626\,0755 \times 10^{-34} \pm 0.60$ ppm according to Handbook of Chemistry and Physics, 70th ed., 1989.

[3]$k = 1.380\,658 \times 10^{-23} \pm 8.5$ ppm according to Handbook of Chemistry and Physics, 70th ed., 1989.

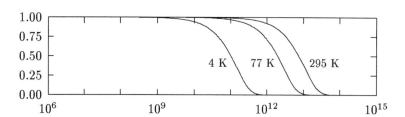

Figure 2.1: Reduction of thermal noise as a function of frequency for temperatures of 295, 77 and 4 K (room temperature and boiling points of nitrogen and helium).

$\sim$ -204 dB$|_{\mathrm{W\,Hz}^{-1}}$. Due to the "nice" numerical value of N' the corresponding temperature, which is close to normal room temperature, is called the standard noise temperature [3,4] and denoted T_0.

$$T_0 \;=\; 290 \quad \mathrm{K} \tag{2.3}$$

It may also be noted that $k\,T_0/q \;=\; 0.0250$ V, where $q =1.602\times10^{-19}$ C is the magnitude of the electronic charge.

Please note that N is used to denote noise power [W] and N' is used for noise power density [W Hz^{-1}].

The noise in a one-port can be represented by either a Thevenin voltage source or a Norton current source as shown in Figure 2.2. Loading these equivalent circuits

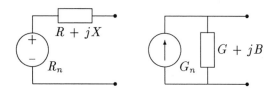

Figure 2.2: One-port equivalent circuits.

with the conjugate of their immittances to obtain power match the source delivers the available power to the load. If the source generates only thermal noise this power is expressed as

$$N \;=\; kT\,\Delta f \;=\; \frac{\langle\,|e|^2\,\rangle}{4\,R} \;=\; \frac{\langle\,|i|^2\,\rangle}{4\,G} \quad \mathrm{[W]}$$

where $\langle\cdots\rangle$ denotes the ensemble average over processes with identical statistical properties.[4] $\langle |e|^2 \rangle$ and $\langle |i|^2 \rangle$ are the magnitude squares of the voltage and current measured in the frequency band Δf with a "true" RMS-meter and R and G are defined in Figure 2.2. From this expression the mean square noise voltage is given by

$$\langle |e|^2 \rangle \;=\; 4\,k\,T\,R\,\Delta f \quad [\mathrm{V}^2] \tag{2.4}$$

and the mean square noise current by

$$\langle |i|^2 \rangle \;=\; 4\,k\,T\,G\,\Delta f \quad [\mathrm{A}^2] \tag{2.5}$$

2.2 Definitions of noise quantities

In noise analysis it is $\langle |e|^2 \rangle$ (and $\langle |i|^2 \rangle$) which is of interest in noise calculations, as $\langle e \rangle$ is equal to zero. As $\langle |e|^2 \rangle$ is dependent on the bandwidth Δf this must also be specified. Sometimes the quantities $\langle |e|^2 \rangle / \Delta f$ $[\mathrm{V}^2\,\mathrm{Hz}^{-1}]$ or $\sqrt{\langle |e|^2 \rangle / \Delta f}$ $[\mathrm{V}\,\mathrm{Hz}^{-\frac{1}{2}}]$ are used[5] but mostly one of the following representations is preferred.

Since noise may have other origins than thermal effects and since it is convenient to have equivalent representations for all types of noise, Equations (2.4) and (2.5) are modified in such a way that they are valid for all kinds of noise. This can be done in two ways. One possibility is to keep the temperature T fixed and then change the values of the resistance R and of the conductance G until the equations are fulfilled. The values for R and G obtained in this way are called the equivalent noise resistance R_n and the equivalent noise conductance G_n respectively. The other possibility is to keep the resistance and the conductance at their physical values and then select the temperature T such that the equations are fulfilled. This change of temperature could also be performed in Equation (2.2). The thus obtained temperature is called the noise temperature of the one-port.

In order to be able to characterize one-ports with negative real parts of their impedance or admittance the available power according to custom is replaced by the exchangeable power[6] as introduced by Haus and Adler [5]. Thus the thermal available noise power is replaced by the exchangeable noise power, N_e, in Equation (2.2).

2.2.1 The equivalent noise resistance

Definition 2.1 The equivalent noise resistance of a one-port R_n is defined as

$$R_n \;\doteq\; \frac{\langle |e|^2 \rangle}{4\,k\,T_0\,\Delta f} \quad [\Omega] \tag{2.6}$$

---

[4]See Appendix A.

[5]The latter is mostly used in dB relative to 1 $\mu\mathrm{V}\,\mathrm{Hz}^{-\frac{1}{2}}$.

[6]See section 3.1.

where $\langle |e|^2 \rangle$ [V^2] is the mean square (open circuited) noise voltage
at the terminals of the one-port in the frequency band Δf [Hz], $k =
1.3807 \times 10^{-23}$ J K^{-1} is Boltzmann's constant and $T_0 = 290$ K is the stan-
dard noise temperature.

It is seen that the chosen fixed temperature in Equation (2.4) is T_0 regardless
of the physical temperature of the one-port. A change in the mean square noise
voltage with temperature determines the variation of R_n with temperature. Other
noise contributions than the thermal noise also change (increase) the equivalent
noise resistance. A variation of R_n with frequency f occurs quite often, and Δf
should be chosen so narrow that $\langle |e|^2 \rangle / \Delta f$ is independent of the value of Δf.[7] The
quantity $\langle |e|^2 \rangle / \Delta f$ can be regarded as the mean square noise voltage density. It is
seen that $R_n > 0$ even if the one-port has a negative real part of its resistance.

A one-port characterized by R_n generates as much noise as a metallic conductor (which only generates thermal noise) with the resistance R_n at standard noise
temperature. R_n does not give any information about the ohmic resistance of the
one-port. The equivalent diagram of the one-port consists of a series connection
of its impedance, a noise voltage generator, the stochastic noise voltage which is
determined by

$$\langle |e|^2 \rangle = 4 k T_0 R_n \Delta f \quad [V^2] \tag{2.7}$$

and perhaps one or more deterministic voltage generators. According to custom a
noise voltage generator in a diagram is shown with the symbol of a normal voltage
generator with R_n beside, as seen in Figure 2.2.

2.2.2 The equivalent noise conductance

Definition 2.2 The equivalent noise conductance of a one-port G_n is defined as

$$G_n \doteq \frac{\langle |i|^2 \rangle}{4 k T_0 \Delta f} \quad [S] \tag{2.8}$$

where $\langle |i|^2 \rangle$ [A^2] is the mean square (short circuited) noise current at
the terminals of the one-port in the frequency band Δf [Hz], $k = 1.3807
\times 10^{-23}$ J K^{-1} is Boltzmann's constant and $T_0 = 290$ K is the standard noise
temperature.

The same remarks as above for the equivalent noise resistance can be added
here if voltage, resistance, impedance and series connection are replaced by current,
conductance, admittance and parallel connection respectively. Similarly, a noise

---

[7]This means that if $\Delta f \to (1+\varepsilon)\Delta f$ then it follows that $\langle |e|^2 \rangle \to (1+\varepsilon)\langle |e|^2 \rangle$ when ε is a small
number.

current generator is shown with the symbol of a normal current generator with a G_n beside and the value of the current is determined by

$$\langle |i|^2 \rangle \quad = \quad 4\,k\,T_0\,G_n\,\Delta f \quad [\text{A}^2] \tag{2.9}$$

2.2.3 The extended noise temperature

The above definitions for R_n and G_n have been chosen in such a way that they are both positive whatever sign the immittance of the one-port has. In the following definition of the noise temperature for a one-port the usual definition [4] has been extended such that the sign of the extended noise temperature shows if the one-port is active[8] or passive [6].

Definition 2.3 The extended noise temperature of a one-port T_{em} is the exchangeable noise power density, N'_e [W/Hz], divided by Boltzmann's constant, $k = 1.3807 \times 10^{-23}$ J K^{-1}.

$$T_{em} \quad \doteq \quad \frac{N'_e}{k} \quad [\text{K}] \tag{2.10}$$

This definition is equivalent to that in Equations (2.4) and (2.5) keeping the values of R and G at their physical values and then adjusting T until the equations are fulfilled. Thus definition 2.3 can be expressed by either

$$T_{em} \quad = \quad \frac{\langle |e|^2 \rangle}{4\,k\,R\,\Delta f} \quad [\text{K}] \tag{2.11}$$

or

$$T_{em} \quad = \quad \frac{\langle |i|^2 \rangle}{4\,k\,G\,\Delta f} \quad [\text{K}] \tag{2.12}$$

The subscript *em* stands for *extended*, which refers to the extension of the noise temperature to active one-ports, and *mono* for one-port.

The extended noise temperature is negative when the one-port is active as N'_e then is negative. This corresponds to a negative R or G in Equations (2.11) and (2.12). Of course T_{em} can not tell anything about the physical temperature except in the case of pure thermal noise (of a one-port which then must be passive) where T_{em} equals the physical temperature. From

$$\langle |e|^2 \rangle \quad = \quad 4\,k\,T_0\,R_n\,\Delta f \quad = \quad 4\,k\,T_{em}\,R\,\Delta f$$

$$\langle |i|^2 \rangle \quad = \quad 4\,k\,T_0\,G_n\,\Delta f \quad = \quad 4\,k\,T_{em}\,G\,\Delta f$$

the following useful relations are easily derived:

$$\frac{T_{em}}{T_0} \quad = \quad \frac{R_n}{R} \quad = \quad \frac{G_n}{G}$$

---

[8]Active means a one-port with negative real part of its immittance.

$$R_n = G_n R^2 \quad \wedge \quad G_n = R_n G^2$$

$$R_n = \frac{T_{em}^2}{T_0^2} \frac{1}{G_n} \quad \wedge \quad G_n = \frac{T_{em}^2}{T_0^2} \frac{1}{R_n}$$

2.3 Calculation with noise quantities

In order to calculate the noise properties of series and parallel connections of uncor-related[9] one-ports it is necessary to develop rules for series connection of one-ports characterized by either R_n or T_{em} and similarly parallel connection of one-ports characterized by either G_n or T_{em}.

Noise generators are represented by stochastic processes, and when those are uncorrelated (independent) the following formulas for the equivalent mean square noise voltage $\langle |e|^2 \rangle$ for series connected noise voltage generators and the equivalent mean square noise current $\langle |i|^2 \rangle$ for parallel connected noise current generators are valid:

$$\langle |e|^2 \rangle = \sum_{i=1}^{I} \sum_{j=1}^{I} \langle e_i\, e_j^* \rangle = \langle |e_1|^2 \rangle + \langle |e_2|^2 \rangle + \cdots + \langle |e_I|^2 \rangle \qquad [\text{V}^2] \ (2.13)$$

$$\langle |i|^2 \rangle = \sum_{i=1}^{I} \sum_{j=1}^{I} \langle i_i\, i_j^* \rangle = \langle |i_1|^2 \rangle + \langle |i_2|^2 \rangle + \cdots + \langle |i_I|^2 \rangle \qquad [\text{A}^2] \ (2.14)$$

2.3.1 Series and parallel connections of one-ports characterized by equivalent noise resistances and conductances

The Thevenin equivalent of a noisy one-port is a series connection of an internal impedance Z (passive or active), a (stochastic) noise voltage generator R_n generating a mean square noise voltage given by Equation (2.4) and perhaps a (deterministic) voltage generator E as shown in Figure 2.3.

From circuit theory the equivalent one-port of I series connected one-ports as shown in Figure 2.3 has $Z = \sum_{i=1}^{I} Z_i$ [Ω] and $E = \sum_{i=1}^{I} E_i$ [V]. As the noise con-tributions from the different one-ports are uncorrelated Equations (2.4) and (2.13) give

$$4\,k\,T_0\,R_n\,\Delta f = \sum_{i=1}^{I} 4\,k\,T_0\,R_{n,i}\,\Delta f \qquad [\text{V}^2]$$

which leads to

$$R_n = \sum_{i=1}^{I} R_{n,i} \qquad [\Omega] \tag{2.15}$$

---

[9] Physically separated one-ports generally have stochastic independent and thus uncorrelated noise generators.

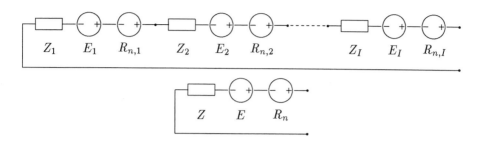

Figure 2.3: Series connection of noisy one-ports and the corresponding equivalent circuit.

In a very similar way, from Equations (2.5) and (2.14) it is found that

$$G_n = \sum_{i=1}^{I} G_{n,i} \quad [\text{S}] \tag{2.16}$$

where I is the number of parallel connected one-ports.

Example 2.1 In the figure a one-port is shown where G_1, G_2 and R_3 are known. Also known are G_2's equivalent noise conductance $G_{n,2}$ and R_3's equivalent noise resistance $R_{n,3}$. When the one-port as such has the noise temperature T_{em}, G_1's equivalent noise conductance $G_{n,1}$ can be

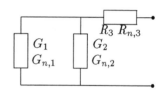

calculated as shown for the following component values: $G_1 = G_2 = 10$ mS, $R_3 = 200$ Ω, $G_{n,2} = 20$ mS, $R_{n,3} = 300$ Ω and $T_{em} = 2\,T_0$.

$$R = \frac{1}{G_1 + G_2} + R_3 = 250 \ \ \Omega \quad \wedge \quad R_n = \frac{T_{em}}{T_0} R = 2\,R = 500 \ \ \Omega$$

$$R_{n,1+2} = R_n - R_{n,3} = 200 \ \ \Omega \quad \wedge \quad G_{n,1+2} = R_{n,1+2}\,(G_1 + G_2)^2 = 80 \ \ \text{mS}$$

$$G_{n,1} = G_{n,1+2} - G_{n,2} = \underline{60 \ \ \text{mS}} \quad \text{or} \quad T_{em,1} = \frac{G_{n,1}}{G_1} T_0 = \underline{6\,T_0}$$

2.3.2 Series and parallel connections of one-ports characterized by extended noise temperatures

Omitting any deterministic voltage sources the series connection of I one-ports characterized by their impedances and extended noise temperatures is shown in Figure 2.4. As the imaginary parts of the impedances do not generate noise only the real parts – the resistances – are shown in Figure 2.4. Inserting Equation (2.4)

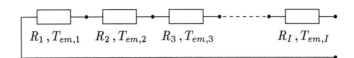

Figure 2.4: Series connection of noisy one-ports characterized by the extended noise temperatures.

into Equation (2.13) gives

$$\langle |e|^2 \rangle \;=\; \sum_{i=1}^{I} 4\,k\,T_{em,i}\,R_i\,\Delta f \;=\; 4\,k\,\Delta f \sum_{i=1}^{I} T_{em,i}\,R_i \qquad [\mathrm{V}^2]$$

From this equation and Equation (2.11) it follows that the extended noise temperature for the series connected one-ports is

$$T_{em} \;=\; \frac{\sum_{i=1}^{I} T_{em,i}\,R_i}{\sum_{i=1}^{I} R_i} \qquad [\mathrm{K}] \qquad\qquad (2.17)$$

As $T_{em,i}$ and R_i have the same sign Equation (2.17) is valid for both active and passive one-ports. The only requirement is that $\sum_{i=1}^{I} R_i \neq 0$.

Similarly, for I parallel connected one-ports one gets

$$T_{em} \;=\; \frac{\sum_{i=1}^{I} T_{em,i}\,G_i}{\sum_{i=1}^{I} G_i} \qquad [\mathrm{K}] \qquad\qquad (2.18)$$

Example 2.2 A tunnel diode circuit is regarded as a one-port with the following data: $G_S = -2$ mS and $T_{em,S} = -7\,T_0$. It is loaded by an admittance with $G_L = 10$ mS and $T_{em,L} = T_0$. The extended noise temperature of the parallel circuit is determined from Equation (2.18):

$$T_{em,t} \;=\; \frac{T_{em,S}\,G_S + T_{em,L}\,G_L}{G_S + G_L} \;=\; \frac{-7\,T_0 \cdot (-2) + T_0 \cdot 10}{-2 + 10} \;=\; \underline{3\,T_0}$$

Example 2.3 In order to isolate an unknown external antenna from an internal ferrite antenna in a receiver the network shown in the figure is used. The nominal antenna impedance $Z_a = 300$ Ω and its noise temperature $T_a = 3\,000$ K. In the network $R_1 = 300$ Ω, $R_2 = 100$ Ω and $R_3 = 62.5$ Ω and they all generate thermal noise at the ambient temperature of 17 °C. Seen from the receiver the antenna and network can be regarded as a one-port with a given impedance, noise

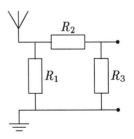

temperature and signal voltage. The solution is found by computing the parallel connection of Z_a and R_1 (Equation (2.16) $-$ R_I, $T_{n,I}$), then the series connection of R_I

and R_2 (Equation (2.15) – R_{II}, $T_{n,II}$) and finally the parallel connection of R_{II} and R_3 (Equation (2.16) – R_{eq}, $T_{n,eq}$).

$$
\begin{array}{lclcccl}
R_I & = & 150 \ \Omega & \wedge & T_{n,I} & = & T_a \frac{G_a}{G_I} + T_0 \frac{G_1}{G_I} & = & 1645 & K \\
R_{II} & = & 250 \ \Omega & \wedge & T_{n,II} & = & T_{n,I} \frac{R}{R_{II}} + T_0 \frac{R_2}{R_{II}} & = & 1103 & K \\
R_{eq} & = & 50 \ \Omega & \wedge & T_{n,eq} & = & T_{n,II} \frac{G_{II}}{G_{eq}} + T_0 \frac{G_3}{G_{eq}} & = & 452.6 & K
\end{array}
$$

The network divides the antenna voltage by 10 (and the power by 16.67) and the antenna noise temperature (and noise power) is divided by 6.6 so the signal to noise ratio is only slightly degraded (≈ 4 dB). At long- and medium-wave bands, where the antenna noise temperature is above $300\,000$ K, the degradation of the signal to noise ratio is hardly noticeable (≈ 0.09 dB), so a lossy network can be used to isolate a receiver from an unknown antenna impedance if the antenna noise temperature is relatively high compared to ambient temperature.

2.4   References

[1] Nyquist, H.: "Thermal agitation of electric charge in conductors", *Physical Review*, vol. 32, pp. 110 – 113, July 1928.

[2] Friis, H. T.: "Noise figures of radio receivers", *Proc. IRE*, vol. 32, pp. 419 – 422, July 1944.

[3] "IRE Standards on Electron Tubes: Definitions of Terms, 1957, 57 IRE 7.S2", *Proc. IRE*, vol. 45, pp. 983 – 1010, July 1957.

[4] "IRE Standards on Electron Tubes: Definitions of Terms, 1962, 62 IRE 7.S2", *Proc. IRE*, vol. 50, pp. 434 – 435, March 1963.

[5] Haus, H. A. & Adler, R. B.: "Circuit theory of linear noisy networks", Technology Press and Wiley, New York, 1959.

[6] Engberg, J. & Larsen, T.: "Extended definitions for noise temperatures of linear noisy one- and two-ports", *IEE Proc. Part H*, vol. 138, pp. 86 – 90, February 1991.

3

Noise characteristics of multi-ports

The multi-ports considered in this chapter can be single response two-ports, multi-response two-ports or multi-ports with one or more responses at each port. In single response two-ports only one single input frequency gives an output at the corresponding output frequency and – of less importance in noise theory – this input frequency leads to no other outputs at other output frequencies. A multi-response port can be considered as many ports as there are responses at the desired output frequency at the output port. This means that multi-response ports are treated as multi-ports with as many ports as the sum of ports times responses requires. It is important to note that only one output port is considered and for spot frequency analysis only a single output frequency is of interest. If more than one output port is of interest each output port is treated separately one by one.

After a short look at some power gain definitions the following noise quantities are introduced: effective input noise temperature, noise factor and operating noise temperature. The noise measure is defined in Chapter 5. Of these the operating noise temperature is intended to be used to describe the noise performance of a system including source and load generated noise. The other definitions are used to describe devices, stages and amplifiers and they are extended to cover negative immittances as well [1,2]. Then the average values of the noise quantities are defined. In a discussion of the noise quantities a definition of an output operating noise temperature is considered.

3.1   Power gains

Two types of power gain are often used in noise theory. One is the available power gain which is extended to include negative immittances and is then called the exchangeable power gain. The other is the transducer gain.

For a passive one-port source the **available power** P_a is defined as **the great-**

est power that can be drawn from the source by arbitrary variation of its
terminal current (or voltage) [2]. This power is the maximum power delivered
to the load when the load immittance is the complex conjugate of the source immittance. Extending the above definition to active sources gives no meaning as the
"maximum" power is infinite and thus not a stationary value (extremum). However, a (negative) stationary value exists in this case so the exchangeable power P_e is
defined as follows.

**Definition 3.1 The exchangeable power of a one-port P_e is the stationary
value (extremum) of the power output from the source, obtained by
arbitrary variation of the terminal current (or voltage).**

For a Thevenin equivalent of a one-port where $Z = R + jX$ and $R \neq 0$ definition
3.1 leads to

$$P_e \;=\; \frac{\langle |e|^2 \rangle}{4R} \;=\; \frac{\langle |e|^2 \rangle}{2(Z + Z^*)} \quad \text{[W]} \qquad \text{for } R \neq 0 \qquad (3.1)$$

It is seen that P_e is negative when the one-port is active ($R < 0$) and positive when
passive ($R > 0$).

The extended version of the available power gain, which is defined as the available
power at the output divided by the available power at the source, is the exchangeable
power gain G_e defined by replacing the available powers by exchangeable powers.

Definition 3.2 The exchangeable power gain G_e is defined as

$$G_e \;\doteq\; \frac{P_{e,o}}{P_{e,S}} \qquad\qquad (3.2)$$

where the exchangeable output power

$$P_{e,o} \;=\; \frac{\langle |e_o|^2 \rangle}{4R_o} \quad \text{[W]}$$

and the exchangeable power at the source

$$P_{e,S} \;=\; \frac{\langle |e_S|^2 \rangle}{4R_S} \quad \text{[W]}$$

**where e_S, e_o, $R_S \neq 0$ and $R_o \neq 0$ are the Thevenin voltage at the source
and output terminals and the corresponding resistances.**

It is seen that

$$G_e > 0 \qquad \text{when} \qquad R_S/R_o > 0$$
$$G_e < 0 \qquad \text{when} \qquad R_S/R_o < 0$$

The transducer gain G_T is defined as the power delivered to the load divided
by the power available from the source. As the available power is always positive
then $G_T > 0$ for passive loads and $G_T < 0$ for active loads. If active sources
are considered $G_T = 0$ as the available power approaches infinity. It is therefore
necessary to extend the definition in the same way as above. This leads to

Definition 3.3 The extended transducer gain G_{eT} is defined as

$$G_{eT} \doteq \frac{P_L}{P_{e,S}} \tag{3.3}$$

where P_L is the power delivered to the load and $P_{e,S}$ is the exchangeable power at the source.

It is seen that

$$G_{eT} > 0 \quad \text{when} \quad R_S/R_L > 0$$

$$G_{eT} < 0 \quad \text{when} \quad R_S/R_L < 0$$

where R_S and R_L are the source and load resistances respectively.

3.2 Definitions of noise quantities

In this section the extended noise temperature, the extended noise factor (and figure) and the operating noise temperature which also is given in an extended version are defined. The extensions of the old definitions include the cases of active sources. This is very convenient as a stage in a cascade of stages as source immittance has the output immittance of the former stage. Often this stage is only stable when loaded which means that the output immittance can well be negative.

The definition of the noise measure will be given in Chapter 5.

Usually the available or exchangeable power gain is used for definitions of spot frequency noise quantities and the transducer or extended transducer gain for average noise quantities and this habit is maintained here. It is, however, not important which type of gain is used as all noise quantities are defined as ratios of powers and a change of e.g. exchangeable output powers in numerator and denominator to powers delivered to the load does not change anything.

3.2.1 The effective noise temperature

The idea behind the definition of the noise temperature is to transfer the noise power generated in the two- or multi-port to the source or sources. As a one-port source with a known exchangeable noise power is characterized by its extended noise temperature this quantity is used to characterize the two- or multi-port.

Definition 3.4 The extended effective (input) noise temperature T_{ee} of a multi-port transducer is defined as the exchangeable output noise power density at a specified output frequency of the transducer with noise free sources, N'_e [W Hz^{-1}] divided by Boltzmann's constant $k = 1.3807 \times 10^{-23}$

J K^{-1} and by the sum of all exchangeable power gains from input responses that give an output at the output port on the specified frequency,

$$T_{ee} \quad \doteq \quad \frac{N'_e}{k \sum_{i=1}^{I} G_{e,i}} \qquad [\text{K}] \qquad\qquad (3.4)$$

where I is the number of responses from all the input ports and $G_{e,i}$ is
the exchangeable power gain from port response i to the output port.

Note 1: There is only one output port and at that port only one frequency is
considered. As the definition uses the exchangeable output noise power density any
noise from the load has no influence on T_{ee}

Note 2: All other ports than the output port are considered input ports and
they should be loaded with any passive or active immittances except short or open
circuits.

Note 3: T_{ee} is a function of the source immittance(s).

The subscript ee stands for **e**xtended $T_{\boldsymbol{e}}$ where T_e is the standard symbol for
the IRE definition of effective input noise temperature [3].

For passive port terminations this definition is equivalent to the IRE definition.
The IRE definition states that the effective (extended) input noise temperature of a
multi-port transducer is the (extended) noise temperature which assigned simultaneously to all input ports of a noise free equivalent of the transducer yields the same
available (exchangeable) output noise power density at a specified output frequency
as the actual transducer with noise free sources.

With all input port terminations active, definition 3.4 gives a value of T_{ee} which
is negative and this corresponds to definition 2.3 of T_{em}. If, however, some input
ports are terminated by passive immittances and some by active immittances T_{ee}
can be of both signs. This situation is discussed below. In these cases with some or
all input terminations active the above IRE definition is equivalent to definition 3.4
when the alterations and the addition indicated in brackets are taken into account.

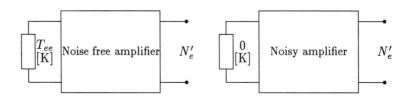

Figure 3.1: Illustration of the extended noise temperature definition for a single response
transducer.

The definition is illustrated in Figure 3.1 for a single response two-port where
at the right the noisy amplifier with a (fictitious) noise free source generates the

exchangeable output noise power density N'_e. The noise free – but otherwise equivalent – amplifier to the left is connected to a source of which the extended noise temperature is varied in such a way that the exchangeable noise power density of this amplifier is also N'_e. Then the extended noise temperature of that source is defined to be the extended noise temperature of the noisy amplifier. As N'_e is a function of frequency the extended noise temperature is also a function of frequency.

As mentioned above the interpretation of T_{ee} with both active and passive input terminations is a little difficult. Consider a three-port transducer with one output port and two input ports. The input ports are loaded with two immittances where one is passive and one active and the output immittance may be either passive or active. As T_{ee} is the same at both input ports and the exchangeable power gains from the two ports have opposite signs, one of the input ports has assigned an extended noise temperature of the opposite sign than the port resistance. It also seems that the noise power density at the output port consists of the difference between two noise power densities. This can be accepted as fictitious reference noise temperatures at the input ports, but it seems more natural if the temperature and resistance had the same sign and that the powers were added. If the input ports were loaded with either only active or only passive immittances, the noise powers at the output port from the input ports would all have the same sign. When this is not the case another – always positive – noise temperature is sometimes useful. It can be expressed from the extended noise temperature by

$$T_{ep} \quad = \quad \frac{\left| T_{ee} \sum_{i=1}^{I} G_{e,i} \right|}{\sum_{i=1}^{I} |G_{e,i}|} \tag{3.5}$$

The idea behind T_{ep} is that if all the input port responses are loaded by immittances with noise temperatures of $\pm T_{ep}$ (+ for a passive load and − for an active load) and the transducer is replaced by a noise free equivalent, then the exchangeable output noise power density is the same as that from the actual transducer with noise free sources. It is seen that if and only if the loads to the input port responses are either all passive or all active then $T_{ep} = |T_{ee}|$. One of the main reasons for choosing definition 3.4 instead of T_{ep} as the definition of the extended noise temperature is that for a single response two-port the function $T_{ee}(Z_S)$, where Z_S is the source impedance, is the quadric surface of a hyperboloid of two sheets.[1] This is a consequence of the simple relations – for single response two-ports – between the extended noise temperature and the extended noise factor defined below.

3.2.2 The noise factor

A much used noise characteristic for two-ports is the noise factor. It is, however, not very practical to use with multi-ports, where the noise temperature is preferable,

---

[1]See Appendix C.

but in some cases more than one response is considered. The noise factor was the first to be "extended" to active sources [4].

Definition 3.5 The extended noise factor of a multi-port transducer, F_e (at a specified input frequency or specified input frequencies which all give an output response at the same output frequency) is defined as the ratio of (1) the total exchangeable noise power density at the output port (and at the corresponding output frequency) when the extended noise temperature of the source (or sources) is/are the standard noise temperature (290 K) at all frequencies (and input ports), to (2) that part of the exchangeable noise power density at the output port which originates from the signal source (or sources) at the input frequency (or frequencies) and at standard noise temperature.

Note 1: F_e is *not* defined for a one-port.[2]

Note 2: For multi-ports with more than one signal response, either an extended noise factor is defined for each signal response, or part (2) in the definition includes noise from those port responses which are used for the input signals.

Note 3: F_e is a function of the source immittance(s).

For two-ports this definition gives the source noise power density $N'_{e,S} = k T_0$. The exchangeable output noise power density $N'_{e,o}$ consists of two parts: the noise power generated by the internal noise sources of the network N'_e and the input noise amplified by the exchangeable power gain of the network $k T_0 G_e$. As definition 3.4 leads to $N'_e = k T_{ee} G_e$ one gets

$$F_e \quad = \quad \frac{N'_{e,o}}{N'_{e,S} G_e} \quad = \quad \frac{k T_0 G_e + k T_{ee} G_e}{k T_0 G_e} \quad = \quad 1 + \frac{T_{ee}}{T_0} \qquad (3.6)$$

$$T_{ee} \quad = \quad (F_e - 1) T_0 \qquad [\text{K}] \qquad (3.7)$$

Please note that Equations (3.6) and (3.7) are only valid for a two-port with a single input and a single output frequency. For multi-ports conversion from T_{ee} to F_e and vice versa is more complicated and is discussed below.

Definition 3.5 is also consistent with H. T. Friis's previous definition [6] when – instead of the exchangeable power gain – the available power gain G_a is used. $G_a = S_o/S_S$ where S_o and S_S are the available signal output and source powers in the frequency range of interest Δf.

$$F \quad = \quad \frac{\frac{S_S}{k T_0 \Delta f}}{\frac{S_o}{N_o}} \quad = \quad \frac{1}{G_a} \frac{N'_o \Delta f}{k T_0 \Delta f}$$

---

[2] Unfortunately the expression $F_e = 1 + T_{em}/T_0$ is sometimes used for a one-port noise factor, which is excluded by the definition.

$$F \quad = \quad \frac{N_o'}{N_S' G_e}$$

From the definition it is seen that

$$F_e > 1 \qquad \text{for} \qquad R_S > 0$$

$$F_e < 1 \qquad \text{for} \qquad R_S < 0$$

The noise factor for $F_e > 1$ is often expressed in dB by

$$F_e \, [dB] \quad = \quad 10 \log F_e \tag{3.8}$$

and then mostly called the noise figure. Also $F_e - 1 = T_{ee}/T_0$ is called the extended excess noise factor.

Example 3.1 In the figure below a two-port consists of three resistors with known resistances and noise temperatures. When the source resistance R_S is known it is possible to determine the noise factor of the two-port by use of definition 3.5.

$$\begin{aligned}
R_S &= 50 \ \Omega \\
R_1 &= 100 \ \Omega \qquad T_{em,1} = T_0 \\
R_2 &= 100 \ \Omega \qquad T_{em,2} = 1.333 \, T_0 \\
R_3 &= 200 \ \Omega \qquad T_{em,3} = 1.875 \, T_0
\end{aligned}$$

By use of Equations (2.18), (2.17) and (2.5) the square of the current in the short-circuited output in a bandwidth of $\Delta f = 1 \, \text{Hz} \, \langle |i_{o,I}|^2 \rangle$ is found as follows:

$$G_a = G_S + G_1 = 30 \quad \text{mS} \qquad \Rightarrow \qquad R_a = 33.3 \quad \Omega$$

$$T_{em,a} = \frac{T_0 \, G_S + T_{em,1} \, G_1}{G_S + G_1} = T_0$$

$$R_b = R_a + R_2 = 133.3 \quad \Omega \qquad \Rightarrow \qquad G_b = 7.5 \quad \text{mS}$$

$$T_{em,b} = \frac{T_{em,a} \, R_a + T_{em,2} \, R_2}{R_a + R_2} = 1.25 \, T_0$$

$$G_c = G_b + G_3 = 12.5 \quad \text{mS}$$

$$T_{em,c} = \frac{T_{em,b} \, G_b + T_{em,3} \, G_3}{G_b + G_3} = 1.50 \, T_0$$

$$\langle |i_{o,I}|^2 \rangle = 4 \, k \, T_{em,c} \, G_c \, \Delta f = 300 \times 10^{-24} \quad \text{A}^2$$

If $\langle |i_{o,I}|^2 \rangle$ is divided by $4 \, G_c$ the total exchangeable noise power density at the output port is determined. To find that part which originates from the source let $T_{em,1} = T_{em,2} = T_{em,3} = 0 \, \text{K}$ and repeat the above calculations. The results are

$$T_{em,a} = \frac{2}{3} T_0 \qquad T_{em,b} = \frac{1}{6} T_0 \qquad T_{em,c} = \frac{1}{10} T_0$$

$$\langle |i_{o,II}|^2 \rangle = 20.0 \times 10^{-24} \quad \text{A}^2$$

$$F_e = \frac{\langle |i_{o,I}|^2 \rangle /(4\,G_c)}{\langle |i_{o,II}|^2 \rangle /(4\,G_c)} = 15$$

Example 3.2 Consider a transmission line made of metallic conductors and therefore generating only thermal noise at the physical temperature of 17 °C ($= T_0$) and a loss of $L = 1/G_e$. In order to find its noise figure the source must have the noise temperature of T_0 also. As a transmission line generates only thermal noise the transmission line and source together can be regarded as a one-port at standard temperature and thus its output noise power density is $k\,T_0$. The noise power density from the source is $k\,T_0$ which is "amplified" by $G_e = 1/L$. The ratio of these two noise power densities determines the noise factor:

$$F_e = \frac{k\,T_0}{k\,T_0/L} = L$$

Now let the transmission line have the physical temperature T_{tl}. To begin with let $T_{tl} = T_0$, then

$$N'_{e\,T_0} + k\,T_0/L = k\,T_0 \qquad \Rightarrow \qquad N'_{e\,T_0} = k\,T_0\,(1 - 1/L)$$

As $N'_{e\,T_{tl}}$ is proportional to T_{tl} the general expression is

$$N'_{e\,T_{tl}} = k\,T_{tl}\,(1 - 1/L)$$

$$F_e = \frac{k\,T_0/L + k\,T_{tl}\,(1 - 1/L)}{k\,T_0/L} = 1 + \frac{T_{tl}}{T_0}(L - 1)$$

Consider a heterodyne system with a normal response and an image response. If it is used for a broadcast receiver the wanted signal is only present at the normal response and the denominator in definition 3.5 includes only noise from the source at the input response frequency, but the numerator contains noise from both the normal and the image responses. If the receiver is for radio astronomy, signals are present at both responses and therefore the denominator in definition 3.5 includes noise from both input responses. Alternatively it may include one response and a separate noise factor derived for each response. The distinction between the two uses of the noise factor is given by calling them single and double sideband noise factors respectively.

In Figure 3.2 the $I + J$ inputs can be loaded by active as well as passive immittances, but signals, which are supposed to be uncorrelated, are applied to only I of them. Suppose the extended noise temperature T_{ee} is given, then the exchangeable output noise density from the transducer itself N'_e can be calculated by

$$N'_e = k\,T_{ee} \sum_{i=1}^{I+J} G_{e,i} \tag{3.9}$$

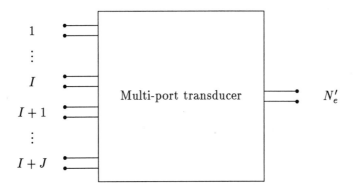

Figure 3.2: A multi-port transducer with I responses with signals and J responses without signals.

where $G_{e,i}$ is the exchangeable power gain from port response i to the output port. If the extended multi-response noise factor is chosen, definition 3.5 leads to

$$F_e = \frac{k T_0 \sum_{i=1}^{I+J} G_{e,i} + N'_e}{k T_0 \sum_{i=1}^{I} G_{e,i}} \tag{3.10}$$

From Equations (3.9) and (3.10) the relations between T_{ee} and the multi-response F_e can be derived. Denoting the spot response factor

$$\mathcal{R} = \frac{\sum_{i=1}^{I+J} G_{e,i}}{\sum_{i=1}^{I} G_{e,i}} \tag{3.11}$$

the relations are

$$F_e = \mathcal{R}\left(1 + \frac{T_{ee}}{T_0}\right) \tag{3.12}$$

$$T_{ee} = \left(\frac{F_e}{\mathcal{R}} - 1\right) T_0 \quad [\text{K}] \tag{3.13}$$

Example 3.3 Consider a three-port transducer with extended noise temperature $T_{ee} = 5 T_0$ and the exchangeable power gain from ports 1 and 2 to the output $G_{e,1} = 15$ and $G_{e,2} = -12$ respectively. From Equation (3.9) and Definition 3.5 the single response extended noise factors are determined:

$$N'_e = k T_{ee} (G_{e,1} + G_{e,2}) = 60.0 \times 10^{-21} \quad \text{W Hz}^{-1}$$

$$F_{e,1} = \frac{k T_0 (G_{e,1} + G_{e,2}) + N'_e}{k T_0 G_{e,1}} = 1.2$$

$$F_{e,2} = \frac{k T_0 (G_{e,1} + G_{e,2}) + N'_e}{k T_0 G_{e,2}} = -1.5$$

The same results are derived from Equation (3.12).

When calculating the multi-response extended noise factor, the spot response factor $\mathcal{R} = 1$ and Definition 3.5 gives

$$F_e \quad = \quad \frac{k\,T_0\,(G_{e,1} + G_{e,2}) + N_e'}{k\,T_0\,(G_{e,1} + G_{e,2})} \quad = \quad 6.0$$

which is the same as F_e derived by Equation (3.12).

Example 3.4 Consider a mixer with exchangeable power gain for the normal response $G_{e,1}$ and for the image response $G_{e,2}$. From definition 3.5

$$F_{e,SSB} \quad = \quad \frac{k\,T_0\,(G_{e,1} + G_{e,2}) + N_e'}{k\,T_0\,G_{e,1}}$$

$$F_{e,DSB} \quad = \quad \frac{k\,T_0\,(G_{e,1} + G_{e,2}) + N_e'}{k\,T_0\,(G_{e,1} + G_{e,2})}$$

This leads to

$$F_{e,SSB} \quad = \quad \frac{G_{e,1} + G_{e,2}}{G_{e,1}}\,F_{e,DSB} \quad = \quad \mathcal{R}\,F_{e,DSB}$$

$$\text{If} \quad G_{e,1} \quad = \quad G_{e,2} \quad = \quad G_e \quad \Rightarrow \quad F_{e,SSB} \quad = \quad 2\,F_{e,DSB}$$

3.2.3   The operating noise temperature

The operating noise temperature T_{op} – sometimes called the system noise temperature – is used to characterize a system under operating conditions and includes all noise contributions which add to the output noise power delivered to the load. Consider the system shown in Figure 3.3 where three sources contribute to the noise power density delivered to the load. One contribution is the noise from the source which could be an antenna with known noise temperature. For the time being only passive sources and loads are considered. The available source noise power amplified by the transducer gain determines the contribution from the source to the noise power density delivered to the load. The next noise source is from the two-port considered (e.g. a receiver). This part is given by the T_{ee} of the two-port and contributes by definition 3.4 as the source at the noise temperature of T_{ee}. The final part is from the load itself. The load generates noise; some of this noise is absorbed by the output port of the two-port (and changes T_{ee}, but often only slightly) and the remaining part is reflected at the output port and then contributes to the noise delivered to the load. This third part is usually very small compared to the other two. While these are amplified by the normally very big transducer gain of the system, the load-generated noise is reduced as it is divided between the load and the output immittance of the transducer.

The IRE definition [5] has been changed in two respects. First it has been extended to cover multi-ports with active as well as passive sources, and secondly

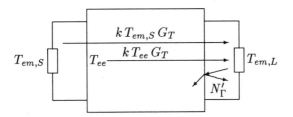

Figure 3.3: Noise contributions to the load.

the noise power density considered at the output is the power delivered to the noise free part of the load circuit. This is because the load noise often flows in the opposite direction to the noise delivered to the load and thus reduce the operating noise temperature [7]. The noise generated in the load and reflected at the output port is included as in the IRE definition.

Definition 3.6 The extended operating noise temperature $T_{e\,op}$ is defined as

$$T_{e\,op} \doteq \frac{N'_L}{k \sum_{i=1}^{I} G_{eT,i}} \quad [\text{K}] \quad (3.14)$$

where N'_L is the total noise power density delivered to the equivalent noise free immittance of the load circuit, $k = 1.3807 \times 10^{-23}\,\text{J K}^{-1}$ is Boltzmann's constant, I is the number of signal responses from all input ports where signals are applied and $G_{eT,i}$ is the extended transducer gain from port response i to the output port.

Note 1: There is only one output port and at that port only one frequency is considered.

Note 2: Noise generated in the load and reflected at the output of the transducer back to the noise free equivalent of the load is part of the numerator in Equation (3.14).

Note 3: All ports other than the output port are considered input ports and they should be loaded with any passive or active immittances except short or open circuits.

Note 4: The denominator of Equation (3.14) includes only gains from input responses where signals are applied.

From

$$N'_L = k\,T_{e\,op} \sum_{i=1}^{I} G_{eT,i}$$

$$= k \sum_{i=1}^{I+J} T_{em,i}\,G_{eT,i} + k\,T_{ee} \sum_{i=1}^{I+J} G_{eT,i} + N'_\Gamma \quad [\text{W Hz}^{-1}]$$

where $N'_\Gamma = |\Gamma|^2 k T_{em\,L}$, I is the number of input responses with signals and $I + J$ is the total number of input responses, it follows that

$$T_{e\,op} = \underbrace{\mathcal{R}\,T_{ee} + \frac{\sum_{i=1}^{I+J} T_{em,i}G_{eT,i}}{\sum_{i=1}^{I} G_{eT,i}}}_{1} + \underbrace{\frac{N'_\Gamma}{k\,\sum_{i=1}^{I} G_{eT,i}}}_{2} \quad [\text{K}] \qquad (3.15)$$

In most cases term 2 in Equation (3.15) is very small compared to term 1 so

$$T_{e\,op} \approx \mathcal{R}\,T_{ee} + \frac{\sum_{i=1}^{I+J} T_{em,i}G_{eT,i}}{\sum_{i=1}^{I} G_{eT,i}} \quad [\text{K}] \qquad (3.16)$$

is a very good approximation. For single response two-ports Equation 3.16 reduces to

$$T_{e\,op} \approx T_{ee} + T_{em} \quad [\text{K}] \qquad (3.17)$$

In order to keep track of the signs a three-port is considered. The two signal input ports are loaded with an active and a passive generator immittance and the possible sign combinations are shown in Table 3.1. If instead only port 1 is applied with a signal the sign combinations shown in Table 3.2 occur.

R_L / ~ / N'_L	R_{S1} / ~ / $T_{em,1}$	R_{S2} / ~ / $T_{em,2}$	$G_{eT,1}$	$G_{eT,2}$	$\sum_{i=1}^{2} G_{eT,i}$	$\mathcal{R}T_{ee}$	$\frac{\sum_{i=1}^{2} T_{em,i}G_{eT,i}}{\sum_{i=1}^{2} G_{eT,i}}$	$T_{e\,op}$ Eq. 3.16	$T_{e\,op}$ Eq. 3.14
+	−	+	−	+	+	+	+	+	+
+	−	+	−	+	−	−	−	−	−
−	−	+	+	−	+	−	−	−	−
−	−	+	+	−	−	+	+	+	+

Table 3.1: Sign combinations regarding $T_{e\,op}$ for a three-port with two signal responses.

R_L / ~ / N'_L	R_{S1} / ~ / $T_{em,1}$	R_{S2} / ~ / $T_{em,2}$	$G_{eT,1}$	$G_{eT,2}$	$\sum_{i=1}^{2} G_{eT,i}$	$\mathcal{R}T_{ee}$	$\frac{\sum_{i=1}^{2} T_{em,i}G_{eT,i}}{G_{eT,1}}$	$T_{e\,op}$ Eq. 3.16	$T_{e\,op}$ Eq. 3.14
+	−	+	−	+	+	−	−	−	−
+	−	+	−	+	−	−	−	−	−
−	−	+	+	−	+	−	−	−	−
−	−	+	+	−	−	−	−	−	−

Table 3.2: Sign combinations regarding $T_{e\,op}$ for a three-port with one signal response.

In general

$$T_{e\,op} > 0 \quad \text{when} \quad \frac{R_L}{\sum_{i=1}^{I} G_{eT,i}} > 0$$

$$T_{e\,op} < 0 \qquad \text{when} \qquad \frac{R_L}{\sum_{i=1}^{I} G_{eT,i}} < 0$$

where I is the number of inputs with signals applied and R_L is the real part of the load impedance.

In most cases where $T_{e\,op}$ is used both the source and load will be passive and most often a single response transducer is considered.

It should be noted that $T_{e\,op}$ is used in the common "figure of merit" G/T, where G is the antenna gain and $T = T_{e\,op}$.

Example 3.5 The importance of only considering the lossless part of the load circuit is illustrated by calculation of the extended operating noise temperature of an attenuator. Consider the one in Example 2.3 and load it with $Z_L = 75 + j\,25\,\Omega$ with a load noise temperature $T_L = 580$ K. Computing the exchangeable power gain of the attenuator one gets $G_e = 60.0 \times 10^{-3}$ and the method used in Example 3.2 determines

$$F_e = 1/G_e = 16.67$$

$$\Rightarrow \quad T_{ee} = (F_e - 1)\,T_0 = 4543 \text{ K}$$

This determines term 1 in Equation (3.15) to be $T_{ee} + T_a = 7543$ K. In order to compute term 2 let the load circuit be substituted by an equivalent noise free impedance Z_L in series with a noise voltage generator whose voltage is determined by

$$\langle |e|^2 \rangle = 4\,k\,T_L\,\text{Re}[Z_L]\,\Delta f$$

The power delivered to the noise free part of the load is

$$N_{L2} = \text{Re}\left\langle \frac{e\,Z_L}{Z_{out} + Z_L}\frac{e^*}{(Z_{out} + Z_L)^*} \right\rangle = \frac{\langle |e|^2 \rangle\,\text{Re}[Z_L]}{|Z_{out} + Z_L|^2}$$

In order to relate this noise power to a part of the operating noise temperature it is divided by $k\,G_{eT}\,\Delta f$ where G_{eT} for $Z_{out} = 50\,\Omega$ is 55.4×10^{-3}:

$$\frac{N_{L2}}{k\,G_{eT}\,\Delta f} = \frac{4\,(\text{Re}[Z_L])^2}{|Z_{out} + Z_L|^2}\frac{T_L}{G_{eT}} = 14500 \text{ K}$$

$$\Rightarrow \quad T_{e\,op} = 22043 \text{ K}$$

If the IRE definition is used the power density flow to the load is the same as above represented by noise temperatures, but the power density flow from the load to the output of the attenuator will then be subtracted. This part of the power density flow is determined by the output terminal voltage U, the current into the load circuit I and $\langle |e|^2 \rangle = 4\,k\,T_L\,\text{Re}[Z_L]\,\Delta f$:

$$N_{L2_{IRE}} = \text{Re}\langle U\,I^* \rangle = \text{Re}\left\langle \frac{e\,Z_{out}}{Z_{out} + Z_L}\frac{-e^*}{(Z_{out} + Z_L)^*} \right\rangle$$

$$= \frac{-\langle |e|^2 \rangle\,\text{Re}[Z_{out}]}{|Z_{out} + Z_L|^2}$$

Referring this power to an input temperature gives

$$\frac{N_{L2_{IRE}}}{k\,G_{eT}\,\Delta f} = \frac{-4\,T_L\,\mathrm{Re}Z_{out}\,\mathrm{Re}Z_L]}{G_{eT}\,|Z_{out}+Z_L|^2}$$

$$= -9667\ \mathrm{K}$$

$$\Rightarrow \qquad T_{e\,op_{IRE}} \quad = \quad -2124\ \mathrm{K}$$

– a somewhat misleading figure. Further information on this is given in [7].

Example 3.6 Let a $12\,\mathrm{GHz}$ satellite-TV receiver require a $G/T = 14\,\mathrm{dB}$. If a 90 cm reflector antenna has a gain of $38\,\mathrm{dB}$ and the noise temperature $T_{em\,a} = 90\ \mathrm{K}$, which noise factor is then the maximum for the receiver?

$$T \quad = \quad T_{e\,op} \quad = \quad (38-14)\,\mathrm{dB\,K} \quad = \quad 24\ \mathrm{dB\,K} \quad \sim \quad 251\ \mathrm{K}$$

$$\Rightarrow \quad T_{ee} \quad \approx \quad T_{e\,op}-T_{em\,a} \quad = \quad 161\ \mathrm{K}$$

$$\Rightarrow \quad F_e \quad = \quad 1+\frac{T_{ee}}{T_0} \quad = \quad 1.56 \quad \sim \quad 1.92\,\mathrm{dB}$$

If the actual receiver has a noise figure of $4\,\mathrm{dB}$ then the antenna diameter D, where the antenna gain is proportional to D^2, is determined by

$$F_e \quad = \quad 2.51 \quad (\sim 4\,\mathrm{dB})$$

$$\Rightarrow \quad T_{ee} \quad = \quad (F_e-1)\,T_0 \quad = \quad 438\ \mathrm{K}$$

$$\Rightarrow \quad T_{e\,op} \quad \approx \quad 528\ \mathrm{K} \quad \sim \quad 27.2\,\mathrm{dB}$$

$$\Rightarrow \quad G \quad = \quad (14+27.2)\,\mathrm{dB} \quad = \quad 41.2\,\mathrm{dB}$$

$$\frac{G}{G_{90\,\mathrm{cm}}} \quad = \quad 10^{\frac{41.2-38}{10}} \quad = \quad 2.10$$

$$\Rightarrow \quad D \quad = \quad 0.9\,\sqrt{2.10} \quad = \quad 1.31\ \mathrm{m}$$

3.3   Average noise quantities and the noise bandwidth

The noise definitions in Section 3.2 above are all defined at a single (a spot) output frequency and therefore often called spot noise quantities. They are functions of the frequency and most useful to describe the noise behaviour of the circuit. Sometimes it is practical to characterize the noise properties in a given frequency band with one number instead of the more useful frequency function and therefore the average noise quantities are introduced. If the gain function of the circuit is not very flat there may be some confusion as to how the reference amplification and the frequency band are defined. These problems are exposed and the equivalent noise bandwidth is defined.

In the IRE definitions of average noise quantities the transducer gain is used. When the definitions are extended to active sources it is necessary to replace it with the extended transducer gain from Equation (3.3), but the really confusing problem is when source immittance or load immittance changes sign in the frequency band of interest. This leads to the possibility of zero exchangeable noise power and gives results which can not be given a physical meaning. If the IRE definitions are going to be extended to active devices it could be more interesting to look at the flow of the noise powers. Then the results may be given an understandable physical meaning, but the equations will be rather complicated. As extended average noise quantities are used rather seldom, it is chosen to keep the mathematics as simple as possible, and thus in cases with both positive and negative noise power flow the values of the noise quantities do not right away give an impression of whether an amplifier has good or bad noise properties.

3.3.1 The average effective noise temperature

The definition of the average extended noise temperature is given as a formula and if all source immittances for all input frequencies have the same sign and the load immittance also has the same sign for all output frequencies a more explanatory equivalent formulation can be given.

Definition 3.7 The average extended effective (input) noise temperature of a multi-port $\overline{T_{ee}}$ is defined as

$$\overline{T_{ee}} \doteq \frac{\sum_{i=1}^{I} \int_0^\infty T_{ee}(f) \, G_{eT,i}(f - f_i) \, df}{\sum_{i=1}^{I} \int_0^\infty G_{eT,i}(f - f_i) \, df} \qquad [\text{K}] \qquad (3.18)$$

where $T_{ee}(f)$ is the extended effective noise temperature of the multiport as a function of the frequency f, and $G_{eT,i}(f - f_i)$ is the extended transducer gain from port response i to the output port at an input frequency $f - f_i$ which originates a corresponding output frequency f.

If $T_{ee}(f)$ for all f have the same sign and also if $G_{eT,i}(f - f_i)$ for all f and i have the same sign, Equation (3.18) is seen as a weighted average of T_{ee}. In words this can be stated as follows:

For a multi-port with a load immittance which has the same sign for all frequencies and with source immittances which all (and for all frequencies) have the same sign, the average extended effective (input) noise temperature $\overline{T_{ee}}$ is the extended noise temperature applied to all input immittances of a noise free equivalent of the multi-port which delivers the same noise power to the load as the noisy multi-port with noise free sources.

3.3.2   The average noise factor

For the average extended noise factor the same comments apply as for the average effective noise temperature.

Definition 3.8 The average extended noise factor of a multi-port $\overline{F_e}$ is defined as

$$\overline{F_e} \;\doteq\; \frac{\sum_{i=1}^{I+J} \int_0^\infty F_e(f)\, G_{eT,i}(f - f_i)\, df}{\sum_{i=1}^{I} \int_0^\infty G_{eT,i}(f - f_i)\, df} \tag{3.19}$$

where $F_e(f)$ is the extended noise factor of the multi-port as a function of the frequency f, $G_{eT,i}(f - f_i)$ is the extended transducer gain from port response i to the output port at an input frequency $f - f_i$ which originates a corresponding output frequency f, I is the number of signal responses and $I+J$ is the total number of responses from sources to load.

If $F_e(f)$ has the same sign for all f and $G_{eT,i}(f - f_i)$ has the same sign for all f and i, Equation (3.19) can be seen as a weighted average of F_e. In words this can be stated as:

For a multi-port with a load immittance which has the same sign for all frequencies and with source immittances which all (and for all frequencies) have the same sign, the average extended noise factor $\overline{F_e}$ is the ratio of (1) the total noise power delivered to the load when the extended noise temperature of the source (or sources) is/are the standard noise temperature (290 K) at all frequencies (and input ports), to (2) that part of the noise power delivered to the load which originates from the signal source (or sources) at standard noise temperature.

Defining the response factor

$$\mathcal{R} \;\doteq\; \frac{\sum_{i=1}^{I+J} \int_0^\infty G_{eT,i}(f - f_i)\, df}{\sum_{i=1}^{I} \int_0^\infty G_{eT,i}(f - f_i)\, df} \tag{3.20}$$

as the ratio of "gain bandwidth product" of all responses to that of the signal responses, the relation corresponding to Equation (3.12) is proved by

$$\overline{F_e} = \mathcal{R}\left(\frac{\overline{T_{ee}}}{T_0} + 1\right) \tag{3.21}$$

$$= \frac{\sum_{i=1}^{I+J} \int_0^\infty G_{eT,i}(f - f_i)\, df}{\sum_{i=1}^{I} \int_0^\infty G_{eT,i}(f - f_i)\, df} \left(\frac{\sum_{i=1}^{I+J} \int_0^\infty T_{ee}(f)\, G_{eT,i}(f - f_i)\, df}{T_0 \sum_{i=1}^{I+J} \int_0^\infty G_{eT,i}(f - f_i)\, df} + 1\right)$$

$$= \frac{\sum_{i=1}^{I+J} \int_0^\infty F_e(f)\, G_{eT,i}(f - f_i)\, df}{\sum_{i=1}^{I} \int_0^\infty G_{eT,i}(f - f_i)\, df} \qquad \text{q. e. d.}$$

From Equation (3.21) $\overline{T_{ee}}$ is found to be

$$\overline{T_{ee}} = \left(\frac{\overline{F_e}}{\mathcal{R}} - 1\right) T_0 \tag{3.22}$$

3.3.3 The average operating noise temperature

The IRE definition [5] uses signal gain which involves the signal bandwidth in the definition of the average operating noise temperature. As many new modulation schemes have been developed in the last decades it seems inconvenient to tie a noise definition to the type of signal. Thus the signal gain is replaced by the extended transducer gain from the signal input ports to the output port in the definition of the average extended operating noise temperature.

Definition 3.9 The average extended operating noise temperature of a multi-port $\overline{T_{e\,op}}$ is defined as

$$\overline{T_{e\,op}} \; \doteq \; \frac{\int_0^\infty N_L'(f)\,df}{k \sum_{i=1}^I \int_0^\infty G_{eT,i}(f-f_i)\,df} \qquad [\text{K}] \qquad (3.23)$$

where $N_L'(f)$ is the total noise power density delivered to the equivalent noise free immittance of the load circuit at the output frequency f, $k = 1.3807 \times 10^{-23}\,\text{J}\,\text{K}^{-1}$ is Boltzmann's constant, I is the number of signal responses from all input ports where signals are applied and $G_{eT,i}(f-f_i)$ is the extended transducer gain from port response i to the output port at an input frequency $f - f_i$ which causes the output frequency f.

Note 1: Noise generated in the load and reflected at the output of the transducer back to the noise free equivalent of the load is part of the numerator in Equation (3.23).

Note 2: The resulting $N_L'(f)$ may consist of parts with opposite signs at a frequency f and also $N_L'(f)$ may change sign with varying frequency. This means that when active immittances occur great care should be taken in interpreting the value of $\overline{T_{e\,op}}$.

Note 3: All ports other than the output port are considered input ports and they should be loaded with any passive or active immittances except short or open circuits and the extended noise temperatures of these loads are the actual ones.

Note 4: The denominator of Equation (3.23) includes only gains from input responses where signals are applied.

From

$$\int_0^\infty N_L'(f)\,df \;=\; k\,\overline{T_{e\,op}}\,\sum_{i=1}^I \int_0^\infty G_{eT,i}(f-f_i)\,df$$

$$= \; k \sum_{i=1}^{I+J} \int_0^\infty T_{ee}(f)\,G_{eT,i}(f-f_i)\,df$$

$$+ \, k \sum_{i=1}^{I+J} \int_0^\infty T_{em}(f)\,G_{eT,i}(f-f_i)\,df \; + \int_0^\infty N_L'(f)\,df$$

follows

$$\overline{T_{e\,op}} \;=\; \mathcal{R}\,\overline{T_{ee}} \;+\; \frac{\sum_{i=1}^{I+J} \int_0^\infty T_{em}(f)\,G_{eT,i}(f-f_i)\,df}{\sum_{i=1}^{I} \int_0^\infty G_{eT,i}(f-f_i)\,df}$$

$$+\; \frac{\int_0^\infty N_L'(f)\,df}{\sum_{i=1}^{I} \int_0^\infty G_{eT,i}(f-f_i)\,df} \qquad\qquad (3.24)$$

As stated in note 2 care should be taken in interpreting the value of $\overline{T_{e\,op}}$ when active immittances occur. Even a very noisy transducer may have almost zero $\overline{T_{e\,op}}$ as the noise power may flow in opposite directions at different frequencies. On the other hand in all those cases where the noise power parts delivered to the load add up with the same sign the average extended noise temperature is a very descriptive quantity.

3.3.4   The equivalent noise bandwidth

For each response a gain function exists. In many expressions in noise theory this gain function is integrated over all f. It is often convenient to replace the integral with the product of a fixed reference gain $G_{eT\,r}$ (often the maximum gain) and a bandwidth which is called the noise bandwidth B_N, when the gain function is the (extended) transducer gain. Even if the extended transducer gain can be used the interpretation is most clear for the common transducer gain function which is possitive for all f.

Definition 3.10 The (extended) equivalent noise bandwidth B_N is defined as

$$B_N \;\doteq\; \frac{\int_0^\infty G_{eT}(f)\,df}{G_{eT\,r}} \qquad [\text{Hz}] \qquad\qquad (3.25)$$

where $G_{eT\,r}$ is the reference value of the extended transducer gain function $G_{eT}(f)$ (at a corresponding reference frequency f_r).

Note 1: Usually $G_{eT\,r}$ is chosen as the maximum value of $G_{eT\,r}(f)$, but it can be chosen freely, so it is important to know which $G_{eT\,r}$ (or f_r) is used when the noise bandwidth is specified.

As seen from Figure 3.4 it is important to specify which $G_{eT\,r}$ or f_r is used when the noise bandwidth is used. It should be noted that all axes in Figure 3.4 are linearly scaled and that the area under the gain function is equal to the area determined by $G_{eT\,r}\,B_N$. The ratio of B_N/B_{3dB} for n LC-circuits with identical Q factor, resonance frequency and without any mutual coupling is given in Table 3.3.

Example 3.7 A tuned amplifier consists of an LC-circuit, a transistor and another LC-circuit with the same Q and resonance frequency as the first, and no coupling exists

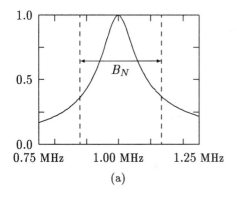
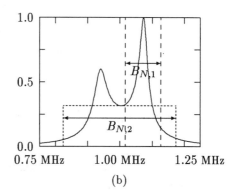

Figure 3.4: Examples of noise bandwidths: (a) a single LC-circuit, (b) a mistuned double LC-circuit.

n:	1	2	3	4	5	6	∞
B_N/B_{3dB}:	1.57	1.221	1.155	1.13	1.11	1.10	1.06

Table 3.3: B_N/B_{3dB} for n uncoupled identical LC-circuits.

between the two resonant circuits. The amplifier shows a $B_{3\,dB}$ of 8 kHz. When the average noise factor of the amplifier is $\overline{F} = 4.0$ and the noise temperature of the source is $T_{em\,S} = 580\,\text{K}\,(\sim 2\,T_0)$ the problem is to find the signal to noise ratio at the load for an input signal power level of 1 pW.

$$\overline{T_{op}} \approx T_{em\,S} + (\overline{F} - 1)\,T_0 = 5\,T_0$$

$$\frac{S_L}{N_L} = \frac{S_i}{k\,\overline{T_{op}}\,B_N} = \frac{S_i}{k\,5\,T_0\,1.221\,B_{3dB}} = 5120 \sim 37.1\,\text{dB}$$

It is supposed that the signal gain $G_S = S_L/S_i$ is equal to the transducer gain $G_T = N_L/(k\,\overline{T_{op}}\,B_N)$, which usually is the case. If the power spectra of S_L and N_L are different due to a special type of modulation and/or very frequency dependent T_{ee} the above calculation may give a wrong result. The IRE definition, which uses G_S in the definition of $\overline{T_{op}}$, will of course give the correct result, but the problem is that the signal spectrum then must be specified together with the $\overline{T_{op}}$ figure to give a full description of the noise in the system and also when measuring the $\overline{T_{op}}$.

The response factor defined in Equation (3.20) can be expressed with use of noise bandwidths for multi-ports which are not loaded with short or open circuits at any frequency as

$$\mathcal{R} \;=\; \frac{\sum_{i=1}^{I+J} G_{eT\,r,i}\, B_{N,i}}{\sum_{i=1}^{I} G_{eT\,r,i}\, B_{N,i}}$$

where $I + J$ is the number of all responses of which I are supplied with signals.

3.4   Discussion of noise quantities

Most often noise quantities are used to describe the noise properties of two-ports. In cascading these, one or more of the two-ports may be only conditionally stable. This means that a source or load for one of the stages is active even if the cascaded circuit is stable. This situation is quite common at higher frequencies. Therefore in extending the gain and noise quantity definitions to include active sources and loads, first priority was given to ensure that the well-known and often used formulae could be used unchanged.

When extending to multi-ports this choice will in rare cases give some values for extended noise quantities which – considered alone – do not give a correct feeling for the noise properties of the device. This is due to the fact that the extended definitions in special cases may have noise powers going in opposite directions and thus being subtracted. It has been decided to accept this rarely occurring calamity in order to preserve the simplicity of the often used formulae.

The same problem occurs for the average noise quantities if either the source, the load or the gain of the device changes sign in the frequency band of interest. At one frequency the noise power flows in one direction and at another in the opposite. Thus the noise power integrated over the frequency band may be a small and rather meaningless figure.

One way to get around these problems is to introduce yet another noise quantity [7]. As all definitions of noise quantities consider only one output port no ambiguity exists on the sign of the load at a given output frequency. By referring a noise temperature to the output terminal many problems are solved. The proposed output noise temperature is called the extended load operating noise temperature and is defined for a multi rosponse transducer as

Definition 3.11 The extended load operating noise temperature $T_{eL\,op}$ of a multi-response transducer is defined as

$$T_{eL\,op} \;=\; \frac{N'_L}{k} \tag{3.26}$$

where N'_L is the noise power density delivered to the equivalent noise free immittance of the load circuit at the output frequency and $k = 1.3807 \times 10^{-23}\,\mathrm{J\,K^{-1}}$ is Boltzmann's constant.

If the extended load operating noise temperature should have an average equivalent and the load immittance should change sign in the frequency band of interest the same problem occurs once again. Two possibilities exist. One is to integrate numerically in order to add all the power flow over the frequency range. The other is just to integrate, which in some rare instances gives meaningless results without further information. But is an average extended load operating noise temperature needed? Perhaps an international, European or American standard committee on noise might be a good idea. If so the number of recommended definitions might be reduced. The first step could be to abolish the average definitions.

3.5 References

[1] Engberg, J. & Larsen, T.: "Extended definitions for noise temperatures of linear noisy one- and two-ports", *IEE Proc. Part H*, vol. 138, pp. 86 – 90, February 1991.

[2] Haus, H. A. & Adler, R. B.: "Circuit theory of linear noisy networks", Technology Press and Wiley, 1959.

[3] "IRE Standards on Electron Tubes: Definitions of Terms, 1957, 57 IRE 7. S2", *Proc. IRE*, vol. 45, pp. 983 – 1010, July 1957.

[4] Haus, H. A. & Adler, R. B.: "An extension of the noise figure definition", *Proc. IRE*, vol. 45, pp. 690 – 691, May 1957.

[5] "IRE Standards on Electron Tubes: Definitions of Terms, 1962, 62 IRE 7. S2", *Proc. IRE*, vol. 50, pp. 434 – 435, March 1963.

[6] Friis, H. T.: "Noise figures of radio receivers", *Proc. IRE*, vol. 32, pp. 419 – 422, July 1944.

[7] Larsen, T.: "Extended source and load operating noise temperatures of nonlinear systems", *IEE Proc. Part H*, vol.139, pp. 121 – 124, April 1992.

4

Noise parameters

Most amplifiers consist of cascaded two-ports. In order to characterize the noisy behaviour of a two-port several sorts of noise parameters have been developed. Noise
parameters can also be developed for multi-ports, but would be more clumsy to
use. Two-ports are conveniently characterized by their small-signal parameters like
H, Y or S parameters. These consist of 8 (4 complex) numbers, which are functions of frequency and bias point in particular and to a minor extent temperature,
radioactive radiation etc. They are, however, used extensively in practical circuit
design. By adding 4 more numbers (2 real and 1 complex) the noise properties of a
two-port can be included as well. Those 4 numbers are called noise parameters and
– in analogy to small-signal parameters – noise parameters exist in many forms.

In this chapter the noise parameters are derived, their use explained, and the
conversion formulae between different types of noise parameters given.

4.1 Noise voltages and currents

In 1955 Rothe and Dahlke [1] introduced noise parameters related to the chain or
$ABCD$ small-signal parameters. These are based on noise voltages and currents. In
a similar way noise power waves related to the scattering parameters have been used
by several authors, e.g. [2,3,4,5,6,7,8,9], but they are still not used as extensively as
the Rothe and Dahlke type of noise parameters.

4.1.1 The equivalent noise two-port

In order to describe a noisy linear two-port, the small-signal equations are enlarged
to cover the noise contributions as well. In Figure 4.1 the noisy two-port and three
examples of equivalent circuits are shown. In Figure 4.1(a) the noisy two-port is
characterized by hatching. In Figures 4.1(b), (c), and (d) the noisy two-port is

replaced by a noise free but otherwise unchanged two-port and two partially correlated noise generators. This is correct as long as the two-port is linear. Physically
the noise sources inside the two-port can contribute to either the input side, the
output side, or both. Mathematically two linear equations exist between the input
and output sides and the noise appears in both equations.

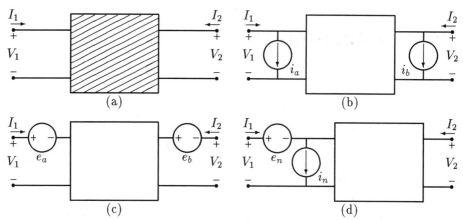

Figure 4.1: Noisy linear two-ports.

The circuit equations corresponding to Figures 4.1(b), (c), and (d), where the
V's and I's are complex Fourier series coefficients of voltages and currents, are

$$
\begin{bmatrix} I_1 \\ I_2 \end{bmatrix} = \begin{bmatrix} Y_{11} & Y_{12} \\ Y_{21} & Y_{22} \end{bmatrix} \cdot \begin{bmatrix} V_1 \\ V_2 \end{bmatrix} + \begin{bmatrix} i_a \\ i_b \end{bmatrix}
$$

$$
\begin{bmatrix} V_1 \\ V_2 \end{bmatrix} = \begin{bmatrix} Z_{11} & Z_{12} \\ Z_{21} & Z_{22} \end{bmatrix} \cdot \begin{bmatrix} I_1 \\ I_2 \end{bmatrix} + \begin{bmatrix} e_a \\ e_b \end{bmatrix}
$$

$$
\begin{bmatrix} V_1 \\ I_1 \end{bmatrix} = \begin{bmatrix} A & B \\ C & D \end{bmatrix} \cdot \begin{bmatrix} V_2 \\ -I_2 \end{bmatrix} + \begin{bmatrix} e_n \\ i_n \end{bmatrix} \tag{4.1}
$$

The noise vectors each represent two noise generators. As the noise is generated
inside the two-port by various physical processes, the contributions to the two noise
generators are more or less correlated. It is therefore necessary to know the correlation between the two generators. The correlation between two stochastic variables
X and Y is determined by the complex correlation coefficient

$$
\gamma = \frac{\langle XY^* \rangle}{\sqrt{\langle |X|^2 \rangle \langle |Y|^2 \rangle}} \tag{4.2}
$$

where $|\gamma| \leq 1$. In Figure 4.1(b) the admittance matrix has been used and in Figure
4.1(c) the impedance matrix has been used. Rothe and Dahlke chose Figure 4.1(d)
corresponding to the chain or $ABCD$ matrix and thus both noise generators are at

the input side which is practical in noise theory as the noise quantities are referred
to the input side.

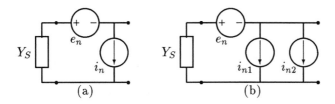

Figure 4.2: Partition of the noise current generator into a correlated and an uncorrelated
part.

As shown in Figure 4.2(b) the noise current generator is partitioned into two
parts, $i_n = i_{n1} + i_{n2}$, where one part (i_{n2}) is partially correlated with the noise
voltage generator and the other part (i_{n1}) is uncorrelated. Thus $\langle i_{n1} i_{n2}^* \rangle = 0$. This
means that i_{n2} is proportional to e_n with the complex proportionality factor Y_γ
which is called the correlation admittance. This leads to

$$i_{n2} = Y_\gamma e_n \tag{4.3}$$

$$\langle |i_n|^2 \rangle = \langle |i_{n1}|^2 \rangle + \langle |i_{n2}|^2 \rangle$$

$$\langle e_n i_n^* \rangle = \langle e_n i_{n1}^* \rangle + \langle e_n i_{n2}^* \rangle = \langle e_n i_{n2}^* \rangle \tag{4.4}$$

Inserting Equation (4.4) into Equation (4.2), it follows that

$$\gamma = \mathrm{Re}\,[\gamma] + j\,\mathrm{Im}\,[\gamma]$$

$$= \frac{\langle e_n i_n^* \rangle}{\sqrt{\langle |e_n|^2 \rangle \langle |i_n|^2 \rangle}} \tag{4.5}$$

$$= \frac{\langle e_n i_{n2}^* \rangle}{\sqrt{\langle |e_n|^2 \rangle \langle |i_n|^2 \rangle}} \tag{4.6}$$

$$\Rightarrow \quad |\gamma|^2 = \frac{\langle |i_{n2}|^2 \rangle}{\langle |i_n|^2 \rangle} \tag{4.7}$$

as

$$\langle e_n i_{n2}^* \rangle \cdot \langle e_n^* i_{n2} \rangle = \langle e_n Y_\gamma^* e_n^* \rangle \cdot \langle e_n^* Y_\gamma e_n \rangle$$

$$= \langle e_n e_n^* \rangle \cdot \langle Y_\gamma^* e_n^* Y_\gamma e_n \rangle$$

$$= \langle |e_n|^2 \rangle \cdot \langle |i_{n2}|^2 \rangle \tag{4.8}$$

Suppose that e_n, i_n and γ are known then

$$\langle |i_{n2}|^2 \rangle \;\; = \;\; |\gamma|^2 \langle |i_n|^2 \rangle$$

$$\langle |i_{n1}|^2 \rangle \;\; = \;\; (1 - |\gamma|^2) \langle |i_n|^2 \rangle$$

Equations (4.3) and (4.6) determine Y_γ by

$$Y_\gamma^* \;\; = \;\; \gamma \sqrt{\frac{\langle |i_n|^2 \rangle}{\langle |e_n|^2 \rangle}} \qquad\qquad (4.9)$$

and thus
$$Y_\gamma \;\; = \;\; G_\gamma + j\, B_\gamma$$

$$\;\; = \;\; \mathrm{Re}\,[\gamma] \sqrt{\frac{\langle |i_n|^2 \rangle}{\langle |e_n|^2 \rangle}} - j\, \mathrm{Im}\,[\gamma] \sqrt{\frac{\langle |i_n|^2 \rangle}{\langle |e_n|^2 \rangle}} \qquad (4.10)$$

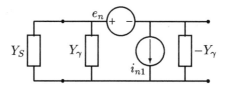

Figure 4.3: Equivalent noise two-port.

The circuit in Figure 4.3 can be replaced by the circuit in Figure 4.4 where Y_γ and $-Y_\gamma$ are noise free which is shown by their noise temperatures of 0 K. The noise generators are conveniently indicated with R_n and G_n by Equations (2.4) and (2.5) replacing the bandwidth dependent e_n and i_{n1}. The two circuits are identical as their open circuit voltages and short circuit currents are the same. The equivalent noise two-port – such as the one in Figure 4.4 – are placed in front of the noise free small signal two-port. When small signal analysis is performed, the noise voltage generator is short circuited, $e_n = 0$, the noise current generator is open circuited, $i_{n1} = 0$, and Y_γ and $-Y_\gamma$ cancel each other, therefore the noise two-port will have no effect on the small signal analysis.

From the two noise generators in Equation (4.1) given by $\langle |e_n|^2 \rangle$, $\langle |i_n|^2 \rangle$ and their correlation coefficient γ, the four noise parameters R_n, G_n and $Y_\gamma = G_\gamma + j\, B_\gamma$ shown in Figure 4.4 are determined.

If the noise voltage generator instead is partitioned – with the noise current generator – into a correlated and an uncorrelated part, the resulting equivalent noise two-port is shown in Figure 4.5. Again the four noise parameters r_n, g_n and $Z_\gamma = R_\gamma + j\, X_\gamma$ are determined by $\langle |e_n|^2 \rangle$, $\langle |i_n|^2 \rangle$ and γ. Note that in general $Z_\gamma \neq 1/Y_\gamma$.

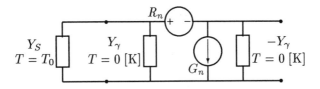

Figure 4.4: Equivalent noise two-port in Π form.

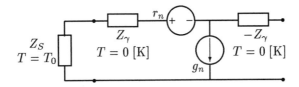

Figure 4.5: Equivalent noise two-port in T form.

Two sets of noise parameters have now been defined. These are R_n, G_n, and Y_γ which are most convenient to use in connection with Y parameters and r_n, g_n, and Z_γ which are convenient with Z parameters.

4.1.2 Y and Z noise parameters

From Figure 4.4 the noise factor can be derived. The amplifier following the noise two-port is noise free so that the noise ratios on the output and input side of the amplifier are the same and therefore the (noise free) amplifier does not contribute to the noise factor. To find the noise factor, the exchangeable noise power densities from the source at T_0 and from the two-port at the output terminals are found. This is done by finding the current in a short circuit over the output terminals from each of the three noise sources G_S, R_n and G_n. The noise power density is then the sum of the three power densities which are each found by squaring the short circuit current and dividing by Δf and by four times the output conductance. This noise power density is also the noise power density of the source admittance at either T_0 (the contribution from the source) or T_{ee} (the contribution from the two-port transferred to the source) both multiplied by the exchangeable power gain of the two-port.

$$i_{R_n} = e_{R_n}\left(Y_S + Y_\gamma\right)$$

$$k\,T_0\,G_e \;=\; \frac{\langle|i^2_{G_S}|\rangle}{4\,(G_S + G_\gamma - G_\gamma)\,\Delta f}$$

$$k\,T_{ee}\,G_e \;=\; \frac{\langle|i^2_{G_n}|\rangle + \langle|i^2_{R_n}|\rangle}{4\,(G_S + G_\gamma - G_\gamma)\,\Delta f}$$

$$=\; \frac{\langle|i^2_{G_n}|\rangle + \langle|e^2_{R_n}|\rangle\,|Y_S + Y_\gamma|^2}{4\,G_S\,\Delta f}$$

Definition 3.5 leads to Equation (3.12). Using Equations (2.7) and (2.9) leads to

$$F_e \;=\; \frac{\langle|i^2_{G_S}|\rangle + \langle|i^2_{G_n}|\rangle + \langle|e^2_{R_n}|\rangle\,|Y_S + Y_\gamma|^2}{\langle|i^2_{G_S}|\rangle}$$

$$=\; 1 + \frac{1}{G_S}\left(G_n + R_n\,|Y_S + Y_\gamma|^2\right) \tag{4.11}$$

In a quite similar way from Figure 4.5 it can be derived that

$$F_e \;=\; 1 + \frac{1}{R_S}\left(r_n + g_n\,|Z_S + Z_\gamma|^2\right) \tag{4.12}$$

Note that $F_e > 1$ if $G_S > 0$ and $F_e < 1$ if $G_S < 0$. The two expressions for F_e, Equations (4.11) and (4.12), show how the extended noise factor depends on the noise parameters and the source immittance.

In order to examine how the noise factor depends on the source immittance Equation (4.11) can be written as

$$\left[G_S - \left(\frac{F_e - 1}{2\,R_n} - G_\gamma\right)\right]^2 + [B_S - (-B_\gamma)]^2 = \frac{(F_e - 1)^2}{4\,R_n} - (F_e - 1)\frac{G_\gamma}{R_n} - \frac{G_n}{R_n} \tag{4.13}$$

which is recognized as the equation for a circle (F_e constant) in the Y_S-plane. The circle has its centre in

$$(G_S, B_S) \;=\; \left(\frac{F_e - 1}{2\,R_n} - G_\gamma,\, -B_\gamma\right) \tag{4.14}$$

and its radius is

$$R_{CF} \;=\; \sqrt{\frac{(F_e - 1)^2}{4\,R_n^2} - (F_e - 1)\frac{G_\gamma}{R_n} - \frac{G_n}{R_n}} \tag{4.15}$$

$F_e(Y_S)$ is symmetrical around $(F_e, G_S, B_S) = (1 + 2\,R_n\,G_\gamma, 0, -B_\gamma)$. Since vanishing of the right-hand side of Equation (4.13) always implies two roots for F_e, two extrema for F_e exist. These are as follows:

$$F_{e\,min} \;=\; 1 + 2\left(R_n G_\gamma + \sqrt{R_n G_n + (R_n G_\gamma)^2}\right) \tag{4.16}$$

for

$$Y_{SOF} \;=\; \sqrt{G_n/R_n + G_\gamma^2} - j\,B_\gamma \tag{4.17}$$

$$F_{e\,max} = 1 + 2\left(R_n G_\gamma - \sqrt{R_n G_n + (R_n G_\gamma)^2}\right) \qquad (4.18)$$

for

$$Y_{SOF'} = -\sqrt{G_n/R_n + G_\gamma^2} - j\,B_\gamma \qquad (4.19)$$

The index SOF stands for **S**ource **O**ptimum with respect to the noise **F**actor and is used for $G_S > 0$. SOF' is used when $G_S < 0$. Note from Equations (4.16) and (4.18) that $F_{min} > F_{max}$. In Figure 4.6 the noise factor contours are shown.

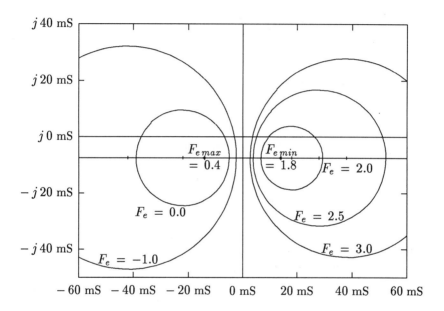

Figure 4.6: Contours for constant extended noise factor in the source admittance plane with data from Example 4.1.

Similarly, from Equation (4.12) a local minimum

$$F_{e\,min} = 1 + 2\left(g_n R_\gamma + \sqrt{g_n r_n + (g_n R_\gamma)^2}\right) \qquad (4.20)$$

for

$$Z_{SOF} = \sqrt{r_n/g_n + R_\gamma^2} - j\,X_\gamma \qquad (4.21)$$

and a local maximum

$$F_{e\,max} = 1 + 2\left(g_n R_\gamma - \sqrt{g_n r_n + (g_n R_\gamma)^2}\right) \qquad (4.22)$$

for

$$Z_{SOF'} = -\sqrt{r_n/g_n + R_\gamma^2} - j\,X_\gamma \qquad (4.23)$$

Equation (4.11), which is illustrated in Figure 4.6, and Equation (4.12) give information on the noise factor as a function of a chosen source immittance. If the minimum noise factor is the design criterion then Equation (4.17) or (4.21) determines the source immittance and Equation (4.16) or (4.20) gives the value of the minimum noise factor. If the source is active the $F_{e\,max}$ gives the smallest noise factor for the stage.

Sometimes the values of $F_{e\,min}$, Y_{SOF}, and R_n are used as noise parameters and similarly with $F_{e\,min}$, Z_{SOF}, and g_n. This leads to

$$F_e \;=\; F_{e\,min} \,+\, \frac{R_n}{G_S} |Y_S - Y_{SOF}|^2 \tag{4.24}$$

$$F_e \;=\; F_{e\,min} \,+\, \frac{g_n}{R_S} |Z_S - Z_{SOF}|^2 \tag{4.25}$$

It should also be noted that the circles for constant F_e, $F_{e\,min}$, and $F_{e\,max}$ can be replaced by T_{ee}, $T_{ee\,min}$, and $T_{ee\,max}$ as seen from Equation (3.13), $T_{ee} = (F_e - 1)\,T_0$.

Example 4.1 In order to draw the noise factor circles as a function of the source admittance for an amplifier with the following noise parameters

$$R_n \;=\; 25\ \Omega$$
$$G_n \;=\; 4.8\ \mathrm{mS}$$
$$Y_\gamma \;=\; 2.0 + j\,7.5\ \mathrm{mS}$$

the first thing to do is to find $F_{e\,min}$ and $F_{e\,max}$ from Equations (4.16) and (4.18) and the corresponding source admittances from Equations (4.17) and (4.19):

$$F_{e\,min} \;=\; 1 + 2\left(R_n G_\gamma + \sqrt{R_n G_n + (R_n G_\gamma)^2} \right)$$

$$=\; 1 + 2\left(0.025 \times 2.0 + \sqrt{0.025 \times 4.8 + (0.025 \times 2.0)^2} \right)$$

$$=\; 1.8 \quad (2.55\ \mathrm{dB})$$

for

$$Y_{SOF} \;=\; \sqrt{\frac{G_n}{R_n} + G_\gamma^2} - j\,B_\gamma$$

$$=\; \sqrt{\frac{4.8}{0.025} + 2.0^2} - j\,7.5 \;=\; 14 - j\,7.5\ \mathrm{mS}$$

and

$$F_{e\,max} \;=\; 1 + 2\left(0.025 \times 2.0 - \sqrt{0.025 \times 4.8 + (0.025 \times 2.0)^2} \right) \;=\; 0.4$$

for

$$Y_{SOF'} \;=\; -\sqrt{\frac{4.8}{0.025} + 2.0^2} - j\,7.5 \;=\; -14 - j\,7.5\ \mathrm{mS}$$

The centres and radii for the noise factor contours are given by Equations (4.14) and (4.15)

$$(G_S, B_S) = \left(\frac{F_e - 1}{2 R_n} - G_\gamma, -B_\gamma \right)$$

$$R_{CF} = \sqrt{ \frac{(F_e - 1)^2}{4 R_n} - (F_e - 1) \frac{G_\gamma}{R_n} - \frac{G_n}{R_n} }$$

Some results are shown in Table 4.1. It is now possible to draw Figure 4.6. If only passive

F_e	$Y_S = G_S + j B_S$ [mS]	R_{CF} [mS]
3.0	$38.0 - j\,7.5$	35.3
2.5	$28.0 - j\,7.5$	24.2
2.0	$18.0 - j\,7.5$	11.3
1.8	$14.0 - j\,7.5$	0.0
0.4	$-14.0 - j\,7.5$	0.0
0.0	$-22.0 - j\,7.5$	17.0
-1.0	$-42.0 - j\,7.5$	39.6

Table 4.1: Centres and radii as function of F_e.

source admittances are possible only the right-hand side of the figure is of interest. From this figure the actual noise factors for different source admittances can be found. However, they can also be found directly from Equation (4.11). If $Y_S = 20 + j\,0$ mS then

$$
\begin{aligned}
F_e &= 1 + \frac{1}{G_S} \left(G_n + R_n |Y_S + Y_\gamma|^2 \right) \\
&= 1 + \frac{1}{20} \left(4.8 + 0.025 \,|20 + 2 + j\,7.5|^2 \right) \\
&= 1.92 \quad (2.82 \text{ dB})
\end{aligned}
$$

The designer has now to decide whether the amplifier should be noise matched, which gives him an extra 0.27 dB, or if a noise figure of 2.82 dB is sufficient. A noise match should be with as little loss as possible, as any losses degrade the noise factor. In this case a loss of a quarter of a dB ruins as much as is gained. With transmission lines an almost lossless match can be performed. If the input frequency is 600 MHz an 83.7 mm shorted 50 Ω stub across the 20 mS source admittance and a 30.3 mm 50 Ω linelength to the amplifier will perform the noise match ($\epsilon_r = 1$). It should be noted that only a narrow band match has been performed.

Example 4.2 An amplifier with the noise parameters $R_n = 20$ Ω, $G_n = 6.4$ mS, and $Y_\gamma = 2 + j\,14$ mS has a source impedance of 50 Ω. By adding a lossless admittance (a susceptance) in parallel with the input of the amplifier the noise factor of the amplifier can be improved. In order to determine by how many decibels, the noise factor without this parallel admittance is computed as

$$F_{e\,old} = 1 + \frac{1}{G_S}\left(G_n + R_n|Y_S + Y_\gamma|^2\right)$$

$$= 1 + \frac{1}{20}\left(6.4 + 0.020\left[(20+2)^2 + (0+14)^2\right]\right)$$

$$= 2.000$$

A lossless admittance in parallel with the input of the amplifier changes the source admittance from $Y_S = 20$ mS to $Y_S = G_S + j\,B_A$ where B_A is the value of the susceptance. If B_A is chosen such that $B_A + B_\gamma = 0$ then the noise factor is at the local minimum. This means that $B_A = -B_\gamma = -14$ mS and the new value of the noise factor is

$$F_{e\,new} = 1 + \frac{1}{20}\left(6.4 + 0.020\left[(20+2)^2 + (-14+14)^2\right]\right) = 1.804$$

From this it follows that the improvement in the noise figure is

$$\Delta F\,[\text{dB}] = 10\log\frac{2.000}{1.804} = 0.448\ \text{dB}$$

If instead of a lossless admittance a complete noise match is performed the $F_{e\,min} = 1.800$ could be obtained. It is, however, a much more complicated solution and the extra 0.010 dB, which theory gives, would almost certainly be lost in the real matching network's internal losses.

4.2   Noise power waves

As Y and Z parameters in many applications have been replaced by S parameters it is natural to look for a set of noise parameters which are also based on the power waves formulation and could work conveniently together with the S parameters.

In order to use the Rothe and Dahlke equivalent from Figure 4.1 – redrawn in Figure 4.7 – in the power wave representation, the T parameters are chosen. The representation is in accordance with Kurokawa [10]. Without noise it follows that

$$\begin{bmatrix} A_1' \\ B_1' \end{bmatrix} = \begin{bmatrix} T_{11} & T_{12} \\ T_{21} & T_{22} \end{bmatrix} \cdot \begin{bmatrix} B_2 \\ A_2 \end{bmatrix} \tag{4.26}$$

The signal power waves are defined as

$$A_1' = \frac{V_1' + Z_1 I_1'}{2\sqrt{|\text{Re}[Z_1]|}} \tag{4.27}$$

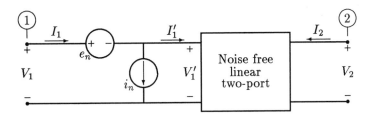

Figure 4.7: The Rothe and Dahlke equivalent.

$$B_1' = \frac{V_1' - Z_1^* I_1'}{2\sqrt{|\mathrm{Re}[Z_1]|}} \qquad (4.28)$$

$$A_2 = \frac{V_2 + Z_2 I_2}{2\sqrt{|\mathrm{Re}[Z_2]|}} \qquad (4.29)$$

$$B_2 = \frac{V_2 - Z_2^* I_2}{2\sqrt{|\mathrm{Re}[Z_2]|}} \qquad (4.30)$$

where Z_1 and Z_2 are the complex reference impedances at ports 1 and 2, which may be complex and also active. The only restrictions are that $\mathrm{Re}[Z_1] \neq 0$ and $\mathrm{Re}[Z_2] \neq 0$.

Including the noise sources in Figure 4.7 gives

$$V_1' = V_1 - e_n \qquad (4.31)$$

$$I_1' = I_1 - i_n \qquad (4.32)$$

Equations (4.26) – (4.32) can be expressed as

$$\frac{V_1 + Z_1 I_1}{2\sqrt{|\mathrm{Re}[Z_1]|}} = T_{11}\frac{V_2 - Z_2^* I_2}{2\sqrt{|\mathrm{Re}[Z_2]|}} + T_{12}\frac{V_2 + Z_2 I_2}{2\sqrt{|\mathrm{Re}[Z_2]|}} + \frac{e_n + Z_1 i_n}{2\sqrt{|\mathrm{Re}[Z_1]|}} \qquad (4.33)$$

$$\frac{V_1 - Z_1^* I_1}{2\sqrt{|\mathrm{Re}[Z_1]|}} = T_{21}\frac{V_2 - Z_2^* I_2}{2\sqrt{|\mathrm{Re}[Z_2]|}} + T_{22}\frac{V_2 + Z_2 I_2}{2\sqrt{|\mathrm{Re}[Z_2]|}} + \frac{e_n - Z_1^* i_n}{2\sqrt{|\mathrm{Re}[Z_1]|}} \qquad (4.34)$$

Here the left-hand side can be defined as power waves A_1 and B_1. The right-hand side consists of a noise free part and a part due to the noise sources. Thus, two noise power waves are defined as

$$a_n = \frac{e_n + Z_1 i_n}{2\sqrt{|\mathrm{Re}[Z_1]|}} \qquad (4.35)$$

$$b_n = \frac{e_n - Z_1^* i_n}{2\sqrt{|\mathrm{Re}[Z_1]|}} \qquad (4.36)$$

As seen from Equations (4.35) and (4.36), the noise power waves are introduced on the input side of the two-port. (It should be noted that Meys [6] uses the same b_n but has changed sign on a_n. Also Meys has the real part of Z_1 positive.)

Equation (4.26) can now be written as

$$\begin{bmatrix} A_1 \\ B_1 \end{bmatrix} = \begin{bmatrix} T_{11} & T_{12} \\ T_{21} & T_{22} \end{bmatrix} \cdot \begin{bmatrix} A_2 \\ B_2 \end{bmatrix} + \begin{bmatrix} a_n \\ b_n \end{bmatrix} \tag{4.37}$$

It is often practical to express Equation (4.37) by using S parameters. Isolating B_1 and B_2 it is found that

$$B_1 = \frac{T_{21}}{T_{11}} A_1 + \frac{T_{11} T_{22} - T_{12} T_{21}}{T_{11}} A_2 - \frac{T_{21}}{T_{11}} a_n + b_n \tag{4.38}$$

$$B_2 = \frac{1}{T_{11}} A_1 - \frac{T_{12}}{T_{11}} A_2 - \frac{1}{T_{11}} a_n \tag{4.39}$$

which in terms of S parameters can be expressed as

$$B_1 = S_{11} A_1 + S_{12} A_2 - S_{11} a_n + b_n \tag{4.40}$$

$$B_2 = S_{21} A_1 + S_{22} A_2 - S_{21} a_n \tag{4.41}$$

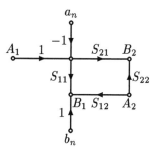

Figure 4.8: S parameter representation of a two-port with noise power waves.

Equations (4.40) and (4.41) are illustrated in the signal flow graph of Figure 4.8. It should be noted that the noise power wave a_n in Figure 4.8 should be treated as an internal noise source together with the S parameter representation.

4.2.1   The extended effective noise temperature

From Appendix B the relationship between the source power wave B_S, and the exchangeable power of the source $P_{e,S}$, which is defined in Equation (3.1), is found to be

$$\langle |B_S|^2 \rangle = p_1 P_{e,S} (1 - |\Gamma_S|^2) \tag{4.42}$$

where $p_i = \mathrm{Re}\,[Z_i]/|\mathrm{Re}\,[Z_i]|$ and i refer to the port number, $\langle |B_S|^2 \rangle$ indicates the ensemble average of $|B_S|^2$, and

$$\Gamma_S = |\Gamma_S|\, e^{j\,\varphi_S} = \frac{Z_S - Z_1}{Z_S + Z_1^*} \qquad (4.43)$$

As shown in Figure 3.1, Definition 3.4 for a linear two-port states that T_{ee} is the noise temperature of the source with a noise free equivalent of the two-port which gives the same exchangeable output noise power as the (noisy) two-port with a noise free source. Figure 4.9 shows the two cases. There the reference impedance of the source is $Z_{rS} = Z_1^*$ and also the reference impedance of the load, $Z_{rL} = Z_2^*$. This is so that an easy connection of source, load and two-port can be achieved as shown in Figure 4.9 (b).

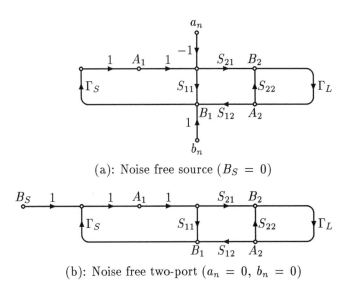

(a): Noise free source $(B_S = 0)$

(b): Noise free two-port $(a_n = 0,\; b_n = 0)$

Figure 4.9: Flow graph illustrating the T_{ee} definition.

Calculating the exchangeable output noise power from Figure 4.9 gives

$$N_{e,\,out}^a = p_2 \frac{|1 - S'_{22}\Gamma_L|^2}{1 - |S'_{22}|^2} \langle |B_2|^2 \rangle$$

$$= p_2 \frac{|1 - S'_{22}\Gamma_L|^2}{1 - |S'_{22}|^2}$$

$$\times \left\langle \left| \frac{S_{21}\Gamma_S b_n - S_{21} a_n}{1 - (S_{11}\Gamma_S + S_{12}S_{21}\Gamma_S\Gamma_L + S_{22}\Gamma_L) + S_{11}S_{22}\Gamma_S\Gamma_L} \right|^2 \right\rangle \quad (4.44)$$

where

$$S'_{22} = S_{22} + \frac{S_{12}S_{21}\Gamma_S}{1 - S_{11}\Gamma_S} \qquad (4.45)$$

is the output reflection coefficient of the two-port, and

$$N^b_{e,\,out} \;=\; p_2 \, \frac{|1 - S'_{22}\Gamma_L|^2}{1 - |S'_{22}|^2} \, \langle\, |B_2|^2 \rangle$$

$$= \; p_2 \, \frac{|1 - S'_{22}\Gamma_L|^2}{1 - |S'_{22}|^2}$$

$$\times \left\langle \left| \frac{S_{21} B_S}{1 - (S_{11}\Gamma_S + S_{12}S_{21}\Gamma_S\Gamma_L + S_{22}\Gamma_L) + S_{11}S_{22}\Gamma_S\Gamma_L} \right|^2 \right\rangle \, (4.46)$$

By definition $N^a_{e,\,out}$ must be identically equal to $N^b_{e,\,out}$, from which

$$B_S \;=\; \Gamma_S\, b_n \,-\, a_n$$

and thus

$$\langle\, |B_S|^2 \rangle \;=\; \langle (\Gamma_S\, b_n - a_n)(\Gamma_S b_n - a_n)^* \rangle$$

$$= \; \langle\, |a_n|^2 \rangle + |\Gamma_S|^2 \langle\, |b_n|^2 \rangle \,-\, 2\,\mathrm{Re}[\Gamma_S \langle a_n^* b_n \rangle] \qquad (4.47)$$

From Equation (4.42), this can be expressed as

$$p_1 N_{e,S}(1 - |\Gamma_S|^2) \;=\; \langle\, |a_n|^2 \rangle + |\Gamma_S|^2 \langle\, |b_n|^2 \rangle \,-\, 2\,\mathrm{Re}[\Gamma_S \langle a_n^* b_n \rangle] \qquad (4.48)$$

The exchangeable source noise power determines the extended source noise temperature T_{ee}. Thus

$$N_{e,S} \;=\; k\, T_{ee} \Delta f \qquad (4.49)$$

where Δf is the noise bandwidth and

$$T_{ee} \;=\; p_1 \, \frac{\langle\, |a_n|^2 \rangle + |\Gamma_S|^2 \langle\, |b_n|^2 \rangle \,-\, 2\,\mathrm{Re}[\Gamma_S \langle a_n^* b_n \rangle]}{k\, \Delta f\, (1 - |\Gamma_S|^2)} \qquad (4.50)$$

From Equations (4.50) and (4.43) the sign of T_{ee} as a function of the signs of the source impedance and the input reference impedance can be found as shown in Table 4.2.

In Equation (4.50), the noise waves are represented as $\langle\, |a_n|^2 \rangle$ and $\langle\, |b_n|^2 \rangle$, which can be regarded as the available power of the ingoing and outgoing power waves at port 1, and as $\langle a_n^* b_n \rangle$, which represents the cross-correlation between the ingoing and outgoing noise power waves. A set of noise parameters, $(T_\alpha, T_\beta, T_\gamma$ and $\varphi_\gamma)$, called T noise parameters, are defined as

$$\langle\, |a_n|^2 \rangle \;=\; k\, T_\alpha \Delta f \qquad (4.51)$$

$$\langle\, |b_n|^2 \rangle \;=\; k\, T_\beta \Delta f \qquad (4.52)$$

$$\langle a_n^* b_n \rangle \;=\; k\, T_\gamma \Delta f e^{j\,\varphi_\gamma} \qquad (4.53)$$

Re[Z_S]	Re[Z_1]	|Γ_S|	T_{ee}
> 0	> 0	< 1	> 0
> 0	< 0	> 1	> 0
< 0	> 0	> 1	< 0
< 0	< 0	< 1	< 0

Table 4.2: Magnitude of $|\Gamma_S|$ and the sign of T_{ee} as functions of the signs of Re[Z_S] and Re[Z_1].

From this definition, the extended effective input noise temperature is given by

$$T_{ee} \quad = \quad p_1 \frac{T_\alpha + |\Gamma_S|^2 T_\beta - 2|\Gamma_S| T_\gamma \cos(\varphi_S + \varphi_\gamma)}{1 - |\Gamma_S|^2} \qquad (4.54)$$

It should be noted that the definitions, Equations (4.51) – (4.53), are the same as Meys's [6,7]. They are, however, used in a different way, as shown in Equations (4.37) and (4.54) (for the real part of Z_1 positive). To distinguish them from each other, the indices have been chosen differently.

The extended noise factor as a function of the T noise parameters and the source reflection coefficient is found from inserting (4.54) into (3.12):

$$F_e \quad = \quad 1 + p_1 \frac{T_\alpha + |\Gamma_S|^2 T_\beta - 2|\Gamma_S| T_\gamma \cos(\varphi_S + \varphi_\gamma)}{T_0 (1 - |\Gamma_S|^2)} \qquad (4.55)$$

From the complex correlation coefficient $\langle a_n^* b_n \rangle / \sqrt{\langle |a_n|^2 \rangle \langle |b_n|^2 \rangle}$ it follows that

$$0 \quad \leq \quad T_\gamma \quad \leq \quad \sqrt{T_\alpha T_\beta} \qquad (4.56)$$

where $T_\gamma = 0$ corresponds to no correlation between the ingoing and outgoing noise power waves and $T_\gamma = \sqrt{T_\alpha T_\beta}$ corresponds to full correlation.

4.2.2 T noise parameters

Another set of noise parameters – the T noise parameters which are T_α, T_β, and $T_\gamma \exp[j \varphi_\gamma]$ – has been defined above and the extended effective noise temperature T_{ee} expressed by the T noise parameters in Equation (4.54). T_{ee} can also be written as circles for constant T_{ee} in the source reflection coefficient plane. Equation (4.54) can be written as

$$\left| \Gamma_S - \frac{T_\gamma e^{-j\varphi_\gamma}}{T_\beta + p_1 T_{ee}} \right|^2 \quad = \quad \frac{T_\gamma^2}{(T_\beta + p_1 T_{ee})^2} - \frac{T_\alpha - p_1 T_{ee}}{T_\beta + p_1 T_{ee}} \qquad (4.57)$$

which, for T_{ee} constant, is an equation for a circle with centre Γ_{CT} and radius R_T given by

$$\Gamma_{CT} \;=\; \frac{T_\gamma}{T_\beta + p_1\,T_{ee}}\, e^{-j\,\Phi_\gamma} \tag{4.58}$$

$$R_T \;=\; \sqrt{\frac{T_\gamma^2}{(T_\beta + p_1\,T_{ee})^2} - \frac{T_\alpha - p_1\,T_{ee}}{T_\beta + p_1\,T_{ee}}} \tag{4.59}$$

When T_{ee} is constant, then from Equation (3.12), $F_e = 1 + T_{ee}/T_0$ is constant. Therefore circles can also be drawn for F_e constant with centre and radius given by substituting $T_{ee} = (F_e - 1)\,T_0$ from Equation (3.13) into Equations (4.58) and (4.59).

If $R_T^2 = 0$ then two extrema for T_{ee} (or F_e) are determined. These and their corresponding source reflection coefficients are

$$T_{ee\,min} \;=\; \frac{1}{2}\left[p_1\,(T_\alpha - T_\beta) + \sqrt{(T_\alpha + T_\beta)^2 - 4\,T_\gamma^2}\,\right] \tag{4.60}$$

$$F_{e\,min} \;=\; 1 + \frac{1}{2\,T_0}\left[p_1\,(T_\alpha - T_\beta) + \sqrt{(T_\alpha + T_\beta)^2 - 4\,T_\gamma^2}\,\right] \tag{4.61}$$

for

$$\Gamma_{SOT} \;=\; \Gamma_{SOF}$$

$$=\; \frac{2\,T_\gamma\,e^{-j\,\varphi_\gamma}}{T_\alpha + T_\beta + p_1\,\sqrt{(T_\alpha + T_\beta)^2 - 4\,T_\gamma^2}} \tag{4.62}$$

$$=\; \frac{T_\alpha + T_\beta - p_1\,\sqrt{(T_\alpha + T_\beta)^2 - 4\,T_\gamma^2}}{2\,T_\gamma}\, e^{-j\,\varphi_\gamma} \tag{4.63}$$

and

$$T_{ee\,max} \;=\; \frac{1}{2}\left[p_1\,(T_\alpha - T_\beta) - \sqrt{(T_\alpha + T_\beta)^2 - 4\,T_\gamma^2}\,\right] \tag{4.64}$$

$$F_{e\,max} \;=\; 1 + \frac{1}{2\,T_0}\left[p_1\,(T_\alpha - T_\beta) - \sqrt{(T_\alpha + T_\beta)^2 - 4\,T_\gamma^2}\,\right] \tag{4.65}$$

for

$$\Gamma_{SOT'} \;=\; \Gamma_{SOF'}$$

$$=\; \frac{2\,T_\gamma\,e^{-j\,\varphi_\gamma}}{T_\alpha + T_\beta - p_1\,\sqrt{(T_\alpha + T_\beta)^2 - 4\,T_\gamma^2}} \tag{4.66}$$

$$=\; \frac{T_\alpha + T_\beta + p_1\,\sqrt{(T_\alpha + T_\beta)^2 - 4\,T_\gamma^2}}{2\,T_\gamma}\, e^{-j\,\varphi_\gamma} \tag{4.67}$$

It should be noted from Equations (4.60) and (4.64) and further from (4.61) and (4.65) that

$$T_{ee\,min} \;\geq\; T_{ee\,max} \qquad \text{and} \qquad F_{e\,min} \;\geq\; F_{e\,max} \tag{4.68}$$

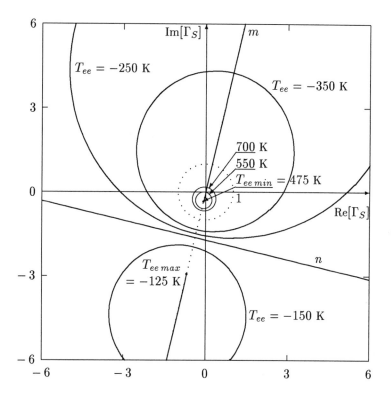

Figure 4.10: Contours for constant T_{ee} in the source reflection coefficient plane for $p_1 = +1$ with data from Example 4.3.

The circles divide into two parts by a singularity when the centre goes towards infinity for $T_\beta + p_1 T_{ee} = 0$. These two parts are separated by a line n, which is determined as

$$n: \quad \text{Im}[\Gamma_{CT}] \quad = \quad \frac{\sin \varphi_\gamma}{\cos \varphi_\gamma} \text{Re}[\Gamma_{CT}] - \frac{T_\alpha + T_\beta}{2 T_\gamma \sin \varphi_\gamma} \qquad (4.69)$$

Before drawing the circles for constant T_{ee} (or constant F_e) it is noteworthy to see from Equation (4.58) that the centres all are located on the line m given by

$$m: \quad \text{Im}[\Gamma_{CT}] \quad = \quad - \frac{\sin \varphi_\gamma}{\cos \varphi_\gamma} \text{Re}[\Gamma_{CT}] \qquad (4.70)$$

The lines m and n intersect at the point

$$\Gamma_{mn} \quad = \quad \frac{1}{2} (\Gamma_{SOT} + \Gamma_{SOT'}) \qquad (4.71)$$

It should also be noted that

$$|\Gamma_{mn}| \quad = \quad \frac{T_\alpha + T_\beta}{2\,T_\gamma} \quad \geq \quad 1 \tag{4.72}$$

T_{ee} as a function of the source reflection coefficient, $\Gamma_S = |\Gamma_S|\,\exp[j\,\varphi_S]$ is shown in Figure 4.10. When the magnitude of this reflection coefficient is kept constant, which corresponds to following a circle around the origin with radius $|\Gamma_S|$ in Figure 4.10, T_{ee} as a function of the phase is

$$T_{ee}(\varphi_S) \quad = \quad p_1 \frac{T_\alpha + |\Gamma_S|^2 T_\beta}{1 - |\Gamma_S|^2} - p_1 \frac{2\,|\Gamma_S|\,T_\gamma}{1 - |\Gamma_S|^2} \cos(\varphi_S + \varphi_\gamma) \tag{4.73}$$

This can be rewritten as

$$T_{ee}(\varphi_S) \quad = \quad T_m - T_a \cos(\varphi_S + \varphi_\gamma) \tag{4.74}$$

where

$$T_m \quad = \quad p_1 \frac{T_\alpha + |\Gamma_S|^2 T_\beta}{1 - |\Gamma_S|^2} \tag{4.75}$$

$$T_a \quad = \quad p_1 \frac{2\,|\Gamma_S|\,T_\gamma}{1 - |\Gamma_S|^2} \tag{4.76}$$

It is seen that $T_{ee}(\varphi_S)$ is sinusoidal with a mean value T_m given by T_α and T_β, and an amplitude T_a given by T_γ and thus by the magnitude of the cross-correlation coefficient of the noise waves. An example of $T_{ee}(\varphi_S)$ is shown in Figure 4.11. If $\varphi_S = -\varphi_\gamma$ then T_{ee} is minimum for a passive source immittance and maximum for an active source immittance.

When looking at T_{ee} as a function of $|\Gamma_S|$ with $\varphi_S = -\varphi_\gamma$, which corresponds to the line m in Figure 4.10, Equation (4.54) can be written as

$$T_{ee}(|\Gamma_S|) \quad = \quad p_1 \frac{T_\alpha + |\Gamma_S|^2\,T_\beta - 2\,|\Gamma_S|\,T_\gamma}{1 - |\Gamma_S|^2} \tag{4.77}$$

and illustrated in Figure 4.12.

From Figure 4.12 or Equation (4.54) it is seen that $T_{ee} \rightarrow \pm\infty$ for $|\Gamma_S| \rightarrow 1$, because $\langle|B_S|^2\rangle \rightarrow 0$ as seen from Equation (4.42). Also

$$\lim_{|\Gamma_S| \rightarrow 0} T_{ee}(|\Gamma_S|) \quad = \quad p_1 T_\alpha$$

$$\lim_{|\Gamma_S| \rightarrow \infty} T_{ee}(|\Gamma_S|) \quad = \quad -p_1 T_\beta$$

Naming the halfplane, which contains the unit circle, as halfplane 1 and the other as halfplane 2,

$$p_1 = +1: \quad \text{Halfplane 1:} \quad T_{ee\,min} \; \leq \; T_{ee} \; < \; \infty \quad \text{(In the unit circle)}$$

$$-\infty \; < \; T_{ee} \; \leq \; -T_\beta$$

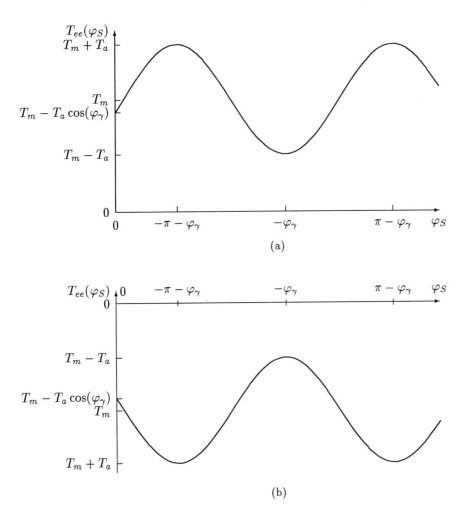

Figure 4.11: An example of the extended effective noise temperature $T_{ee}(\varphi_S)$ as a function of the phase of the source reflection coefficient.
(a) Passive source $(\text{Re}[Z_1] > 0, |\Gamma_S| < 1)$ or $(\text{Re}[Z_1] < 0, |\Gamma_S| > 1)$.
(b) Active source $(\text{Re}[Z_1] > 0, |\Gamma_S| > 1)$ or $(\text{Re}[Z_1] < 0, |\Gamma_S| < 1)$.

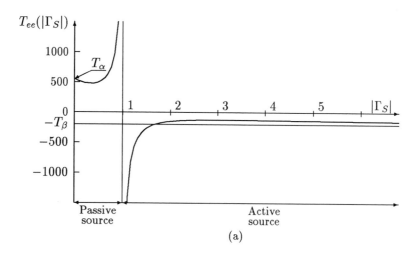

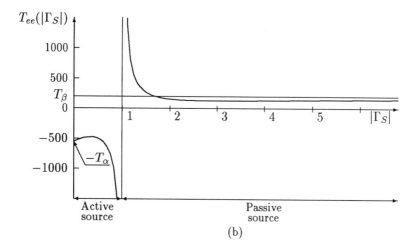

Figure 4.12: An example of the extended effective noise temperature $T_{ee}(|\Gamma_S|)$ as a function of the magnitude of the source reflection coefficient for $\varphi_S = -\varphi_\gamma$.
(a) $\mathrm{Re}(Z_1) > 0$.
(b) $\mathrm{Re}(Z_1) < 0$.

Halfplane 2: $\quad -T_\beta \leq T_{ee} \leq T_{ee\,max}$

$p_1 = -1:$ Halfplane 1: $\quad -\infty < T_{ee} \leq T_{ee\,max}$ (In the unit circle)

$$T_\beta \leq T_{ee} < \infty$$

Halfplane 2: $\quad T_{ee\,min} \leq T_{ee} \leq T_\beta$

Example 4.3 In Figure 4.10 the circles for constant T_{ee} in the source reflection plane are drawn. These are constructed for a transistor with the following T noise parameters:

$$T_\alpha \;=\; 550\text{ K} \qquad T_\beta \;=\; 200\text{ K} \qquad T_\gamma e^{j\,\varphi_\gamma} \;=\; 225\,e^{-j\,1.8}\text{ K}$$

First $T_{ee\,min}$ and $T_{ee\,max}$ are found from Equations (4.60) and (4.64) to be 475 K and -125 K respectively. Then for $T_{ee} \geq 475$ K and $T_{ee} \leq -125$ K the centres and radii of the circles of constant T_{ee} are found from Equations (4.58) and (4.59). Some of the results are shown in Table 4.3. The two lines n and m are found as n goes through the origin and Γ_{mn} given by

$$\Gamma_{mn} \;=\; \frac{1}{2}(\Gamma_{SOT} + \Gamma_{SOT'})$$

$$=\; \frac{1}{2}(0.333 + 3.000)\,e^{-j\,1.8} \;=\; 1.667\,e^{-j\,1.8}$$

and the line m is perpendicular to n in Γ_{mn}.

T_{ee}	Γ_{CT}	R_T
700	$0.250\,e^{-j\,1.8}$	0.479
550	$0.300\,e^{-j\,1.8}$	0.300
475	$0.333\,e^{-j\,1.8}$	0.000
-125	$3.000\,e^{-j\,1.8}$	0.000
-150	$4.500\,e^{-j\,1.8}$	2.500
-250	$-4.500\,e^{-j\,1.8}$	6.021
-350	$-1.500\,e^{-j\,1.8}$	2.872

Table 4.3: Centres and radii as functions of noise temperature.

4.3 Transformations between sets of noise parameters

There are almost an infinite number of ways in which a set of noise parameters can be defined. So far in this book five sets have been defined. One further set of noise parameters was introduced by Bächtold and Strutt [11] and it uses $F_{e\,min}$ and Γ_S. From Equation (4.24) one has

$$F_e = F_{e\,min} + \frac{R_n}{G_S} |Y_S - Y_{SOF}|^2 \qquad (4.78)$$

As

$$\Gamma_S = \frac{Z_S - Z_1}{Z_S + Z_1^*} = \frac{1 - Y_S Z_1}{1 + Y_S Z_1^*}$$

where Z_S is the source impedance and Z_1 is the reference impedance for the input port, one gets

$$Y_S = \frac{1 - \Gamma_S}{\Gamma_S Z_1^* + Z_1}$$

and

$$|Y_S - Y_{SOF}|^2 = \frac{4\,(\mathrm{Re}[Z_1])^2\,|\Gamma_{SOF} - \Gamma_S|^2}{|\Gamma_S\,Z_1^* + Z_1|^2\,|\Gamma_{SOF}\,Z_1^* + Z_1|^2} \qquad (4.79)$$

$$G_S = \frac{1}{2}(Y_S + Y_S^*) = \frac{1}{2}\left(\frac{1 - \Gamma_S}{\Gamma_S\,Z_1^* + Z_1} + \frac{1 - \Gamma_S^*}{\Gamma_S^*\,Z_1 + Z_1^*}\right)$$

From this it follows that

$$G_S = \frac{\mathrm{Re}[Z_1]\,(1 - |\Gamma_S|^2)}{|\Gamma_S\,Z_1^* + Z_1|^2} \qquad (4.80)$$

Equations (4.79), (4.80), and (4.78) lead to

$$F_e = F_{e\,min} + \frac{4\,\mathrm{Re}[Z_1]\,R_n}{|\Gamma_{SOF}\,Z_1^* + Z_1|^2} \cdot \frac{|\Gamma_S - \Gamma_{SOF}|^2}{1 - |\Gamma_S|^2} \qquad (4.81)$$

Here F_e is expressed by $F_{e\,min}$, Γ_{SOF}, and R_n. It is, however, convenient to introduce a new noise parameter, Q_{nc} as

$$Q_{nc} = \frac{4\,\mathrm{Re}[Z_1]\,R_n}{|\Gamma_{SOF}\,Z_1^* + Z_1|^2} \qquad (4.82)$$

and F_e can be written as

$$F_e = F_{e\,min} + Q_{nc}\,\frac{|\Gamma_S - \Gamma_{SOF}|^2}{1 - |\Gamma_S|^2} \qquad (4.83)$$

4.3.1 Transformation formulae

From G_n, R_n, and Y_γ

To $F_{e\,min}$, R_n, and Y_{SOF}:

$$F_{e\,min} = 1 + 2\left(R_n G_\gamma + \sqrt{R_n G_n + (R_n G_\gamma)^2}\right) \qquad (4.84)$$

$$R_n = R_n \qquad (4.85)$$

$$Y_{SOF} = \sqrt{G_n/R_n + G_\gamma^2} - j\,B_\gamma \qquad (4.86)$$

To r_n, g_n, and Z_γ:

$$r_n = \frac{G_n}{G_n/R_n + |Y_\gamma|^2} \tag{4.87}$$

$$g_n = G_n + R_n|Y_\gamma|^2 \tag{4.88}$$

$$Z_\gamma = \frac{Y_\gamma^*}{G_n/R_n + |Y_\gamma|^2} \tag{4.89}$$

To $F_{e\,min}$, g_n, and Z_{SOF}:

$$F_{e\,min} = 1 + 2\left(R_n G_\gamma + \sqrt{R_n G_n + (R_n G_\gamma)^2}\right) \tag{4.90}$$

$$g_n = G_n + R_n|Y_\gamma|^2 \tag{4.91}$$

$$Z_{SOF} = \frac{\sqrt{G_n/R_n + G_\gamma^2} + j\,B_\gamma}{G_n/R_n + |Y_\gamma|^2} \tag{4.92}$$

To $F_{e\,min}$, Q_{nc}, and Γ_{SOF}:[1]

$$F_{e\,min} = 1 + 2\left(R_n G_\gamma + \sqrt{R_n G_n + (R_n G_\gamma)^2}\right) \tag{4.93}$$

$$Q_{nc} = \frac{4\,\mathrm{Re}[Z_1]\,R_n}{\left|\frac{1-\left(\sqrt{G_n/R_n+G_\gamma^2}-j\,B_\gamma\right)Z_1}{1+\left(\sqrt{G_n/R_n+G_\gamma^2}-j\,B_\gamma\right)Z_1^*}\,Z_1^* + Z_1\right|^2} \tag{4.94}$$

$$\Gamma_{SOF} = \frac{1 - \left(\sqrt{G_n/R_n + G_\gamma^2} - j\,B_\gamma\right)Z_1}{1 + \left(\sqrt{G_n/R_n + G_\gamma^2} - j\,B_\gamma\right)Z_1^*} \tag{4.95}$$

To T_α, T_β, and $T_\gamma\,e^{j\,\varphi_\gamma}$:

$$T_\alpha = \frac{T_0}{|R_1|}\left(|Z_1|^2 G_n + R_n\left[1 + |Z_1|^2|Y_\gamma|^2 + 2\,(R_1 G_\gamma - X_1 B_\gamma)\right]\right) \tag{4.96}$$

$$T_\beta = \frac{T_0}{|R_1|}\left(|Z_1|^2 G_n + R_n\left[1 + |Z_1|^2|Y_\gamma|^2 - 2\,(R_1 G_\gamma + X_1 B_\gamma)\right]\right) \tag{4.97}$$

$$T_\gamma\,e^{j\,\varphi_\gamma} = \frac{T_0}{|R_1|}\left(\left[X_1^2 - R_1^2\right]G_n + R_n\left[1 - |Z_1|^2|Y_\gamma|^2 + 2\,X_1(X_1|Y_\gamma|^2 - B_\gamma)\right]\right)$$
$$+ j\,2\,p_1\,T_0\left[X_1\,G_n + R_n\,(X_1\,|Y_\gamma|^2 - B_\gamma)\right] \tag{4.98}$$

---

[1] $Z_1 = R_1 + j\,X_1$ is the complex reference impedance at the input port.

From $F_{e\,min}$, R_n, and Y_{SOF}

To G_n, R_n, and Y_γ:

$$G_n = G_{SOF}(F_{e\,min} - 1) - \frac{(F_{e\,min} - 1)^2}{4\,R_n} \tag{4.99}$$

$$R_n = R_n \tag{4.100}$$

$$Y_\gamma = \frac{F_{e\,min} - 1}{2\,R_n} - G_{SOF} - j\,B_{SOF} \tag{4.101}$$

To r_n, g_n, and Z_γ:

$$r_n = \frac{(4\,R_n G_{SOF} - F_{e\,min} + 1)(F_{e\,min} - 1)}{4\,R_n\,|Y_{SOF}|^2} \tag{4.102}$$

$$g_n = R_n\,|Y_{SOF}|^2 \tag{4.103}$$

$$Z_\gamma = \frac{F_{e\,min} - 1 - 2\,R_n G_{SOF} + j\,2\,R_n B_{SOF}}{2\,R_n\,|T_{SOF}|^2} \tag{4.104}$$

To $F_{e\,min}$, g_n, and Z_{SOF}:

$$F_{e\,min} = F_{e\,min} \tag{4.105}$$

$$g_n = R_n\,|Y_{SOF}|^2 \tag{4.106}$$

$$Z_{SOF} = \frac{Y_{SOF}^*}{|Y_{SOF}|^2} \tag{4.107}$$

To $F_{e\,min}$, Q_{nc}, and Γ_{SOF}:

$$F_{e\,min} = F_{e\,min} \tag{4.108}$$

$$Q_{nc} = \frac{4\,\mathrm{Re}[Z_1]\,R_n}{\left|\frac{1 - Y_{SOF}Z_1}{1 + Y_{SOF}Z_1^*}\,Z_1^* + Z_1\right|^2} \tag{4.109}$$

$$\Gamma_{SOF} = \frac{1 - Y_{SOF}Z_1}{1 + Y_{SOF}Z_1^*} \tag{4.110}$$

To T_α, T_β, and $T_\gamma\,e^{j\,\varphi_\gamma}$:

$$T_\alpha = \frac{T_0}{|R_1|}\Big(R_n(1 + |Z_1|^2|Y_{SOF}|^2) + R_1(F_{e\,min} - 1)$$

$$- 2\,R_n(R_1 G_{SOF} - X_1 B_{SOF})\Big) \tag{4.111}$$

$$T_\beta = \frac{T_0}{|R_1|}\Big(R_n(1 + |Z_1|^2|Y_{SOF}|^2) - R_1(F_{e\,min} - 1)$$

$$+ 2\,R_n(R_1 G_{SOF} + X_1 B_{SOF})\Big) \tag{4.112}$$

$$T_\gamma\,e^{j\,\varphi_\gamma} \;=\; \frac{T_0}{|R_1|}\left(R_n\left[1 - |Y_{SOF}|^2(R_1^2 - X_1^2)\right] + 2\,R_nX_1B_{SOF}\right)$$
$$+\,j\,2\,p_1T_0R_n(B_{SOF} + X_1|Y_{SOF}|^2) \tag{4.113}$$

From r_n, g_n, and Z_γ

To G_n, R_n, and Y_γ:

$$G_n \;=\; \frac{r_n}{r_n/g_n + |Z_\gamma|^2} \tag{4.114}$$

$$R_n \;=\; r_n + g_n|Z_\gamma|^2 \tag{4.115}$$

$$Y_\gamma \;=\; \frac{Z_\gamma^*}{r_n/g_n + |Z_\gamma|^2} \tag{4.116}$$

To $F_{e\,min}$, R_n, and Y_{SOF}:

$$F_{e\,min} \;=\; 1 + 2\left(g_nR_\gamma + \sqrt{g_nr/n + (g_nR_\gamma)^2}\right) \tag{4.117}$$

$$R_n \;=\; r_n + g_n|Z_\gamma|^2 \tag{4.118}$$

$$Y_{SOF} \;=\; \frac{\sqrt{r_n/g_n + R_\gamma^2} + j\,X_\gamma}{r_n/g_n + |Z_\gamma|^2} \tag{4.119}$$

To $F_{e\,min}$, g_n, and Z_{SOF}:

$$F_{e\,min} \;=\; 1 + 2\left(g_nR_\gamma + \sqrt{g_nr/n + (g_nR_\gamma)^2}\right) \tag{4.120}$$

$$g_n \;=\; g_n \tag{4.121}$$

$$Z_{SOF} \;=\; \sqrt{r_n/g_n + R_\gamma^2} - j\,X_\gamma \tag{4.122}$$

To $F_{e\,min}$, Q_{nc}, and Γ_{SOF}:

$$F_{e\,min} \;=\; 1 + 2\left(g_nR_\gamma + \sqrt{g_nr_n + (g_nR_\gamma)^2}\right) \tag{4.123}$$

$$Q_{nc} \;=\; \frac{4\,\mathrm{Re}[Z_1]\,(r_n + g_n\,|Z_\gamma|^2)}{\left|\dfrac{\sqrt{r_n/g_n + R_\gamma^2} - j\,X_\gamma - Z_1}{\sqrt{r_n/g_n + R_\gamma^2} - j\,X_\gamma + Z_1^*}\,Z_1^* + Z_1\right|^2} \tag{4.124}$$

$$\Gamma_{SOF} \;=\; \frac{\sqrt{r_n/g_n + R_\gamma^2} - j\,X_\gamma - Z_1}{\sqrt{r_n/g_n + R_\gamma^2} - j\,X_\gamma + Z_1^*} \tag{4.125}$$

To T_α, T_β, and $T_\gamma\, e^{j\,\varphi_\gamma}$:

$$T_\alpha = \frac{T_0}{|R_1|}\left(r_n + g_n\left[(R_\gamma + R_1)^2 + (X_\gamma + X_1)^2\right]\right) \quad (4.126)$$

$$T_\beta = \frac{T_0}{|R_1|}\left(r_n + g_n\left[(R_\gamma - R_1)^2 + (X_\gamma + X_1)^2\right]\right) \quad (4.127)$$

$$T_\gamma\, e^{j\,\varphi_\gamma} = \frac{T_0}{|R_1|}\left(r_n + g_n\left[(R_\gamma^2 - R_1^2) + (X_\gamma + X_1)^2\right]\right)$$
$$+ j\,2\,p_1 T_0\, g_n(X_\gamma + X_1) \quad (4.128)$$

From $F_{e\,min}$, g_n, and Z_{SOF}

To G_n, R_n, and Y_γ:

$$G_n = \frac{(4\,g_n R_{SOF} - F_{e\,min} + 1)(F_{e\,min} - 1)}{4\,g_n|Z_{SOF}|^2} \quad (4.129)$$

$$R_n = g_n\,|Z_{SOF}|^2 \quad (4.130)$$

$$Y_\gamma = \frac{F_{SOF} - 1 - 2\,g_n R_{SOF} + j\,2\,g_n X_{SOF}}{2\,g_n|Z_{SOF}|^2} \quad (4.131)$$

To $F_{e\,min}$, R_n, and Y_{SOF}:

$$F_{e\,min} = F_{e\,min} \quad (4.132)$$

$$R_n = g_n\,|Z_{SOF}|^2 \quad (4.133)$$

$$Y_{SOF} = \frac{Z_{SOF}*}{|Z_{SOF}|^2} \quad (4.134)$$

To r_n, g_n, and Z_γ:

$$r_n = R_{SOF}(F_{e\,min} - 1) - \frac{(F_{e\,min} - 1)^2}{4\,g_n} \quad (4.135)$$

$$g_n = g_n \quad (4.136)$$

$$Z_\gamma = \frac{F_{e\,min} - 1}{2\,g_n} - R_{SOF} - j\,X_{SOF} \quad (4.137)$$

To $F_{e\,min}$, Q_{nc}, and Γ_{SOF}:

$$F_{e\,min} = F_{e\,min} \quad (4.138)$$

$$Q_{nc} = \frac{4\,\mathrm{Re}[Z_1]\,g_n\,|Z_{SOF}|^2}{\left|\frac{Z_{SOF} - Z_1}{Z_{SOF} + Z_1^*}\,Z_1^* + Z_1\right|^2} \quad (4.139)$$

$$\Gamma_{SOF} = \frac{Z_{SOF} - Z_1}{Z_{SOF} + Z_1^*} \quad (4.140)$$

To T_α, T_β, and $T_\gamma\, e^{j\,\varphi_\gamma}$:

$$T_\alpha = \frac{T_0}{|R_1|}\left(g_n(|Z_{SOF}|^2 + |Z_1|^2) + R_1(F_{e\,min} - 1)\right.$$

$$\left. - 2\,g_n(R_{SOF}R_1 + X_{SOF}X_1)\right) \tag{4.141}$$

$$T_\beta = \frac{T_0}{|R_1|}\left(g_n(|Z_{SOF}|^2 + |Z_1|^2) - R_1(F_{e\,min} - 1)\right.$$

$$\left. + 2\,g_n(R_{SOF}R_1 - X_{SOF}X_1)\right) \tag{4.142}$$

$$T_\gamma\, e^{j\,\varphi_\gamma} = \frac{T_0\,g_n}{|R_1|}\left(|Z_{SOF}|^2 - (R_1^2 - X_1^2) - 2\,X_{SOF}X_1\right)$$

$$+ j\,2\,p_1\,T_0\,g_n\,(x_1 - X_{SOF}) \tag{4.143}$$

From $F_{e\,min}$, Q_{nc}, and Γ_{SOF}

To G_n, R_n, and Y_γ:

$$G_n = \mathrm{Re}[Z_1]\,(F_{e\,min} - 1)\frac{Q_{nc}(1 - |\Gamma_{SOF}|^2) - (F_{e\,min} - 1)^2}{Q_{nc}|\Gamma_{SOF}Z_1^* + Z_1|^2} \tag{4.144}$$

$$R_n = \frac{Q_{nc}|\Gamma_{sof}Z_1^* + Z_1|^2}{4\,\mathrm{Re}[Z_1]} \tag{4.145}$$

$$Y_\gamma = \frac{2\,\mathrm{Re}[Z_1]\,(F_{e\,min} - 1) + Q_{nc}(|\Gamma_{SOF}|^2 Z_1 - Z_1^*) + 2\,j\,Q_{nc}\,\mathrm{Im}[\Gamma_{SOF}^* Z_1]}{Q_{nc}\,|\Gamma_{SOF}Z_1^* + Z_1|^2}$$

$$\tag{4.146}$$

To $F_{e\,min}$, R_n, and Y_{SOF}:

$$F_{e\,min} = F_{e\,min} \tag{4.147}$$

$$R_n = \frac{Q_{nc}|\Gamma_{sof}Z_1^* + Z_1|^2}{4\,\mathrm{Re}[Z_1]} \tag{4.148}$$

$$Y_{SOF} = \frac{1 - \Gamma_{SOF}}{\Gamma_{SOF}Z_1^* + Z_1} \tag{4.149}$$

To r_n, g_n, and Z_γ:

$$r_n = \mathrm{Re}[Z_1]\,(F_{e\,min} - 1)\frac{Q_{nc}\,(1 - |\Gamma_{SOF}|^2) - (F_{e\,min} - 1)}{Q_{nc}\,|1 - \Gamma_{SOF}|^2} \tag{4.150}$$

$$g_n = \frac{Q_{nc}\,|1 - \Gamma_{SOF}|^2}{4\,\mathrm{Re}[Z_1]} \tag{4.151}$$

$$Z_\gamma = \frac{2\,\text{Re}[Z_1]\,(F_{e\,min} - 1) + Q_{nc}\,(|\Gamma_{SOF}|^2\,Z_1^* - Z_1 - 2\,j\,\text{Im}[\Gamma_{SOF}^*\,Z_1])}{Q_{nc}\,|1 - \Gamma_{SOF}|^2} \tag{4.152}$$

To $F_{e\,min}$, g_n, and Z_{SOF}:

$$F_{e\,min} = F_{e\,min} \tag{4.153}$$

$$g_n = \frac{Q_{nc}\,|1 - \Gamma_{SOF}|^2}{4\,\text{Re}[Z_1]} \tag{4.154}$$

$$Z_{SOF} = \frac{\Gamma_{SOF}\,Z_1^* + Z_1}{1 - \Gamma_{SOF}} \tag{4.155}$$

To T_α, T_β, and $T_\gamma\,e^{j\,\varphi_\gamma}$:

$$T_\alpha = p_1 T_0 (F_{e\,min} - 1 + Q_{nc}|\Gamma_{SOF}|^2) \tag{4.156}$$

$$T_\beta = p_1 T_0 (1 - F_{e\,min} + Q_{nc}) \tag{4.157}$$

$$T_\gamma\,e^{j\,\varphi_\gamma} = p_1\,T_0\,Q_{nc}\,|\Gamma_{SOF}|\,e^{-j\,\varphi_{SOF}} \tag{4.158}$$

From T_α, T_β, and $T_\gamma\,e^{j\,\varphi_\gamma}$

To G_n, R_n, and Y_γ:

$$G_n = \frac{|R_1|}{T_0}$$

$$\times \frac{T_\alpha T_\beta - T_\gamma^2}{|Z_1|^2(T_\alpha + T_\beta + 2\,T_\gamma\cos\varphi_\gamma) - 4\,X_1\,T_\gamma(X_1\cos\varphi_\gamma + R_1\sin\varphi_\gamma)} \tag{4.159}$$

$$R_n = \frac{1}{4\,T_0\,|R_1|}$$

$$\times \Big(|Z_1|^2(T_\alpha + T_\beta + 2\,T_\gamma\cos\varphi_\gamma) - 4\,X_1\,T_\gamma(X_1\cos\varphi_\gamma + R_1\sin\varphi_\gamma)\Big) \tag{4.160}$$

$$Y_\gamma =$$

$$\frac{R_1(T_\alpha - T_\beta) + j\,[X_1(T_\alpha + T_\beta - 2\,T_\gamma\cos\varphi_\gamma) - 2\,R_1 T_\gamma\sin\varphi_\gamma]}{|Z_1|^2(T_\alpha + T_\beta + 2\,T_\gamma\cos\varphi_\gamma) - 4\,X_1 T_\gamma(X_1\cos\varphi_\gamma + R_1\sin\varphi_\gamma)} \tag{4.161}$$

To $F_{e\,min}$, R_n, and Y_{SOF}:

$$F_{e\,min} = 1 + \frac{1}{2\,T_0}\left(p_1(T_\alpha - T_\beta) + \sqrt{(T_\alpha + T_\beta)^2 - 4\,T_\gamma^2}\right) \tag{4.162}$$

$$R_n = \frac{1}{4\,T_0|R_1|}\Big(|Z_1|^2(T_\alpha + T_\beta + 2\,T_\gamma\cos\varphi_\gamma)$$

$$- 4\,X_1 T_\gamma(X_1\cos\varphi_\gamma + R_1\sin\varphi_\gamma)\Big) \tag{4.163}$$

$$Y_{SOF} = \frac{|R_1|\sqrt{(T_\alpha + T_\beta)^2 - 4T_\gamma^2}}{|Z_1|^2(T_\alpha + T_\beta + 2T_\gamma \cos\varphi_\gamma) - 4X_1 T_\gamma(X_1 \cos\varphi_\gamma + R_1 \sin\varphi_\gamma)}$$

$$+ j \frac{2R_1 T_\gamma \sin\varphi_\gamma - X_1(T_\alpha + T_\beta + 2T_\gamma \cos\varphi_\gamma)}{|Z_1|^2(T_\alpha + T_\beta + 2T_\gamma \cos\varphi_\gamma) - 4X_1 T_\gamma(X_1 \cos\varphi_\gamma + R_1 \sin\varphi_\gamma)} \quad (4.164)$$

To r_n, g_n, and Z_γ:

$$r_n = \frac{|R_1|(T_\alpha T_\beta - T_\gamma^2)}{T_0(T_\alpha + T_\beta - 2T_\gamma \cos\varphi_\gamma)} \quad (4.165)$$

$$g_n = \frac{T_\alpha + T_\beta - 2T_\gamma \cos\varphi_\gamma}{4T_0|R_1|} \quad (4.166)$$

$$Z_\gamma = \frac{R_1(T_\alpha - T_\beta)}{T_\alpha + T_\beta - 2T_\gamma \cos\varphi_\gamma}$$

$$j\left(\frac{2T_\gamma R_1 \sin\varphi_\gamma}{T_\alpha + T_\beta - 2T_\gamma \cos\varphi_\gamma} - X_1\right) \quad (4.167)$$

To $F_{e\,min}$, g_n, and Z_{SOF}:

$$F_{e\,min} = 1 + \frac{1}{2T_0}\left(p_1(T_\alpha - T_\beta) + \sqrt{(T_\alpha + T_\beta)^2 - 4T_\gamma^2}\right) \quad (4.168)$$

$$g_n = \frac{T_\alpha + T_\beta - 2T_\gamma \cos\varphi_\gamma}{4T_0|R_1|} \quad (4.169)$$

$$Z_{SOF} = |R_1|\frac{\sqrt{(T_\alpha + T_\beta)^2 - 4T_\gamma^2}}{T_\alpha + T_\beta - 2T_\gamma \cos\varphi_\gamma}$$

$$+ j\left(X_1 - \frac{2T_\gamma R_1 \sin\varphi_\gamma}{T_\alpha + T_\beta - 2T_\gamma \cos\varphi_\gamma}\right) \quad (4.170)$$

To $F_{e\,min}$, Q_{nc}, and Γ_{SOF}:

$$F_{e\,min} = 1 + \frac{1}{2T_0}\left(p_1(T_\alpha - T_\beta) + \sqrt{(T_\alpha + T_\beta)^2 - 4T_\gamma^2}\right) \quad (4.171)$$

$$Q_{nc} = \frac{1}{2T_0}\left(p_1(T_\alpha + T_\beta) + \sqrt{(T_\alpha + T_\beta)^2 - 4T_\gamma^2}\right) \quad (4.172)$$

$$\Gamma_{SOF} = \frac{1}{2T_\gamma}\left((T_\alpha + T_\beta) - p_1\sqrt{(T_\alpha + T_\beta)^2 - 4T_\gamma^2}\right) \quad (4.173)$$

4.4   References

[1] Rothe, H. & Dahlke, W.: "Theory of noisy fourpoles", *Proc. IRE*, vol. 44, pp. 811 – 818, June 1956.

[2] Bauer, H. & Rothe, H.: "Der äquivalente Rauschvierpol als Wellenvierpol", *Archiv der elektrischen Übertragung*, vol. 10, pp. 241 – 252, 1956. *Proc. IRE*, vol. 44, pp. 811 – 818, June 1956.

[3] Penfield, P.: "Noise in negative-resistance amplifiers", *IRE Trans. on Circuit Theory*, vol. CT-7, pp. 166 – 170, June 1960.

[4] Penfield, P.: "Wave representation of amplifier noise", *IRE Trans. on Circuit Theory*, vol. CT-9, pp. 84 – 86, March 1962.

[5] Engen, G. F.: "A new method of characterizing amplifier noise performance", *IEEE Trans. on Instrumentation and Measurement*, vol. IM–19, pp. 344 – 349, Nov. 1970.

[6] Meys, R.: "A wave approach to the noise properties of linear microwave devices", *IEEE Trans. on Microwave Theory and Techniques*, vol. MTT-26, pp. 34 – 37, January 1978.

[7] Meys, R. & Milecan, M.: "Accurate experimental noise characterization of GaAs FET's at 18 and 20 GHz through the use of the noise wave model", *Proc. 11th European Microwave Conference*, Amsterdam, 1981, pp. 177 – 182.

[8] Dobrowolski, J. A.: "A CAD-oriented method for noise figure computation of two-ports with any internal topology", *IEEE Trans. on Microwave Theory and Techniques*, vol. MTT-37, pp. 15 – 20, January 1989.

[9] Engberg, J. & Larsen, T.: "Extended definitions for noise temperatures of linear noisy one- and two-ports", *IEE Proc. Part H*, vol. 138, pp. 86 – 90, February 1991.

[10] Kurokawa, K.: "Power waves and the scattering matrix", *IEEE Trans. on Microwave Theory and Techniques*, vol. MTT-13, pp. 194 – 202, March 1965.

[11] Bächtold, W. & Strutt, M. J. O.: "Darstellung der Rauschzahl und der verfübaren Verstärkung in der Ebene des komplexen Quellenreflexionsfactors", *A. E. Ü.* Band 21, pp. 631 – 633, 1967.

5

Noise measure and graphic representations

In this chapter the extended effective noise temperature and the extended noise factor for cascaded single response two-ports are derived. Then the extended noise measure of a single response two-port is introduced. The behaviour of the extended gain and especially the extended noise measure in both the source admittance plane and the source reflection coefficient plane are treated in the last two sections.

5.1   Two-ports in cascade

Often amplifiers consist of two-ports in cascade. This section computes the extended noise factor and the extended noise temperature of two-ports in cascade.

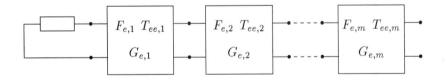

Figure 5.1: Noisy two-ports in cascade.

The definition of extended effective noise temperature (Equation (3.4)) uses the exchangeable output noise power density for the amplifier alone (without noise from the source). Single response amplifiers mean that each two-port has only one exchangeable gain at the appropriate frequency, thus

$$N'_e \quad = \quad T_{ee}G_e \quad = \quad T_{ee}G_{e,1}G_{e,2}\cdots G_{e,m} \quad \text{[W Hz}^{-1}\text{]} \quad (5.1)$$

The output noise power density from Figure 5.1 is

$$N_e' = k\left(T_{ee,1}G_{e,1}G_{e,2}\cdots G_{e,m} + T_{ee,2}G_{e,2}\cdots G_{e,m} + \cdots + T_{ee,m}G_{e,m}\right) \quad (5.2)$$

These two equations lead to

$$T_{ee} = T_{ee,1} + \frac{T_{ee,2}}{G_{e,1}} + \frac{T_{ee,3}}{G_{e,1}G_{e,2}} + \cdots + \frac{T_{ee,m}}{G_{e,1}G_{e,2}\cdots G_{e,m-1}} \quad \text{[K]} \quad (5.3)$$

$$= T_{ee,1} + \sum_{i=2}^{m} \frac{T_{ee,i}}{\prod_{j=1}^{i-1} G_{e,j}} \quad \text{[K]} \quad (5.4)$$

The extended noise factor is given by Equation (3.12) which for cascaded two-ports looks like $F_e = 1 + T_{ee}/T_0$. Together with Equations (5.3) and (5.4) this leads to

$$F_e = 1 + \frac{T_{ee,1}}{T_0} + \frac{T_{ee,2}}{G_{e,1}T_0} + \frac{T_{ee,3}}{G_{e,1}G_{e,2}T_0} + \cdots + \frac{T_{ee,m}}{G_{e,1}G_{e,2}\cdots G_{e,m-1}T_0}$$

$$= F_{e,1} + \frac{F_{e,2} - 1}{G_{e,1}} + \frac{F_{e,3} - 1}{G_{e,1}G_{e,2}} + \cdots + \frac{F_{e,m} - 1}{G_{e,1}G_{e,2}\cdots G_{e,m-1}} \quad (5.5)$$

$$= F_{e,1} + \sum_{i=2}^{m} \frac{F_{e,i} - 1}{\prod_{j=1}^{i-1} G_{e,j}} \quad (5.6)$$

In order to see how the signs behave the combination of two stages is examined. From Table 4.2 it is seen that T_{ee} has the same sign as the source resistance R_S. The formula for T_{ee} of two stages in cascade is (Equation 5.3):

$$T_{ee} = T_{ee,1} + \frac{T_{ee,2}}{G_{e,1}}$$

This gives the signs as shown in Table 5.1.

$T_{ee,1}$	$G_{e,1}$	$T_{ee,2}$	T_{ee}
> 0	> 0	> 0	> 0
> 0	< 0	< 0	> 0
< 0	> 0	< 0	< 0
< 0	< 0	> 0	< 0

Table 5.1: Signs for T_{ee} for two two-ports in cascade.

It is seen from Table 5.1 that the sign for T_{ee} for a two-stage amplifier is the same as the sign of the source immittance. Taking these two two-ports as a single two-port and then adding another two-port, the combination of three two-ports also follows the sign of the source immittance. This procedure can be repeated as often as necessary, so

$$\text{for} \quad R_S > 0 \quad \text{then} \quad T_{ee} > 0 \quad \text{and} \quad F_e - 1 > 0$$

$$\text{and for} \quad R_S < 0 \quad \text{then} \quad T_{ee} < 0 \quad \text{and} \quad F_e - 1 < 0$$

Example 5.1 An antenna amplifier consists of two identical stages each with the following data:

$$
\begin{array}{llll}
Y_{11} & = & 10 + j\,2.1 \text{ mS} & \quad Y_{12} & = & 0.50 - j\,0.86 \text{ mS} \\
Y_{21} & = & 19 - j\,30 \text{ mS} & \quad Y_{22} & = & 1.0 + j\,3.0\text{mS} \\
R_n & = & 25\;\Omega & \quad G_n & = & 4.8 \text{ mS} \\
Y_\gamma & = & 2.0 + j\,7.5 \text{ mS} & & &
\end{array}
$$

The source admittance is the antenna admittance which is $Y_S = 20$ mS.

The noise factor of the antenna amplifier is computed in the following way.

$$
F_{e,1} = 1 + \frac{1}{G_S}\left(G_n + R_n\,|Y_S + Y_\gamma|^2\right)
$$

$$
= 1 + \frac{1}{20}\left(4.8 + 0.025\,|20 + 2 + j\,7.5|^2\right) = 1.92
$$

$$
G_{e,1} = \frac{|Y_{21}|^2\,G_S}{\text{Re}\,[\{(Y_{11} + Y_S)Y_{22} - Y_{12}Y_{21}\}(Y_{11} + Y_S)^*]} =
$$

$$
\frac{(19^2 + 30^2)\,20}{\text{Re}\,[\{(10 + j\,2.1 + 20)(1.0 + j\,3.0) - (0.50 - j\,0.86)(19 - j\,30)\}(10 + j\,2.1 + 20)^*]}
$$

$$
= 17.3
$$

$$
Y_{out,1} = Y_{22} - \frac{Y_{12}\,Y_{21}}{Y_{11} + Y_S}
$$

$$
= 1.0 + j\,3.0 - \frac{(0.50 - j\,0.86)(19 - j\,30)}{10 + j\,2.1 + 20}
$$

$$
= 1.61 + j\,4.00 \text{ mS}
$$

$$
F_{e,2} = 1 + \frac{1}{G_{out,1}}\left(G_n + R_n\,|Y_{out,1} + Y_\gamma|^2\right)
$$

$$
= 1 + \frac{1}{1.61}\left(4.8 + 0.025\,|1.61 + j\,4.00 + 2.0 + j\,7.5|^2\right) = 6.23
$$

$$
F_e = F_{e,1} + \frac{F_{e,2} - 1}{G_{e,1}} = 1.92 + \frac{6.23 - 1}{17.3} = 2.22
$$

It is seen that the second stage has a rather high noise factor, but because of the also rather high gain of the first stage it does not spoil the combined noise factor very much. If a lossless matching network was added between the two amplifier stages in order to transform the output admittance of the first stage to the optimum source admittance for minimum noise factor for the second stage the combined noise factor can be determined. Instead of using $F_{e,2}$ above the $F_{e,2\,min}$ is used:

$$
\begin{aligned}
F_{e,2\,min} &= 1 + 2\left(R_n G_\gamma + \sqrt{R_n G_n + (R_n G_\gamma)^2}\right) \\
&= 1 + 2\left(0.025 \times 2.0 + \sqrt{0.025 \times 4.8 + (0.025 \times 2.0)^2}\right) = 1.8
\end{aligned}
$$

The second stage source admittance is computed by Equation (4.17) to $14 - j\,7.5$ mS. As the exchangeable power gain is unchanged

$$
F_e = 1.92 + \frac{1.8 - 1}{17.3} = 1.96
$$

Example 5.2 An antenna with impedance 50 Ω and noise temperature $T_a = 2900$ K is connected to an amplifier via a 50 Ω cable. This cable has a loss of 1.76 dB and its temperature can be taken to be the standard temperature, T_0 (17 °C). The noise factor of the amplifier is 1.8.

The 50 Ω cable has a loss of 1.76 dB and as its temperature is T_0 [K] it is seen from Example 3.2 that its noise factor is $1.50 \sim 1.76$ dB and the gain is 1/1.50. For the combination of cable and amplifier, the result is

$$
F_e = F_{e,cable} + \frac{F_{e,amp} - 1}{G_{e,cable}} = 1.50 + \frac{1.8 - 1}{\frac{1}{1.50}} = 2.7
$$

$$
T_{ee} = (F_e - 1)T_0 = (2.7 - 1)\,290 = 493 \text{ K}
$$

From this the operating noise temperature for the antenna – cable – amplifier combination referred to the antenna terminal is

$$
T_{e\,op} = T_a + T_{ee} = 2900 + 493 = 3393 \sim 3400 \text{ K}
$$

It should be noted that a rise in noise factor from 1.5 to 2.7 – which often should be avoided – is of no importance, when another noise source – here the antenna noise – is dominating.

5.2   The noise measure

Take two two-ports with extended noise factors $F_{e,1}$ and $F_{e,2}$ and exchangeable power gains $G_{e,1}$ and $G_{e,2}$ and find out which cascade combination is best – the one with two-port one first or the one with two-port two first. Let the result be that

$F_{e12} < F_{e21}$. From Equation (5.5) it follows, for a passive source and $G_{e,1} < 0$ or $G_{e,1} > 1$ and $G_{e,2} < 0$ or $G_{e,2} > 1$, that

$$F_{e,1} + \frac{F_{e,2} - 1}{G_{e,1}} < F_{e,2} + \frac{F_{e,1} - 1}{G_{e,2}}$$

$$(F_{e,1} - 1)\left(1 - \frac{1}{G_{e,2}}\right) < (F_{e,2} - 1)\left(1 - \frac{1}{G_{e,1}}\right)$$

$$\frac{F_{e,1} - 1}{1 - \frac{1}{G_{e,1}}} < \frac{F_{e,2} - 1}{1 - \frac{1}{G_{e,2}}}$$

It is seen that for two two-ports with a positive source immittance and an amplification which is either negative or greater than one, the choice should be the two-port with the smaller extended noise measure, M_e, first, where the extended noise measure is defined by

$$M_e \doteq \frac{F_e - 1}{1 - \frac{1}{G_e}} \tag{5.7}$$

If the source is chosen to be negative the same procedure shows that the negative noise measure closest to zero is the best. Haus and Adler [1] have extended the definition to the general case where F_e and G_e may differ from the conventional F and G_a and denoted this extended definition of noise measure by M_e.

In the discussion of the noise performance of amplifiers, it turns out to be important to bear in mind the algebraic sign that M_e assumes under various physical conditions. These are summarized in Table 5.2.

R_S	R_o	G_e	F_e	$|G_e|$	M_e
> 0	> 0	> 0	> 1	> 1	> 0
> 0	> 0	> 0	> 1	< 1	< 0
> 0	< 0	< 0	> 1	≠ 1	> 0
< 0	> 0	< 0	< 1	≠ 1	< 0
< 0	< 0	> 0	< 1	> 1	< 0
< 0	< 0	> 0	< 1	< 1	> 0

Table 5.2: Signs for M_e under various conditions.

It should be noted that, for a passive source, M_e is positive for $G_e > 1$, which corresponds to a normal amplifier, and for $G_e < 0$, which holds for an amplifier with negative real part of the output immittance. Also M_e is negative when $0 < G_e < 1$ which corresponds to an attenuator.

When m two-ports are going to be used for an amplifier the order of the two-ports for minimum noise is

$$0 < M_{e,1} \leq M_{e,2} \leq M_{e,3} \leq \cdots \leq M_{e,m} \tag{5.8}$$

In contrast, when a postamplifier is specified and a single preamplifier is required to precede the postamplifier, then one preamplifier must be chosen from several possibilities [2]. Here the minimum noise measure does not give enough information for the choice. This is demonstrated by writing the expression for the noise factor of the cascaded arrangement as follows:

$$F_{e,12} = F_{e,1} + \frac{F_{e,2} - 1}{G_{e,1}}$$

$$= 1 + \left(\frac{F_{e,1} - 1}{F_{e,2} - 1} + \frac{1}{G_{e,1}} \right) (F_{e,2} - 1)$$

$$= 1 + \left[1 - \left(1 - \frac{1}{G_{e,1}} - \frac{F_{e,1} - 1}{F_{e,2} - 1} \right) \right] (F_{e,2} - 1)$$

$$F_{e,12} = 1 + \left[1 - \left(1 - \frac{1}{G_{e,1}} - \frac{M_{e,1}}{F_{e,2} - 1} \right) \right] (F_{e,2} - 1)$$

Suppose that a second stage is given. The second stage noise factor can be kept constant at the minimum value of $F_{e,2}$, for example, by keeping the source immittance presented to stage two constant. Next, consider a candidate preamplifier with the real part of its output immittance positive. A lossless, passive two-port can be used to transform preamplifier output immittance to the value of stage two source immittance for minimum $F_{e,2}$. In subsequent comparisons of preamplifier candidates, let a lossless two-port transforming network, such as discussed above, be included with each preamplifier, with the net result that $F_{e,2}$ stays constant at its minimum value. It is seen that, given a particular second stage (the postamplifier), the first stage should be chosen such that the expression

$$\left(1 - \frac{1}{G_{e,1}} \right) \left(1 - \frac{M_{e,1}}{F_{e,2} - 1} \right) \tag{5.9}$$

is maximized.

$G_{e,1}$	$M_{e,1}$	$\left(1 - \frac{1}{G_{e,1}}\right)\left(1 - \frac{M_{e,1}}{F_{e,2}-1}\right)$	$F_{e,12}$
$0 < G_{e,1} < 1$	$M_{e,1} < 1$	< 0	$> F_{e,2}$
$G_{e,1} > 1$	$M_{e,1} > F_{e,2} - 1$	< 0	$> F_{e,2}$
$G_{e,1} > 1$	$0 < M_{e,1} > F_{e,2} - 1$	> 0	$< F_{e,2}$

Table 5.3: Three common cases of preamplifier performances.

Table 5.3 shows the noise factor of a cascaded amplifier for three cases of exchangeable gain of the first two-port. Only the third case gives a lower noise factor for the cascaded amplifier. The first case in Table 5.3 corresponds to an attenuator;

its noise measure is negative and the cascaded noise factor greater than $F_{e,2}$. In the second case, the preamplifier has a noise measure greater than the excess noise factor (the noise factor minus one) of the second stage, and again the cascaded noise factor is greater than $F_{e,2}$. The third case is the normal case, where the first stage noise measure is less than the excess noise factor of the second stage. The cascaded noise factor is lower than $F_{e,2}$. Here it is interesting to note that, from preamplifiers with the same noise measure, the one with the highest gain should be chosen.

The noise measure for m cascaded two-ports is derived from Equations (5.7), (5.5), and (5.6) as

$$M_e = M_{e,1}\frac{G_{e,1}-1}{G_e-1}G_{e,2}G_{e,3}\cdots G_{e,m}$$
$$+ M_{e,2}\frac{G_{e,2}-1}{G_e-1}G_{e,3}\cdots G_{e,m} + \cdots + M_{e,m}\frac{G_{e,m}-1}{G_e-1} \quad (5.10)$$

or

$$M_e = \sum_{i=1}^{m}\left(M_{e,i}\frac{G_{e,i}-1}{G_e-1}\prod_{j=i+1}^{m}G_{e,j}\right) \quad (5.11)$$

It is, however, normally much simpler to find the cascaded noise measure by computing the noise factor and the exchangeable gain of the cascaded two-ports and using the definition of the noise measure (Equation (5.7)).

When looking at this definition it is seen that

$$\lim_{G_e\to\infty} M_e = F_e - 1$$

This is also the case when cascading an infinite number of identical two-ports. Let these all have the noise factor $F_e > 1$ ($\sim$ a passive source) and gain $G_e > 1$, then

$$F_{e,tot} = F_e + \frac{F_e-1}{G_e} + \frac{F_e-1}{G_e^2} + \frac{F_e-1}{G_e^3} + \cdots$$
$$= 1 + (F_e-1) + (F_e-1)\left(\frac{1}{G_e}\right) + (F_e-1)\left(\frac{1}{G_e}\right)^2 + \cdots$$
$$= 1 + \lim_{n\to\infty}\left((F_e-1)\frac{1-\frac{1}{G_e^n}}{1-\frac{1}{G_e}}\right)$$
$$= 1 + \frac{F_e-1}{1-\frac{1}{G_e}} = 1 + M_e$$

It should be noted that like the noise factor, the exchangeable gain, and other noise characteristics, the noise measure is also a function of the source immittance.

Example 5.3 This example illustrates the importance of the noise measure and the effect of matching. Consider three amplifiers having the following noise parameters:

$$R_{n,1} \;=\; 25\,\Omega \qquad G_{n,1} \;=\; 4.8\text{ mS} \qquad Y_{\gamma,1} \;=\; 2.0 + j\,12\text{ mS}$$

$$R_{n,2} \;=\; 24\,\Omega \qquad G_{n,2} \;=\; 4.8\text{ mS} \qquad Y_{\gamma,2} \;=\; 5.0 + j\,9\text{ mS}$$

$$R_{n,3} \;=\; 6.25\,\Omega \qquad G_{n,3} \;=\; 9.6\text{ mS} \qquad Y_{\gamma,3} \;=\; 8.0 + j\,14\text{ mS}$$

The amplifiers are unilateralized which means that their output admittances are constant whatever the source admittance might be. Let all output admittances be 10 mS and let that be the value of the source admittance as well. The exchangeable power gains of the amplifiers are

$$G_{e,1} \;=\; 4.0 \qquad G_{e,2} \;=\; 10 \qquad G_{e,3} \;=\; 10$$

When all three amplifiers are used in cascade find the order giving the lowest overall noise factor. In order to find this the noise measure for each amplifier is computed.

$$F_e \;=\; 1 + \frac{1}{G_S}\left(G_n + R_n|Y_S + Y_\gamma|^2\right)$$

$$F_{e,1} \;=\; 1 + \frac{1}{10}\left(4.8 + 0.025|10 + 2 + j\,12|^2\right) \;=\; 2.200$$

$$F_{e,2} \;=\; 1 + \frac{1}{10}\left(4.8 + 0.024|10 + 5 + j\,9|^2\right) \;=\; 2.214$$

$$F_{e,3} \;=\; 1 + \frac{1}{10}\left(9.6 + 0.00625|10 + 8 + j\,14|^2\right) \;=\; 2.285$$

$$M_e \;=\; \frac{F_e - 1}{1 - \frac{1}{G_e}}$$

$$M_{e,1} \;=\; \frac{2.200 - 1}{1 - \frac{1}{4.0}} \;=\; 1.600$$

$$M_{e,2} \;=\; \frac{2.214 - 1}{1 - \frac{1}{10}} \;=\; 1.349$$

$$M_{e,3} \;=\; \frac{2.285 - 1}{1 - \frac{1}{10}} \;=\; 1.428$$

As $0 < M_{e,2} < M_{e,3} < M_{e,1}$ the best amplifier is the one which consists of amplifier stages with the numbers $2 - 3 - 1$.

If each amplifier in front of its input terminal has a network which transforms the 10 mS of the source or the output admittance of the former amplifier to the optimum admittance for minimum noise factor then the order of the amplifiers is changed. This is demonstrated in the following:

$$F_{e\,min} \;=\; 1 + 2\left(R_n G_\gamma + \sqrt{R_n G_n + (R_n G_\gamma)^2}\right)$$

$$F_{e,1\,min} \;=\; 1 + 2\left(0.025 \times 2.0 + \sqrt{0.025 \times 4.8 + (0.025 \times 2.0)^2}\right) \;=\; 1.80$$

$$F_{e,2\,min} = 1 + 2\left(0.024 \times 5.0 + \sqrt{0.024 \times 4.8 + (0.024 \times 5.0)^2}\right) = 1.96$$

$$F_{e,3\,min} = 1 + 2\left(0.00625 \times 8.0 + \sqrt{0.00625 \times 9.6 + (0.00625 \times 8.0)^2}\right)$$

$$= 1.60$$

$$M_{e\,min,1} = \frac{1.80 - 1}{1 - \frac{1}{4.0}} = 1.067$$

$$M_{e\,min,2} = \frac{1.96 - 1}{1 - \frac{1}{10}} = 1.067$$

$$M_{e\,min,3} = \frac{1.60 - 1}{1 - \frac{1}{10}} = 0.667$$

As $0 < M_{e\,min,3} < M_{e\,min,1} = M_{e\,min,2}$ then one of the two equal combinations 3 – 1 – 2 or 3 – 2 – 1 gives the best combined amplifier.

The two examined amplifier chains can be compared by computing their overall noise factors or noise temperatures. The results are:

$$F_{e,2\,3\,1} = F_{e,2} + \frac{F_{e,3} - 1}{G_{e,2}} + \frac{F_{e,1} - 1}{G_{e,2}G_{e,3}}$$

$$= 2.214 + \frac{2.285 - 1}{10} + \frac{2.200 - 1}{10 \times 10} = 2.355$$

$$T_{ee,2\,3\,1} = (F_{e,2\,3\,1} - 1)\,T_0 = (2.355 - 1)\,290 = 393\,K$$

$$F_{e,3\,1\,2} = F_{e,3\,2\,1} = 1.60 + \frac{1.96 - 1}{10} + \frac{1.80 - 1}{10 \times 10} = 1.704$$

$$T_{ee,3\,1\,2} = T_{ee,3\,2\,1} = (1.704 - 1)\,290 = 204\,K$$

It is interesting to note that the combination with matching networks generates about half as much noise power as the combination without matching networks.

5.3 Graphic representation in the admittance plane

In order to examine graphically the behaviour of the extended noise measure it is necessary to look at the involved parts, the extended noise factor and the exchangeable power gain. The extended noise factor is shown in Appendix C to be a hyperboloid of two sheets. The exchangeable power gain is examined before the extended noise measure is investigated. Fukui [3] was the first to be interested in this subject and the present work is an enlargement of Fukui's work.

5.3.1   Graphic representation of the exchangeable power gain

The exchangeable power gain for a two-port can be expressed as a function of the two-port's Y parameters and the source admittance as:

$$G_e = \frac{|Y_{21}|^2 G_S}{\text{Re}\left[(\Delta_Y + Y_{22}Y_S)(Y_{11} + Y_S)^*\right]} \tag{5.12}$$

where $\Delta_Y = Y_{11}Y_{22} - Y_{12}Y_{21}$ and for any index the admittance $Y = G + jB$. Since the extended noise measure equation contains the exchangeable power gain as $1/G_e$ the contours for $1/G_e$ as a function of Y_S are found by rewriting Equation (5.12) as follows:

$$\left[G_S - \left(\frac{|Y_{21}|^2}{2\,G_e G_{22}} + \frac{\text{Re}[Y_{12}Y_{21}]}{2\,G_{22}} - G_{11}\right)\right]^2$$

$$+ \left[B_S - \left(\frac{\text{Im}[Y_{12}Y_{21}]}{2\,G_{22}} - B_{11}\right)\right]^2 = R_G^2 \tag{5.13}$$

where

$$R_G^2 = \frac{|Y_{12}Y_{21}|^2}{2\,G_{22}^2} + \frac{|Y_{21}|^2}{G_e G_{22}}\left(\frac{|Y_{21}|^2}{4\,G_e G_{22}} + \frac{\text{Re}[Y_{12}Y_{21}]}{2\,G_{22}} - G_{11}\right) \tag{5.14}$$

and it is seen that when $R_G^2 > 0$, the contours for constant $1/G_e$ (and thus constant G_e) are circles in the Y_S-plane. Extrema exist for $1/G_e$ if $R_G^2 = 0$, or when

$$\frac{1}{G_e} = \frac{1}{|Y_{21}|^2}\left(2\,G_{11}G_{22} - \text{Re}\,[Y_{12}Y_{21}]\right.$$

$$\left.\pm \sqrt{4\,G_{11}^2 G_{22}^2 - 4\,G_{11}G_{22}\text{Re}[Y_{12}Y_{21}] - (\text{Im}[Y_{12}Y_{21}])^2}\right) \tag{5.15}$$

is real. The latter requires that

$$4\,G_{11}^2 G_{22}^2 - 4\,G_{11}G_{22}\text{Re}[Y_{12}Y_{21}] - (\text{Im}[Y_{12}Y_{21}])^2 \geq 0 \tag{5.16}$$

This inequality is equivalent to $k^2 \geq 1$, where $k = 1/C$ ($C \sim$ Linvill's factor) is a common stability factor

$$k = \frac{2\,G_{11}G_{22} - \text{Re}[Y_{12}Y_{21}]}{|Y_{12}Y_{21}|} \tag{5.17}$$

(Unconditional stability requires $k \geq 1$.) When $k^2 \geq 1$, extrema given by Equation (5.15) exist for

$$Y_{SOG} = \pm\frac{1}{2\,G_{22}}\sqrt{4\,G_{11}^2 G_{22}^2 - 4\,G_{11}G_{22}\text{Re}[Y_{12}Y_{21}] - (\text{Im}[Y_{12}Y_{21}])^2}$$

$$+ j\left(\frac{\text{Im}[Y_{12}Y_{21}]}{2\,G_{22}} - B_{11}\right) \tag{5.18}$$

Here the index SOG stands for **S**ource **O**ptimum with respect to the exchangeable **G**ain. It is easy to see that, in the expression for the extrema of $1/G_e$, the two extreme values have the same sign (which is the sign of k). The proof is:

$$\sqrt{4\,G_{11}^2 G_{22}^2 - 4\,G_{11}G_{22}\mathrm{Re}[Y_{12}Y_{21}] - (\mathrm{Im}[Y_{12}Y_{21}])^2}$$

$$< \quad |2\,G_{11}G_{22} - \mathrm{Re}[Y_{12}Y_{21}]|$$

or $\qquad\qquad -(\mathrm{Im}[Y_{12}Y_{21}])^2 \quad < \quad (\mathrm{Re}[Y_{12}Y_{21}])^2 \qquad$ q. e. d.

When $k^2 < 1$, $R_G^2 > 0$ for all values of $1/G_e$; so there are no extrema for $1/G_e$, which thus is unlimited. The circles for constant $1/G_e$ given by Equation (5.13) all cross each other at the two points:

$$Y_S \;=\; 0 + j\left(\frac{\mathrm{Im}[Y_{12}Y_{21}]}{2\,G_{22}} - B_{11} \pm \sqrt{\frac{|Y_{12}Y_{21}|^2}{4\,G_{22}^2} - \left(G_{11} - \frac{\mathrm{Re}[Y_{12}Y_{21}]}{2\,G_{22}}\right)^2} \right)$$

$$(5.19)$$

It should be noted that $1/G_e$ is not defined for $G_S = 0$ where the two points are situated. Figure 5.2 shows the contours for $1/G_e$ as a function of Y_S (a) with extrema, and (b) without extrema.

$1/G_e$ as a function of Y_S is symmetrical around

$$\left(\frac{1}{G_e}, G_S, B_S \right) \;=\; \left(\frac{2\,G_{11}G_{22} - \mathrm{Re}[Y_{12}Y_{21}]}{|Y_{21}|^2}, 0, \frac{\mathrm{Im}[Y_{12}Y_{21}]}{2\,G_{22}} - B_{11} \right) \quad (5.20)$$

Keeping $B_S = \mathrm{Im}[Y_{12}Y_{21}]/(2\,G_{22}) - B_{11} = B_{CG}$, which is the B_S value for all the centres of the contour circles, then $\frac{1}{G_e}(G_S)$ can be given in terms of the value of G_{CG} for the centre of a contour circle and the radius, R_G, of that circle can also be expressed in terms of G_{CG}. The centres satisfy the equation for a straight line:

$$\frac{1}{G_e} \;=\; \frac{2\,G_{22}}{|Y_{21}|^2}\,G_{CG} + \frac{2\,G_{11}G_{22} - \mathrm{Re}[Y_{12}Y_{21}]}{|Y_{21}|^2} \qquad (5.21)$$

and the radii are expressed as follows:

$$R_G \;=\; \sqrt{G_{CG}^2 + \frac{|Y_{12}Y_{21}|^2}{4\,G_{22}^2} - \left(G_{11} - \frac{\mathrm{Re}[Y_{12}Y_{21}]}{2\,G_{22}}\right)^2} \qquad (5.22)$$

Figure 5.3 shows three cases of $\frac{1}{G_e}(G_S)\Big|_{B_S = B_{CG}}$, which, together with Figure 5.2, illustrates the behaviour of $\frac{1}{G_e}(Y_S)$.

5.3.2 Graphic representation of the extended noise measure

The extended noise measure was defined in Equation (5.7). Inserting F_e from E-quation (4.11) and G_e from Equation (5.12) into Equation (5.7) leads to

$$M_e \;=\; \frac{|Y_{21}|^2\,(G_n + R_n|Y_S + Y_\gamma|^2)}{|Y_{21}|^2 G_S - \mathrm{Re}[(Y_{11} + Y_S)^*(\Delta_Y + Y_{22}Y_S)]} \qquad (5.23)$$

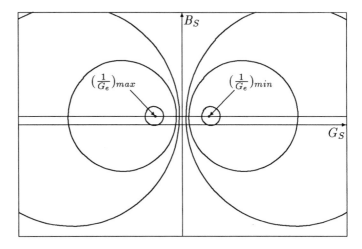

(a) $4G_{11}^2 G_{22}^2 - 4G_{11}G_{22}\text{Re}[Y_{12}Y_{21}] - (\text{Im}[Y_{12}Y_{21}])^2 > 0$

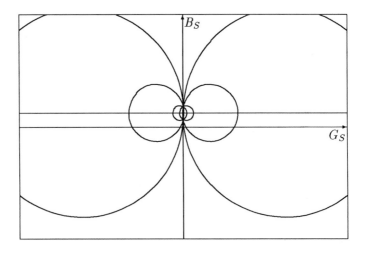

(b) $4G_{11}^2 G_{22}^2 - 4G_{11}G_{22}\text{Re}[Y_{12}Y_{21}] - (\text{Im}[Y_{12}Y_{21}])^2 < 0$

Figure 5.2: Contours for constant $1/G_e$ as a function of (a) with extrema, and (b) without extrema.

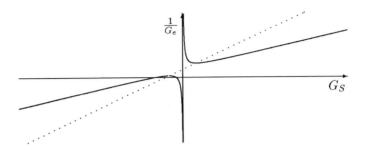

Condition (a) from Figure 5.2 and $2\,G_{11}G_{22} - \mathrm{Re}[Y_{12}Y_{21}] > 0$

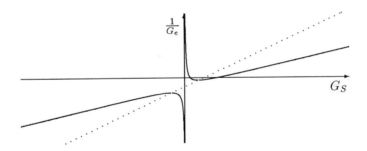

Condition (a) from Figure 5.2 and $2\,G_{11}G_{22} - \mathrm{Re}[Y_{12}Y_{21}] < 0$

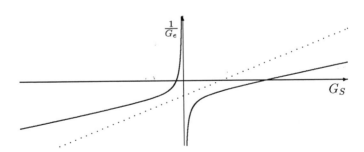

Condition (b) from Figure 5.2 and $2\,G_{11}G_{22} - \mathrm{Re}[Y_{12}Y_{21}] < 0$

Figure 5.3: $1/G_e$ as a function of G_S for $B_S = \mathrm{Im}[Y_{12}Y_{21}]/(2\,G_{22}) - B_{11}$ and $G_{22} > 0$.

Extrema of $M_e(Y_S)$ can be found by setting the partial derivatives with respect to G_S and B_S equal to zero. This yields two equations of fourth order in G_S and B_S with no explicit analytical solutions. As the expressions for both F_e and $1/G_e$ are represented by circles, the theory of linear transformations of analytic functions shows that Equation (5.23) must also represent circles and can be written as

$$(G_S - G_{CM})^2 + (B_S - B_{CM})^2 \;=\; R_M^2 \tag{5.24}$$

where

$$G_{CM} \;=\; \frac{-1}{|Y_{21}|^2 R_n + M_e G_{22}}$$

$$\times \left\{ |Y_{21}|^2 G_\gamma R_n - \frac{1}{2} M_e |Y_{21}|^2 \right.$$

$$\left. + M_e G_{11} G_{22} - \frac{1}{2} M_e \mathrm{Re}[Y_{12} Y_{21}] \right\} \tag{5.25}$$

$$B_{CM} \;=\; \frac{-1}{|Y_{21}|^2 R_n + M_e G_{22}}$$

$$\times \left\{ |Y_{21}|^2 B_\gamma R_n + M_e B_{11} G_{22} - \frac{1}{2} M_e \mathrm{Im}[Y_{12} Y_{21}] \right\} \tag{5.26}$$

$$R_M^2 \;=\; \frac{|Y_{21}|^2}{(|Y_{21}|^2 R_n + M_e G_{22})^2}$$

$$\times \left\{ M_e^2 \frac{1}{4} \left(|Y_{12}|^2 + |Y_{21}|^2 - 4 G_{11} G_{22} + 2\,\mathrm{Re}[Y_{12} Y_{21}] \right) \right.$$

$$+ M_e \left(R_n \mathrm{Re}[Y_{12} Y_{21}(Y_{11} - Y_\gamma)^*] - G_{22} R_n |Y_{11} - Y_\gamma|^2 \right.$$

$$\left. - G_{22} G_n - |Y_{21}|^2 G_\gamma R_n \right)$$

$$\left. - |Y_{21}|^2 G_n R_n \right\} \tag{5.27}$$

Equation (5.24) is valid when $R_M^2 \geq 0$. By examining Equations (5.24) – (5.27) and by setting the expression in curly brackets in Equation (5.27) equal to $A\,M_e^2 + B\,M_e + C$ where $C < 0$, the following conclusions can be drawn:

1. $(G_{CM}(M_e), B_{CM}(M_e))$ is on a straight line in the Y_S plane:

$$B_{CM} \;=\; \frac{1}{|Y_{21}|^2 + \mathrm{Re}[Y_{12} Y_{21}] + 2\,G_{22}(G_\gamma - G_{11})}$$

$$\times \left\{ (2\,G_{22}(B_\gamma - B_{11}) + \mathrm{Im}[Y_{12} Y_{21}])\,G_{CM} + G_\gamma \mathrm{Im}[Y_{12} Y_{21}] \right.$$

$$\left. + 2\,G_{22} \mathrm{Im}[Y_{11}^* Y_\gamma] - B_\gamma \left(|Y_{21}|^2 + \mathrm{Re}[Y_{12} Y_{21}] \right) \right\} \tag{5.28}$$

2. From Equation (5.25), or the expression:

$$M_e(G_{CM}) = \frac{-1}{G_{22}}|Y_{21}|^2 R_n + \frac{1}{G_{22}}|Y_{21}|^2 R_n$$

$$\times \frac{|Y_{21}|^2 - 2\,G_{11}G_{22} + \text{Re}[Y_{12}Y_{21}] + 2\,G_\gamma G_{22}}{|Y_{21}|^2 - 2\,G_{11}G_{22} + \text{Re}[Y_{12}Y_{21}] - 2\,G_{CM}G_{22}} \quad (5.29)$$

it is seen that $M_e(G_{CM})$ is monotonic with one pole.

3. Since $C < 0$, then for $A > 0$ or

$$|Y_{12}|^2 + |Y_{21}|^2 - 4\,G_{11}G_{22} + 2\,\text{Re}[Y_{12}Y_{21}] \quad > \quad 0 \quad (5.30)$$

extrema will exist for M_e. The signs for the extrema are different because $-4\,AC > 0$ and thus $|B| < \sqrt{B^2 - 4\,AC}$. Note that the condition of Equation (5.30) is the same as that for $R_G^2 > 0$ when $1/G_e = 1$.

4. When $A < 0$, $|B| > \sqrt{B^2 - 4\,AC}$ whenever $B^2 - 4\,AC > 0$, so that the signs for the extrema are the same.

5. Calling the extrema M_{e1} and M_{e2} where $M_{e1} > M_{e2}$, the range for M_e is

$$M_e \geq M_{e1} > 0 \text{ and } M_e \leq M_{e2} < 0 \text{ for } A > 0$$

$$M_{e2} \leq M_e \leq M_{e1} \text{ where } \text{sgn}[M_{e1}] = \text{sgn}[M_{e2}] \text{ for } A < 0$$

Figures 5.4 and 5.5 indicate the behaviour of the noise neasure. The extreme values are found from Equation (5.27) by setting $R_M^2 = 0$, and the corresponding optimum source admittance, Y_{SOM}, from Equations (5.25) and (5.26).

5.4 Graphic representation in the reflection plane

As in the former section the exchangeable gain and the extended noise measure can be illustrated as functions of the source reflection coefficient. As these quantities are all traced as circles in the source admittance plane it follows from the theory of linear transformations of analytic functions that they should also be generalized circles (a circle, a straight line, or a point) in the plane of the source reflection coefficient. This section examines the behaviour of G_e and M_e as functions of the source reflection coefficient.

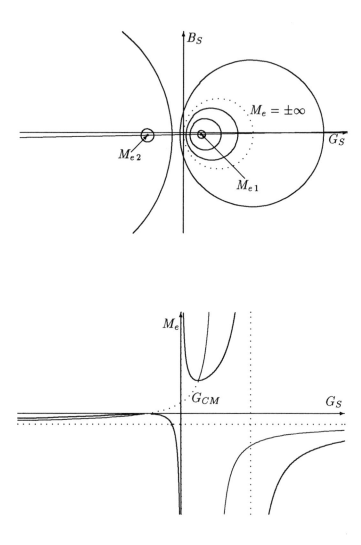

Figure 5.4: Contours for constant extended noise measure in the source admittanced plane for $A > 0$ and the corresponding function of extended noise measure versus source conductance when $B_S = B_{CM}$.

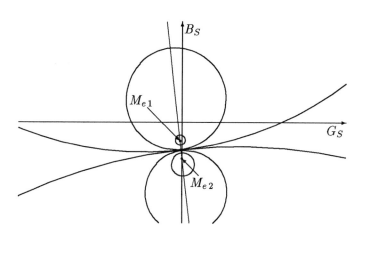

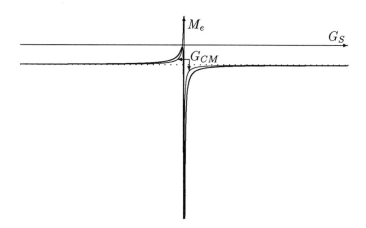

Figure 5.5: Contours for constant extended noise measure in the source admittanced plane for $A < 0$ and the corresponding function of extended noise measure versus source conductance when $B_S = B_{CM}$.

5.4.1   Graphic representation of the exchangeable power gain

The exchangeable power gain is defined as the ratio of the exchangeable power at the output to the exchangeable power from the source by Equation (3.2) where

$$P_{eo} = p_2 \frac{|1 - S'_{22}\Gamma_L|^2}{1 - |S'_{22}|^2} \langle |B_2|^2 \rangle \qquad (5.31)$$

$$P_{eS} = p_1 \frac{1}{1 - |\Gamma_S|^2} \langle |B_S|^2 \rangle \qquad (5.32)$$

Here again

$$p_i = \frac{\mathrm{Re}[Z_i]}{|\mathrm{Re}[Z_i]|}$$

where i refers to the port number;

$$S'_{22} = \frac{S_{12}S_{21}\Gamma_S}{1 - S_{11}\Gamma_S}$$

is the output reflection coefficient of the two-port, where Γ_S is the source reflection coefficient; and

$$\frac{\langle |B_2|^2 \rangle}{\langle |B_S|^2 \rangle} = \frac{|S_{21}|^2}{|1 - S_{11}\Gamma_S|^2 |1 - S'_{22}\Gamma_L|^2}$$

where Γ_L is the load reflection coefficient.

These equations give the following expression for the exchangeable power gain:

$$G_e = \frac{p_1 p_2 |S_{21}|^2 (1 - |\Gamma_S|^2)}{1 - |S_{22}|^2 + |\Gamma_S|^2(|S_{11}|^2 - |\Delta_S|^2) - 2\,\mathrm{Re}[\Gamma_S(S_{11} - S_{22}^*\Delta_S)]} \qquad (5.33)$$

As in the former section the contours for constant $1/G_e$ (and thus constant G_e) are found to be circles, as Equation (5.33) can be rewritten as:

$$\left| \Gamma_S - \frac{S_{11}^* - S_{22}\Delta_S^*}{|S_{11}|^2 - |\Delta_S|^2 + p_1 p_2 |S_{21}|^2 \frac{1}{G_e}} \right|^2 = R_G^2 \qquad (5.34)$$

where

$$R_G^2 =$$

$$\frac{|S_{11}^* - S_{22}\Delta_S^*|^2 + (|S_{11}|^2 - |\Delta_S|^2 + p_1 p_2 |S_{21}|^2 \frac{1}{G_e})(p_1 p_2 |S_{21}|^2 \frac{1}{G_e} - 1 + |S_{22}|^2)}{(|S_{11}|^2 - |\Delta_S|^2 + p_1 p_2 |S_{21}|^2 \frac{1}{G_e})^2}$$

$$(5.35)$$

Equation (5.34) shows that for $1/G_e$ constant the contours for $1/G_e$ are circles with radii equal to the square root of the expression in Equation (5.35) and with centres, Γ_{CG}, given by

$$\Gamma_{CG} = \frac{S_{11}^* - S_{22}\Delta_S^*}{|S_{11}|^2 - |\Delta_S|^2 + p_1 p_2 |S_{21}|^2 \frac{1}{G_e}} \qquad (5.36)$$

Extrema for $1/G_e$ are found for $R_G^2 = 0$. The numerator in Equation (5.35) can be written as

$$|S_{12}|^2 |S_{21}|^2 + p_1 p_2 |S_{21}|^2 \left[|S_{11}|^2 + |S_{22}|^2 - 1 - |\Delta_S|^2\right] \frac{1}{G_e} + |S_{21}|^4 \frac{1}{G_e^2} = 0$$

When using the stability factor k, which here is

$$k = \frac{1 + |\Delta_S|^2 - |S_{11}|^2 - |S_{22}|^2}{2|S_{12}S_{21}|} \qquad (5.37)$$

the extrema for $1/G_e$ are given by

$$\frac{1}{G_e} = \frac{|S_{12}|}{|S_{21}|} \left(p_1 p_2 k \pm \sqrt{k^2 - 1}\right) \qquad (5.38)$$

It is seen that extrema only exist when $k^2 > 1$. For $k^2 \le 1$ $1/G_e$ is unlimited as all the circles pass through one point (for $k^2 = 1$) or two points (for $k^2 < 1$) on the unit circle.

For $k^2 > 1$ it follows that

$$\min\left[\frac{1}{G_e}\right] = \frac{1}{G_{e\,max}} = \frac{|S_{12}|}{|S_{21}|} \left(p_1 p_2 k + \sqrt{k^2 - 1}\right) \qquad (5.39)$$

which gives Γ_{SOG} (for **s**ource **o**ptimum **g**ain) by inserting Equation (5.39) into Equation (5.35):

$$\Gamma_{SOG} = \frac{2(S_{11}^* - S_{22}\Delta_S^*)}{1 + |S_{11}|^2 - |S_{22}|^2 - |\Delta_S|^2 - 2p_1 p_2 |S_{12}S_{21}|\sqrt{k^2 - 1}} \qquad (5.40)$$

$$= \frac{1 + |S_{11}|^2 - |S_{22}|^2 - |\Delta_S|^2 + 2p_1 p_2 |S_{12}S_{21}|\sqrt{k^2 - 1}}{2|S_{11}^* - S_{22}\Delta_S^*|^2}$$

$$\times (S_{11}^* - S_{22}\Delta_S^*) \qquad (5.41)$$

and that

$$\max\left[\frac{1}{G_e}\right] = \frac{1}{G_{e\,min}} = \frac{|S_{12}|}{|S_{21}|} \left(p_1 p_2 k - \sqrt{k^2 - 1}\right) \qquad (5.42)$$

which gives $\Gamma_{SOG'}$ by inserting Equation (5.42) into Equation (5.36):

$$\Gamma_{SOG'} = \frac{2\left(S_{11}^* - S_{22}\Delta_S^*\right)}{1 + |S_{11}|^2 - |S_{22}|^2 - |\Delta_S|^2 + 2\,p_1p_2|S_{12}S_{21}|\sqrt{k^2 - 1}} \tag{5.43}$$

$$= \frac{1 + |S_{11}|^2 - |S_{22}|^2 - |\Delta_S|^2 - 2\,p_1p_2|S_{12}S_{21}|\sqrt{k^2 - 1}}{2\,|S_{11}^* - S_{22}\Delta_S^*|^2}$$

$$\times \left(S_{11}^* - S_{22}\Delta_S^*\right) \tag{5.44}$$

From Equations (5.39) and (5.42) it is seen that

$$G_{e\,max} \ \leq \ G_{e\,min} \tag{5.45}$$

This inequality is valid whatever the sign of p_1p_2 or k. It is also easy to see that the sign of $G_{e\,max}$ and $G_{e\,min}$ is the same. Thus the signs shown in Table 5.4 are obtained.

p_1p_2	k	$\mathrm{sgn}[G_{e\,max}] = \mathrm{sgn}[G_{e\,min}]$
$+1$	$> +1$	$+$
$+1$	< -1	$-$
-1	$> +1$	$-$
-1	< -1	$+$

Table 5.4: Sign for $G_{e\,max}$ and $G_{e\,min}$ as a function of p_1p_2 and k.

When examining Γ_{SOG} and $\Gamma_{SOG'}$ it is seen from Equations (5.40) and (5.44) that

$$\Gamma_{SOG}\Gamma_{SOG'}^* \ = \ 1 \ \Longrightarrow \ |\Gamma_{SOG}||\Gamma_{SOG'}| \ = \ 1 \tag{5.46}$$

This means that either Γ_{SOG} or $\Gamma_{SOG'}$ is passive while the other is active. The special case, when $|\Gamma_{SOG}| = |\Gamma_{SOG'}| = 1$, requires that $\mathrm{Re}[Z_S] = 0$. As the exchangeable gain is defined only for $\mathrm{Re}[Z_S] \neq 0$ the unit circle is excluded from the region of definition. It is also seen that

$$\Gamma_{SOG}|_{p_1p_2=+1} \ = \ \Gamma_{SOG'}|_{p_1p_2=-1} \tag{5.47}$$

$$\Gamma_{SOG}|_{p_1p_2=-1} \ = \ \Gamma_{SOG'}|_{p_1p_2=+1} \tag{5.48}$$

From Equation 5.36 it is seen that the centres are located on the straight line r through the origin:

$$\mathrm{Im}[\Gamma_{CG}] \ = \ \frac{\mathrm{Im}[S_{11}^* - S_{22}\Delta_S^*]}{\mathrm{Re}[S_{11}^* - S_{22}\Delta_S^*]}\mathrm{Re}[\Gamma_{CG}] \tag{5.49}$$

Another line of interest is the line q which divides the circles into two parts. This corresponds to a singularity when the denominator in the expressions for Γ_{CG}

and R_G (Equations (5.36) and (5.35)) equals zero. In order to locate the line q the quantity $\Gamma_{CG} - R_G$, where R_G is on the line r, is investigated when the singularity is approached. The singularity is at

$$G_e = \frac{p_1 p_2 |S_{21}|^2}{|\Delta_S|^2 - |S_{11}|^2} \tag{5.50}$$

Let $G_e \leq 0$ from which it follows that $|S_{11}|^2 - |\Delta_S|^2 \geq 0 \wedge p_1 p_2 = +1$ or $|S_{11}|^2 - |\Delta_S|^2 < 0 \wedge p_1 p_2 = -1$. Coming from one side the point Γ_{qr} on the line r is determined by

$$\Gamma_{qr} = \lim_{(|S_{11}|^2 - |\Delta_S|^2)G_e + p_1 p_2 |S_{21}|^2 \to 0+} (\Gamma_{CG} + R_G e^{j \varphi_\sigma})$$

where

$$\sigma = S_{11}^* - S_{22}\Delta_S^* = |\sigma| e^{j \varphi_\sigma}$$

$$\Gamma_{CG} + R_G e^{j \varphi_\sigma} = \frac{|S_{11}^* - S_{22}\Delta_S^*|}{|S_{11}|^2 - |\Delta_S|^2 + p_1 p_2 |S_{21}|^2 \frac{1}{G_e}} e^{j \varphi_\sigma}$$

$$\times \left(1 + \sqrt{\left[1 + \frac{(|S_{11}|^2 - |\Delta_S|^2 + p_1 p_2 |S_{21}|^2 \frac{1}{G_e})(p_1 p_2 |S_{21}|^2 \frac{1}{G_e} - 1 + |S_{22}|^2)}{|S_{11}^* - S_{22}\Delta_S^*|^2} \right]} \right)$$

In approaching the limit the Taylor approximation $\sqrt{1 + x} \approx 1 + \frac{1}{2} x$ for small x is used and

$$\Gamma_{qr} = \lim_{(|S_{11}|^2 - |\Delta_S|^2)G_e + p_1 p_2 |S_{21}|^2 \to 0+} \left(\frac{|S_{11}^* - S_{22}\Delta_S^*|}{|S_{11}|^2 - |\Delta_S|^2 + p_1 p_2 |S_{21}|^2 \frac{1}{G_e}} \left[1 - 1 \right. \right.$$

$$\left. \left. - \frac{1}{2} \frac{(|S_{11}|^2 - |\Delta_S|^2 + p_1 p_2 |S_{21}|^2 \frac{1}{G_e})(p_1 p_2 |S_{21}|^2 \frac{1}{G_e} - 1 + |S_{22}|^2)}{|S_{11}^* - S_{22}\Delta_S^*|^2} \right] e^{j \varphi_\sigma} \right)$$

$$= \lim_{(|S_{11}|^2 - |\Delta_S|^2)G_e + p_1 p_2 |S_{21}|^2 \to 0+} \left(-\frac{1}{2} \frac{(p_1 p_2 |S_{21}|^2 \frac{1}{G_e} - 1 + |S_{22}|^2)}{|S_{11}^* - S_{22}\Delta_S^*|^2} e^{j \varphi_\sigma} \right)$$

$$= -\frac{1}{2} \frac{(p_1 p_2 |S_{21}|^2 \frac{|\Delta_S|^2 - |S_{11}|^2}{p_1 p_2 |S_{21}|^2} - 1 + |S_{22}|^2)}{|S_{11}^* - S_{22}\Delta_S^*|^2} e^{j \varphi_\sigma}$$

$$= \frac{1 + |S_{11}|^2 - |S_{22}|^2 - |\Delta_S|^2}{2 |S_{11}^* - S_{22}\Delta_S^*|^2} (S_{11}^* - S_{22}\Delta_S^*)$$

Approaching the point Γ_{qr} from the other side and keeping $G_e \leq 0$ the same result as above is obtained. Also when $|S_{11}|^2 - |\Delta_S|^2 < 0 \wedge p_1 p_2 = +1$ or $|S_{11}|^2 - |\Delta_S|^2 \geq 0 \wedge p_1 p_2 = -1$ the result is the same when the limit is approached through positive as well as negative values.

One important thing about the location of Γ_{qr} is that it is located just midway between Γ_{SOG} and $\Gamma_{SOG'}$ which is seen from Equations (5.40) and (5.44):

$$\Gamma_{qr} = \tfrac{1}{2}(\Gamma_{SOG} + \Gamma_{SOG'})$$

$$= \frac{1 + |S_{11}|^2 - |S_{22}|^2 - |\Delta_S|^2}{2|S_{11}^* - S_{22}\Delta_S^*|^2}(S_{11}^* - S_{22}\Delta_S^*) \qquad (5.51)$$

Another characteristic of Γ_{qr} is that it is always located on or outside the unit circle. This is seen from

$$|\Gamma_{qr}| > 1$$

$$|\Gamma_{qr}|^2 > 1$$

$$(1 - |S_{22}|^2 + |S_{11}|^2 - |\Delta_S|^2)^2 > 4|S_{11}^* - S_{22}\Delta_S^*|^2$$

$$(1 - |S_{22}|^2)^2 + (|S_{11}|^2 - |\Delta_S|^2)^2$$
$$+ 2(1 - |S_{22}|^2)(|S_{11}|^2 - |\Delta_S|^2) > 4(1 - |S_{22}|^2)(|S_{11}|^2 - |\Delta_S|^2)$$
$$+ 4|S_{12}S_{21}|^2$$

$$(1 - |S_{22}|^2 - |S_{11}|^2 + |\Delta_S|^2)^2 > 4|S_{12}S_{21}|^2$$

$$k^2 > 1 \qquad\qquad \text{q. e. d.}$$

For $k^2 \le 1$ it follows that $R_G \ge 0$ for all values of G_e. It can be shown that different circles corresponding to different constant G_e values all intersect each other in two points on the unit circle for $k^2 < 1$ and in one point on the unit circle for $k^2 = 1$. In these points G_e is not defined as G_e is not defined on the unit circle. The two (one) points are given by

$$\Gamma_S = \exp\left[j\left(\varphi_\sigma \pm \arccos\frac{1 + |S_{11}|^2 - |S_{22}|^2 - |\Delta_S|^2}{2|S_{11}^* - S_{22}\Delta_S^*|}\right)\right] \qquad (5.52)$$

where

$$S_{11}^* - S_{22}\Delta_S^* = |S_{11}^* - S_{22}\Delta_S^*|\, e^{j\varphi_\sigma}$$

It remains to be shown that the numerical value of the argument to the arccos expression in Equation (5.52) is less than or equal to one.

$$\left|\frac{1 + |S_{11}|^2 - |S_{22}|^2 - |\Delta_S|^2}{2|S_{11}^* - S_{22}\Delta_S^*|}\right| \le 1$$

$$(1 - |S_{22}|^2 + |S_{11}|^2 - |\Delta_S|^2)^2 \le 4|S_{11}^* - S_{22}\Delta_S^*|^2$$

$$(1 - |S_{22}|^2)^2 + (|S_{11}|^2 - |\Delta_S|^2)^2$$
$$+ 2(1 - |S_{22}|^2)(|S_{11}|^2 - |\Delta_S|^2) \le 4(1 - |S_{22}|^2)(|S_{11}|^2 - |\Delta_S|^2)$$
$$+ 4|S_{12}S_{21}|^2$$

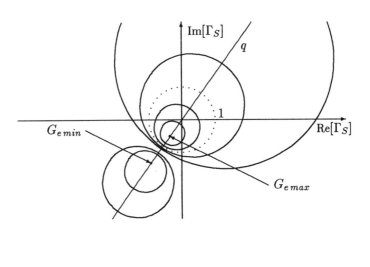

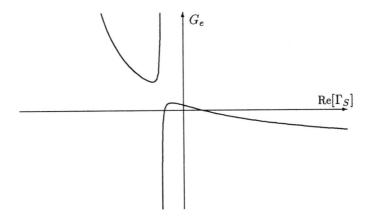

Figure 5.6: Contours for constant exchangeable power gain in the source reflection coefficient plane for $k^2 > 1$ and the corresponding function of exchangeable power gain along the line q. A third part of the curve (bottom, left) is outside the figure – compare with Figure 5.7.

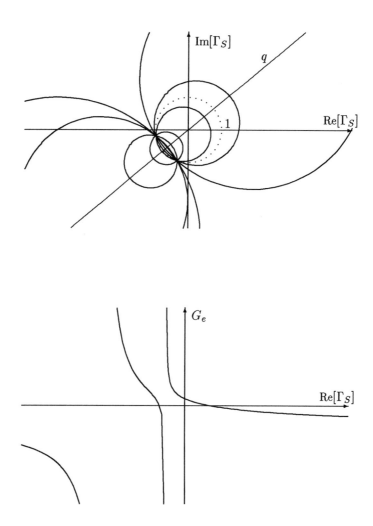

Figure 5.7: Contours for constant exchangeable power gain in the source reflection coefficient plane for $k^2 < 1$ and the corresponding function of exchangeable power gain along the line q.

$$(1 - |S_{22}|^2 - |S_{11}|^2 + |\Delta_S|^2)^2 \leq 4|S_{12}S_{21}|^2$$

$$k^2 \leq 1 \qquad \text{q. e. d.}$$

The equal sign is valid, when $k^2 = 1$, and that corresponds to one common point on the unit circle where all constant G_e circles intersect. When $k^2 < 1$ the G_e circles intersect in two points on the unit circle.

5.4.2 Graphic representation of the extended noise measure

Inserting F_e from Equation (4.55) and G_e from Equation (5.12) into Equation (5.7) the expression for the noise measure is

$$M_e = \frac{p_1|S_{21}|^2 (T_\alpha + |\Gamma_S|^2 T_\beta - 2T_\gamma \mathrm{Re}[\Gamma_S e^{j\varphi_\gamma}])}{D} \qquad (5.53)$$

where

$$D = T_0 \left(\{|S_{21}|^2 - p_1 p_2(1 - |S_{22}|^2)\} \right.$$
$$\left. - |\Gamma_S|^2 \{|S_{21}|^2 + p_1 p_2(|S_{11}|^2 - |\Delta_S|^2)\} + 2 p_1 p_2 \mathrm{Re}[\Gamma_S^*(S_{11}^* - S_{22}\Delta_S^*)] \right)$$

As M_e can be illustrated as circles for constant M_e in the admittance plane it follows from the theory of analytic functions that M_e also can be illustrated as circles in the source reflection coefficient plane. Equation (5.53) can be written as a set of circles:

$$|\Gamma_S - \Gamma_{CM}|^2 = R_M^2 \qquad (5.54)$$

where

$$\Gamma_{CM} = \frac{p_1|S_{21}|^2 T_\gamma e^{-j\varphi_\gamma} + p_1 p_2 M_e T_0(S_{11}^* - S_{22}\Delta_S^*)}{p_1|S_{21}|^2 T_\beta + M_e T_0[|S_{21}|^2 + p_1 p_2(|S_{11}|^2 - |\Delta_S|^2)]} \qquad (5.55)$$

$$R_M^2 = \frac{|p_1|S_{21}|^2 T_\gamma e^{-j\varphi_\gamma} + p_1 p_2 M_e T_0(S_{11}^* - S_{22}\Delta_S^*)|^2}{\{p_1|S_{21}|^2 T_\beta + M_e T_0[|S_{21}|^2 + p_1 p_2(|S_{11}|^2 - |\Delta_S|^2)]\}^2}$$
$$+ \frac{M_e T_0[|S_{21}|^2 - p_1 p_2(1 - |S_{22}|^2)] - p_1|S_{21}|^2 T_\alpha}{p_1|S_{21}|^2 T_\beta + M_e T_0[|S_{21}|^2 + p_1 p_2(|S_{11}|^2 - |\Delta_S|^2)]} \qquad (5.56)$$

The extrema are found by setting $R_M^2 = 0$ in Equation (5.56). Doing this and rearranging the equation the following expression is obtained:

$$A M_e^2 + B M_e + C = 0 \qquad (5.57)$$

where

$$A = p_1 p_2 T_0^2 [|S_{11}|^2 + |S_{22}|^2 + p_1 p_2(|S_{12}|^2 + |S_{21}|^2) - |\Delta_S|^2 - 1] \quad (5.58)$$

$$B = -p_2 T_0 \{ T_\alpha(|S_{11}|^2 - |\Delta_S|^2) + T_\beta(1 - |S_{22}|^2)$$

$$+ p_1 p_2 |S_{21}|^2 (T_\alpha - T_\beta) - 2 T_\gamma \mathrm{Re}[(S_{11}^* - S_{22}\Delta_S^*) e^{j\varphi_\gamma}] \} \tag{5.59}$$

$$C = -|S_{21}|^2 (T_\alpha T_\beta - T_\gamma^2) \tag{5.60}$$

From Equation (4.56) it is seen that $C \leq 0$ which is important to remember in the following:

1. $A > 0$

When $A > 0$ then $B^2 - 4AC \geq 0$. This means that there exist two extrema for M_e. It is also seen that $|B| \leq \sqrt{B^2 - 4AC}$ from which it follows that the signs of the extrema are different. As A is positive it is seen from Equation (5.57) that R_M^2 is negative between the two roots and thus $M_{e\,min} \geq 0$ and $M_{e\,max} \leq 0$. This yields

$$M_{e\,min} = \frac{-B + \sqrt{B^2 - 4AC}}{2A} \geq 0 \tag{5.61}$$

$$M_{e\,max} = \frac{-B - \sqrt{B^2 - 4AC}}{2A} \leq 0 \tag{5.62}$$

2. $A < 0 \wedge B^2 - 4AC \geq 0$

Here it is seen that two extrema exist as $B^2 - 4AC \geq 0$. As $A < 0$ and $C \leq 0$ it follows that $|B| \geq \sqrt{B^2 - 4AC}$ and thus the signs for the extrema are the same. It is also seen that the range for M_e is between the extrema. Therefore

$$M_{e\,max} = \frac{-B - \sqrt{B^2 - 4AC}}{2A} \tag{5.63}$$

$$M_{e\,min} = \frac{-B + \sqrt{B^2 - 4AC}}{2A} \tag{5.64}$$

3. $A < 0 \wedge B^2 - 4AC < 0$

There are no real solutions for M_e.

The centres are located on a straight line, v, whose equation is found from the complex Equation (5.55) by eliminating M_e. The result is

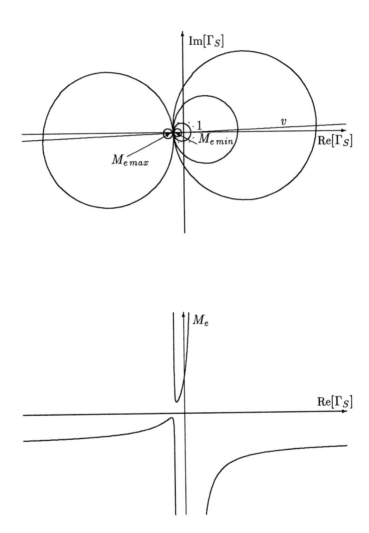

Figure 5.8: Contours for constant extended noise measure in the source reflection coefficient plane when $A > 0$ and the corresponding function of extended noise measure along the line v.

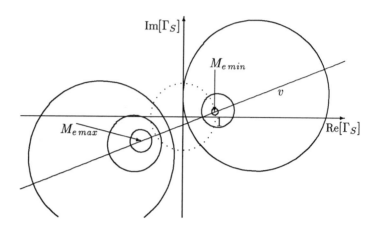

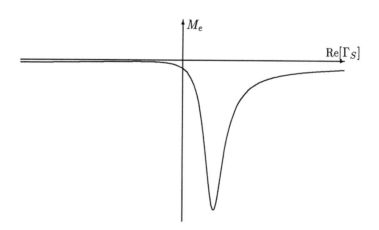

Figure 5.9: Contours for constant extended noise measure in the source reflection coefficient plane when $A < 0$ and the corresponding function of extended noise measure along the line v.

$$\{p_1 p_2 T_\gamma (|S_{21}|^2 + p_1 p_2 [|S_{11}|^2 - |\Delta_S|^2]) \sin\varphi_\gamma + T_\beta \operatorname{Im}[S_{11}^* - S_{22}\Delta_S^*]\} \operatorname{Re}[\Gamma_{CM}]$$

$$+ \{p_1 p_2 T_\gamma (|S_{21}|^2 + p_1 p_2 [|S_{11}|^2 - |\Delta_S|^2]) \cos\varphi_\gamma - T_\beta \operatorname{Re}[S_{11}^* - S_{22}\Delta_S^*]\} \operatorname{Im}[\Gamma_{CM}]$$

$$= T_\gamma (\operatorname{Im}[S_{11}^* - S_{22}\Delta_S^*] \cos\varphi_\gamma + \operatorname{Re}[S_{11}^* - S_{22}\Delta_S^*] \sin\varphi_\gamma) \tag{5.65}$$

It should be noted that this line generally does not go through the origin. Figures 5.8 and 5.9 show two examples of constant extended noise measure circles, one corresponding to $A > 0$ and the other corresponding to $A < 0$.

5.5 References

[1] Haus, H. A. & Adler, R. B.: "Circuit theory of linear noisy networks", Technology Press and Wiley, 1959.

[2] Engberg, J. & Gawler, G.: "Significance of the noise measure for cascaded stages", *Proc. IEEE Trans. on Circuit Theory*, vol. CT-16, pp. 259 – 260, May 1969.

[3] Fukui, H.: "Available power gain, noise figure and noise measure of two-ports and their graphical representations", *Proc. IEEE Trans. on Circuit Theory*, vol. CT-13, pp. 137 – 142, June 1966.

[4] Engberg, J.: "Simultaneous input power match and noise optimization using feedback", R69ELS-79, Electronics Laboratory, General Electric, Syracuse, NY, 1969.

6

Noise of embedded networks

Noise parameters of a transistor or another active element are often given in datasheets, but what happens if bias network, feedback elements, or other external elements are added to the original network? This is the subject of this chapter where lumped elements, distributed elements, and combinations of two-ports are analyzed. Then an example is given of a two-port with feedback via two transformers in such a way that the rules analyzed in section 6.1 do not apply. Only three-poles are considered, as most active elements have three poles which are the common reference terminal, the input and the output terminals. Therefore formulae for transformation of noise parameters from common emitter or source to common collector or drain and to common base or gate are given. Finally, some thoughts on computer aided design of linear circuits with noise are expressed.

6.1 Lumped embedding

Consider a three-pole network as in Figure 6.1 to which linear and lumped one-ports such as resistors, capacitors, and inductors are added by a number of successive parallel and series connections. It will be convenient to define "embedded" and "embedding" networks. The original network is called the embedded network (denoted by 1, 2, and 3 in Figure 6.1), and the linear one-ports are called the embedding network. The resulting network is also a three-pole network (denoted by 1', 2', and 3'). A practical way to obtain the new signal parameters of such a network can be explained as follows. Add the first set of parallel admittances to the Y parameters of the embedded network. Transform the resulting Y parameters to Z parameters and add the first set of series impedances. Then transform the resulting Z parameters back again to Y parameters and add the second set of parallel elements. Continue this procedure until complete [1,2].

The noise parameters of the composite network are computed in a similar way.

101

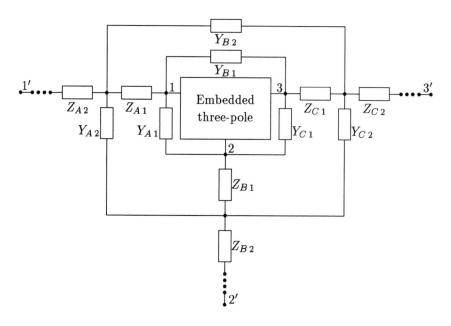

Figure 6.1: Three-pole embedded in lumped one-ports.

Like signal parameters, noise parameters have several forms as shown in Chapter 4. The most useful forms of this application are the equivalent Π and T noise two-ports introduced by Rothe and Dahlke [3] and examined in Chapter 4. The transformation procedure is as follows. First, compute the noise parameters (in Π form which corresponds to Y parameters) of the network consisting of the embedded network (with its noise parameters in Π form) and the first set of parallel elements. Transform the resulting noise parameters to T form, and combine the noise parameters (in Y form) of the resulting network with those of the first set of added series elements. Then transform the resulting noise parameters back to Π form and add the effect of the second set of parallel elements, continuing this procedure until complete.

Since the admittance of a parallel element and the impedance of a series element may be zero, it is possible to compute the signal and noise parameters of a rather general class of embedding networks (feedback, biasing, stray elements etc.).

The formulae, which are derived in [1], are given below. The unprimed entities are the small signal and noise parameters of the three-port without new embedding, and the primed entities are the parameters with a set of embedding elements.

Parallel embedding:

$$Y'_{11} = Y_{11} + Y_A + Y_B \tag{6.1}$$

$$Y'_{12} = Y_{12} - Y_B \tag{6.2}$$

$$Y'_{21} = Y_{21} - Y_B \tag{6.3}$$

$$Y'_{22} = Y_{22} + Y_B + Y_C \tag{6.4}$$

$$R'_n = \frac{E_y}{D_y} \tag{6.5}$$

$$Y'_\gamma = \frac{H_y}{E_y} \tag{6.6}$$

$$G'_n = \frac{L_y}{D_y} - \frac{|H_y|^2}{D_y E_y} \tag{6.7}$$

where

$$D_y = |Y_B - Y_{21}|^2$$

$$E_y = G_B + G_C + |Y_{21}|^2 R_n$$

$$H_y = (Y_{11} + Y_A + Y_{21}) G_B + (Y_{11} + Y_A + Y_B) G_C$$
$$+ (Y_{11} - Y_\gamma) Y_B Y_{21}^* R_n + (Y_A + Y_B + Y_\gamma) |Y_{21}|^2 R_n$$

$$L_y = |Y_B - Y_{21}|^2 (G_A + G_n) + |Y_{11} + Y_A + Y_{21}|^2 G_B$$
$$+ |Y_{11} + Y_A + Y_B|^2 G_C + |(Y_{11} - Y_\gamma) Y_B + (Y_A + Y_B + Y_\gamma) Y_{21}|^2 R_n$$

Series embedding:

$$Z'_{11} = Z_{11} + Z_A + Z_B \tag{6.8}$$

$$Z'_{12} = Z_{12} + Z_B \tag{6.9}$$

$$Z'_{21} = Z_{21} + Z_B \tag{6.10}$$

$$Z'_{22} = Z_{22} + Z_B + Z_C \tag{6.11}$$

$$g'_n = \frac{E_z}{D_z} \tag{6.12}$$

$$Z'_\gamma = \frac{H_z}{E_z} \tag{6.13}$$

$$r'_n = \frac{L_z}{D_z} - \frac{|H_z|^2}{D_z E_z} \tag{6.14}$$

where

$$D_z = |Z_B + Z_{21}|^2$$

$$E_z = R_B + R_C + |Z_{21}|^2 g_n$$

$$H_z = (Z_{11} + Z_A - Z_{21}) R_B + (Z_{11} + Z_A + Z_B) R_C$$
$$- (Z_{11} - Z_\gamma) Z_B Z_{21}^* g_n + (Z_A + Z_B + Z_\gamma) |Z_{21}|^2 g_n$$

$$L_z = |Z_B + Z_{21}|^2(R_A + r_n) + |Z_{11} + Z_A - Z_{21}|^2 R_B$$
$$+ |Z_{11} + Z_A + Z_B|^2 R_C + |(Z_\gamma - Z_{11}) Z_B + (Z_A + Z_B + Z_\gamma) Z_{21}|^2 g_n .$$

Transformation from Z to Y parameters:

$$Y_{11} = \frac{Z_{22}}{\Delta_Z} \tag{6.15}$$

$$Y_{12} = \frac{-Z_{12}}{\Delta_Z} \tag{6.16}$$

$$Y_{21} = \frac{-Z_{21}}{\Delta_Z} \tag{6.17}$$

$$Y_{22} = \frac{Z_{11}}{\Delta_Z} \tag{6.18}$$

$$R_n = r_n + g_n|Z_\gamma|^2 \tag{6.19}$$

$$G_n = \frac{r_n}{|Z_\gamma|^2 + r_n/g_n} \tag{6.20}$$

$$Y_\gamma = \frac{Z_\gamma^*}{|Z_\gamma|^2 + r_n/g_n} \tag{6.21}$$

$$\text{where} \quad \Delta_Z = Z_{11} Z_{22} - Z_{12} Z_{21}$$

Transformation from Y to Z parameters:

$$Z_{11} = \frac{Y_{22}}{\Delta_Y} \tag{6.22}$$

$$Z_{12} = \frac{-Y_{12}}{\Delta_Y} \tag{6.23}$$

$$Z_{21} = \frac{-Y_{21}}{\Delta_Y} \tag{6.24}$$

$$Z_{22} = \frac{Y_{11}}{\Delta_Y} \tag{6.25}$$

$$g_n = G_n + R_n|Y_\gamma|^2 \tag{6.26}$$

$$r_n = \frac{G_n}{|Y_\gamma|^2 + G_n/R_n} \tag{6.27}$$

$$Z_\gamma = \frac{Y_\gamma^*}{|Y_\gamma|^2 + G_n/R_n} \tag{6.28}$$

$$\text{where} \quad \Delta_Y = Y_{11} Y_{22} - Y_{12} Y_{21}$$

It is interesting to note that the formulae work both ways: embedding and deembedding. As an example take an encapsulated transistor whose small signal parameters and noise parameters are measured. The effect of (the known) stray capacitors and stray inductors can be removed by adding the negative of the stray elements. This is done such that the stray elements closest to the terminals of the encapsulated transistor are removed first and then the second set of stray elements until finally the small signal and noise parameters of the transistor chip are obtained.

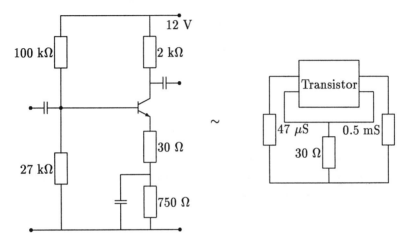

Figure 6.2: Transistor with bias and feedback and equivalent circuit. Capacitors are considered as shorts at the working frequency.

Example 6.1 A transistor amplifier stage as seen in Figure 6.2 consists of a transistor with known Y and noise parameters and also bias and load resistors and finally a series feedback resistor. The data for the transistor are

$$Y_{11} = 10 + j\,2.1 \quad \text{mS}$$

$$Y_{12} = 0.5 - j\,0.866 \quad \text{mS}$$

$$Y_{21} = 125 - j\,29 \quad \text{mS}$$

$$Y_{22} = 1 + j\,3 \quad \text{mS}$$

$$R_n = 25 \quad \Omega$$

$$G_n = 4.8 \quad \text{mS}$$

$$Y_\gamma = 2 + j\,7.5 \quad \text{mS}$$

These data give $F_{min} = 1.8$ and $Y_{SOF} = 14 - j\,7.5$mS. In order to compute the small signal and noise parameters of the stage, one has to apply Equations (6.22) – (6.28) to get the transistor data into Z form. Then Equations (6.8) – (6.14) add the series feedback resistor (while $Z_A = Z_C = 0 + j\,0$). Equations (6.15) – (6.21) transform to the Y parameters and Equations (6.1) – (6.7) add the bias and collector resistor (with $Y_B = 0$). The final result, computed with a Fortran program where the equations used are shown as subroutines in Appendix D, is:

$$Y_{11} = 1.586 + j\,1.548 \quad \text{mS}$$

$$Y_{12} = 0.4448 - j\,1.017 \quad \text{mS}$$

$$Y_{21} = 25.16 - j\,2.935 \quad \text{mS}$$

$$Y_{22} = 0.2837 + j\,1.469 \quad \text{mS}$$

$$R_n = 62.81 \quad \Omega$$

$$G_n = 4.735 \quad \text{mS}$$

$$Y_\gamma = 3.442 + j\,3.432 \quad \text{mS}$$

This corresponds to $F_{min} = 2.606$ and $Y_{SOF} = 9.340 - 3.432\,\text{mS}$. The main part of this noise degregation is due to the series feedback. The bias and collector resistors raise the noise factor only slightly – from 1.800 to 1.807.

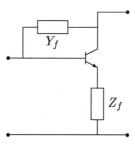

Figure 6.3: Feedback elements for simultaneous input power match and noise optimization.

Example 6.2 An optimization procedure for simultaneous input power match and noise optimization using only one series and one shunt feedback element as shown in Figure 6.3 has been developed. Since lossless feedback does not change the value of the minimum noise measure (does not add noise), only lossless elements are considered. The fact that $M_{e\,min}$ is constant for lossless feedback is a convenient check on results of computer programs. What does change, of course, is the value of the source admittance, Y_{SOM}, for minimum noise measure.

For each pair of lossless feedback elements a computer program – using the subroutines in Appendix D – gives the values for the primed signal and noise parameters from Equations (6.1) – (6.28). The minimum noise measure, $M_{e\,min}$, and the corresponding source admittance, Y_{SOM}, from Equations (5.27), (5.25) and (5.26) are also computed.

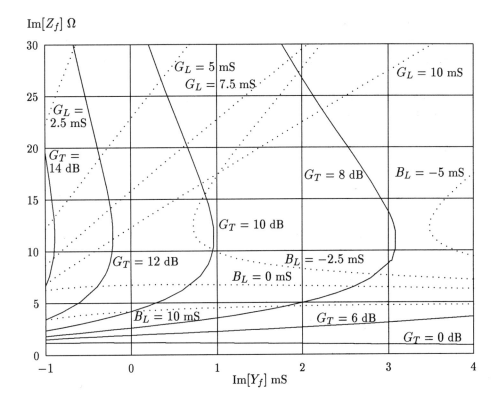

Figure 6.4: Transducer gain G_T (full lines) and load admittance G_L, B_L (dotted lines) versus lossless feedback for unity VSWR and minimum noise measure.

When the source admittance, $Y_S = Y_{SOM}$ is determined, input power matching occurs when the load admittance, Y_L, is chosen to satisfy the relation:

$$Y_L = -Y_{22} + \frac{Y_{12}Y_{21}}{y_{11} - Y_S^*}$$

This equation allows choosing the load admittance for input power match and minimum noise measure (Y_{LOM}). Then, with the source and load admittances equal to Y_{SOM} and Y_{LOM} respectively, the transducer gain, G_T, of the stage is computed.

From these results various parameter curves are drawn with the values of the lossless feedback elements as coordinates. Such a graph is shown in Figure 6.4, where the real and imaginary parts of the load admittance and the transducer gain for that load admittance are the parameters. The graph is to be read as follows.

Each point in the graph assigns a load admittance (G_L, B_L) to a feedback pair (B_f, X_f) such that input matching and minimum noise measure are attained simultaneously. The corresponding transducer gain can also be read from Figure 6.4.

It should be noted that simultaneous input power match and minimum noise measure without feedback can only be obtained with a transducer gain less than one. With only series feedback $(X_f = 12\ \Omega)$ the optimization is obtained with a transducer gain of about 11.5 dB.

Lehmann and Heston [4] have constructed an integrated low noise amplifier on X-band using this technique.

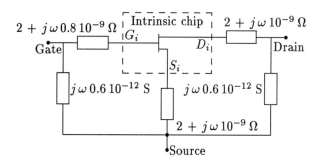

Figure 6.5: Transistor with lumped encapsulation.

Example 6.3 Consider an encapsulated transistor as shown in Figure 6.5. Data for the encapsulated transistor are

$$Y_{11} = 10 + j\,2.1\ \ \text{mS}$$
$$Y_{12} = 0.5 - j\,0.866\ \ \text{mS}$$
$$Y_{21} = 19.5 - j\,29\ \ \text{mS}$$
$$Y_{22} = 1 + j\,3\ \ \text{mS}$$
$$R_n = 25\ \ \Omega$$
$$G_n = 4.8\ \ \text{mS}$$
$$Y_\gamma = 2 + j\,7.5\ \ \text{mS}$$

and the data for the encapsulation are seen in Figure 6.5. By adding the negative admittance of the two capacitors in parallel with the input and output terminals they disappear. This is done by applying Equations (6.1) – (6.7). Using Equations (6.22) – (6.28) the Z parameters appear. Then the resistances and inductances in series with the intrinsic transistor are removed by adding the negative amount in Equations (6.8) – (6.14). Finally conversion to the Y parameters is performed by the Equations (6.15)

– (6.21). The results at 1 GHz – computed with a Fortran program[1] – are as follows:

$$Y_{11} = 14.41 + j\,0.8113 \quad \text{mS}$$

$$Y_{12} = 0.787 - j\,0.7576 \quad \text{mS}$$

$$Y_{21} = 36.23 - j\,31.19 \quad \text{mS}$$

$$Y_{22} = 1.948 - j\,1.114 \quad \text{mS}$$

$$R_n = 23.37 \quad \Omega$$

$$G_n = 4.269 \quad \text{mS}$$

$$Y_\gamma = 1.260 + j\,6.552 \quad \text{mS}$$

6.2 Reference plane transformation of noise parameters

When working with noise parameters it is often convenient to know them at a reference plane different from that where the noise parameters are actually measured. This is a kind of embedding as a piece of transmission line in front of a transistor can be considered as an embedding element. The result leads to the noise parameters of a transmission line as a two-port. In the next section this two-port can be an embedding element in parallel or in cascade with other two-ports. In this section, formulae are presented for transformation of noise parameters along a transmission line with known characteristic impedance and known attenuation and phase constants. It is assumed, however, that the temperature of the transmission line is the standard noise temperature of 290 K [5].

When calculating the noise performance of an active element embedded in passive circuit elements, it is necessary, at microwave frequencies, to include distributed elements as well as lumped elements. The most significant distributed element with respect to noise performance is the transmission line leading to the first active element. This section considers the noise parameters of a network consisting of a transmission line with known constants preceding an active two-port with known noise parameters. The result can be used to transform the noise parameters measured in one reference plane to another along the known transmission line. The transformation works both ways; it adds the additional noise contribution when the transmission line is made longer and subtracts noise when the reference plane is moved closer to the active two-port. This feature is useful when correcting measured noise data of an active two-port and is carried out by adding a "negative" length of transmission line.

---

[1]See Appendix D.

With transmission line constants Z_0 (real and positive) and $\gamma = \alpha + j\beta$ and with l the electrical length of the transmission line in front of the known two-port with noise parameters R_n, G_n and Y_γ the new (primed) noise parameters are presented below as

$$
\begin{aligned}
R'_n = {} & \frac{Z_0}{4} \Big\{ e^{2\alpha l} - e^{-2\alpha l} + Z_0 G_n \Big[e^{2\alpha l} + e^{-2\alpha l} - 2\cos 2\beta l \Big] \\
& + \frac{R_n}{Z_0} \Big[Z_0^2 \left(G_\gamma^2 + B_\gamma^2 \right) \left(e^{2\alpha l} + e^{-2\alpha l} - 2\cos 2\beta l \right) \\
& + 2 Z_0 G_\gamma \left(e^{2\alpha l} - e^{-2\alpha l} \right) - 4 Z_0 B_\gamma \sin 2\beta l \\
& + e^{2\alpha l} + e^{-2\alpha l} - 2\cos 2\beta l \Big] \Big\}
\end{aligned}
\tag{6.29}
$$

$$
\begin{aligned}
G'_\gamma = {} & \frac{1}{4R'_n} \Big\{ e^{2\alpha l} + e^{-2\alpha l} - 2 + Z_0 G_n \Big[e^{2\alpha l} - e^{-2\alpha l} \Big] \\
& + R_n \Big[Z_0 \left(G_\gamma^2 + B_\gamma^2 + Z_0^{-2} \right) \left(e^{2\alpha l} - e^{-2\alpha l} \right) \\
& + 2 G_\gamma \left(e^{2\alpha l} - e^{-2\alpha l} \right) \Big] \Big\}
\end{aligned}
\tag{6.30}
$$

$$
\begin{aligned}
B'_\gamma = {} & \frac{-1}{2R'_n} \Big\{ Z_0 G_n \sin 2\beta l + R_n \Big[Z_0 \left(G_\gamma^2 + B_\gamma^2 - Z_0^{-2} \right) \sin 2\beta l \\
& - 2 B_\gamma \cos 2\beta l \Big] \Big\}
\end{aligned}
\tag{6.31}
$$

$$
\begin{aligned}
G'_n = {} & \frac{1}{4Z_0} \Big\{ e^{2\alpha l} - e^{-2\alpha l} + Z_0 G_n \Big[e^{2\alpha l} + e^{-2\alpha l} + 2\cos 2\beta l \Big] \\
& + \frac{R_n}{Z_0} \Big[Z_0^2 \left(G_\gamma^2 + B_\gamma^2 \right) \left(e^{2\alpha l} + e^{-2\alpha l} + 2\cos 2\beta l \right) \\
& + 2 Z_0 G_\gamma \left(e^{2\alpha l} - e^{-2\alpha l} \right) + 4 Z_0 B_\gamma \sin 2\beta l \\
& + e^{2\alpha l} + e^{-2\alpha l} - 2\cos 2\beta l \Big] \Big\} - \frac{R'_n}{Z_0} \left(G'_\gamma + B'_\gamma \right)
\end{aligned}
\tag{6.32}
$$

Transformation of noise parameters through a lossy transmission line

Equations (6.29) – (6.32) are derived in the following. With reference to Figure 6.6 the problem can be stated as follows: find the (new) noise parameters at reference plane A expressed by the (old and known) noise parameters at reference plane B and the transmission line constants l, Z_0 (real and positive) and $\gamma = \alpha + j\beta$. First, all immittances are normalized to Z_0 or $Y_0 = 1/Z_0$:

$$
r_n = \frac{R_n}{Z_0}
$$

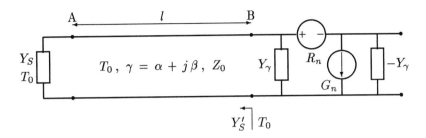

Figure 6.6: Noise two-port with preceding transmission line.

$$g_n = \frac{G_n}{Y_0}$$

$$y_\gamma = g_\gamma + j b_\gamma = \frac{Y_\gamma}{Y_0}$$

$$y_S = g_S + j b_S = \frac{Y_S}{Y_0}$$

The unknown normalized noise parameters are primed:

$$r'_n, \quad g'_n \quad \text{and} \quad y'_\gamma = g'_\gamma + j b'_\gamma$$

The unknown noise factor at reference plane A can be expressed by

$$F = 1 + \frac{1}{g_S}\left(g'_n + r'_n \left|y_S + y'_\gamma\right|^2\right) \tag{6.33}$$

Using the definition of the extended noise factor as the total exchangeable noise power at the output divided by that part of it which originates from the source at standard temperature T_0, the noise factor at reference plane B can be expressed as

$$F = \frac{g'_S + g_n + r_n|y'_S + y_\gamma|^2}{g''_S} \tag{6.34}$$

Here use is made of the fact that the one-port to the left of B in Figure 6.6, consisting of a source admittance and a transmission line at the temperature T_0, generates noise as the real part of Y'_S at T_0. The denominator g''_S represents that part of the exchangeable noise power at the output which is generated in the source.

$$y'_S = g'_S + j b'_S = \frac{y_S \cosh \gamma l + \sinh \gamma l}{\cosh \gamma l + y_S \sinh \gamma l} =$$

$$\frac{[(g_S+1)^2 + b_S^2]e^{2\alpha l} - [(g_S-1)^2 + b_S^2]e^{-2\alpha l} + j\,[4b_S \cos 2\beta l - 2(g_S^2 + b_S^2 - 1)\sin 2\beta l]}{[(g_S+1)^2 + b_S^2]e^{2\alpha l} + [(g_S-1)^2 + b_S^2]e^{-2\alpha l} - 4b_S \sin 2\beta l - 2(g_S^2 + b_S^2 - 1)\cos 2\beta l} \tag{6.35}$$

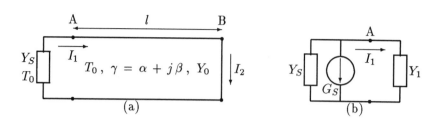

Figure 6.7: Generator with shorted transmission line and equivalent circuit.

The only remaining unknown in Equation (6.34) is now g_S''. Some of the noise power generated in the source is dissipated in the lossy transmission line, and as one only has to consider the noise generated in the source all other noise sources are eliminated. The relation between the short circuit noise current at B (see Figure 6.7) and g_S'' is

$$\langle |I_2|^2\rangle \quad = \quad 4\,k\,T_0\,\Delta f\,g_S''\,Y_0 \tag{6.36}$$

where $k = 1.38 \times 10^{-23}$ J K^{-1} (Boltzmann's constant) and Δf is a frequency increment in Hz. The currents I_1 and I_2 are related by

$$I_1 \quad = \quad I_2 \cosh \gamma l \tag{6.37}$$

Now I_1 can be calculated from Figure 6.7(b), where Y_1 represents the admittance of the shorted transmission line looking into it at A:

$$y_1 \quad = \quad \frac{Y_1}{Y_0} \quad = \quad \coth \gamma l$$

$$= \quad \frac{e^{2\alpha l} - e^{-2\alpha l} - j\,2\sin 2\beta l}{e^{2\alpha l} + e^{-2\alpha l} - 2\cos 2\beta l} \tag{6.38}$$

$$\langle |I_1|^2\rangle \quad = \quad \frac{|y_1|^2}{|y_S + y_1|^2}\,4\,k\,T_0\,\Delta f\,g_S\,Y_0 \tag{6.39}$$

From Equations (6.36) – (6.39) the following expression for g_S'' can be calculated:

$$g_S'' \quad = \quad \frac{4}{e^{2\alpha l} + e^{-2\alpha l} + 2\cos 2\beta l}\,\frac{|y_1|^2}{|y_S + y_1|^2}\,g_S \tag{6.40}$$

where

$$\frac{|y_1|^2}{|y_S + y_1|^2} \quad =$$

$$\frac{(e^{2\alpha l} - e^{-2\alpha l})^2 + 4\sin^2 2\beta l}{[g_S(e^{2\alpha l} + e^{-2\alpha l} - 2\cos 2\beta l) + e^{2\alpha l} - e^{-2\alpha l}]^2 + [b_S(e^{2\alpha l} + e^{-2\alpha l} - 2\cos 2\beta l) - 2\sin 2\beta l]^2}$$

Equation (6.34) is now completely determined by the known noise parameters r_n, g_n and y_γ and the transmission line constants αl and βl, as y_S' and its real part g_S' are expressed in Equation (6.35), and g_S'' is expressed in Equation (6.40), in terms of only αl, βl and y_S.

With Equations (6.35) and (6.40) this expression for the noise factor must, for all $y_S = g_S + j\,b_S$, be identically equal with the noise factor expression in Equation (6.33), and it is thus possible to determine the new noise parameters in terms of the old ones and the transmission line data by choosing four values of y_S and then solving the four identity equations for the new normalized noise parameters. The results are:

$$
\begin{aligned}
r_n' = {} & \frac{1}{4}\Big\{ e^{2\alpha l} - e^{-2\alpha l} + g_n \left[e^{2\alpha l} + e^{-2\alpha l} - 2\cos 2\beta l \right] \\
& + r_n \left[\left(g_\gamma^2 + b_\gamma^2 \right) \left(e^{2\alpha l} + e^{-2\alpha l} - 2\cos 2\beta l \right) \right. \\
& + 2\,g_\gamma \left(e^{2\alpha l} - e^{-2\alpha l} \right) - 4\,b_\gamma \sin 2\beta l \\
& \left. + e^{2\alpha l} + e^{-2\alpha l} - 2\cos 2\beta l \right] \Big\}
\end{aligned}
\tag{6.41}
$$

$$
\begin{aligned}
g_\gamma' = {} & \frac{1}{4r_n'}\Big\{ e^{2\alpha l} + e^{-2\alpha l} - 2 + g_n \left[e^{2\alpha l} - e^{-2\alpha l} \right] \\
& + r_n \left[\left(g_\gamma^2 + b_\gamma^2 + 1 \right) \left(e^{2\alpha l} - e^{-2\alpha l} \right) \right. \\
& \left. + 2g_\gamma \left(e^{2\alpha l} - e^{-2\alpha l} \right) \right] \Big\}
\end{aligned}
\tag{6.42}
$$

$$
\begin{aligned}
b_\gamma' = {} & \frac{-1}{2R_n'}\Big\{ g_n \sin 2\beta l + r_n \left[\left(g_\gamma^2 + b_\gamma^2 - 1 \right) \sin 2\beta l \right. \\
& \left. - 2b_\gamma \cos 2\beta l \right] \Big\}
\end{aligned}
\tag{6.43}
$$

$$
\begin{aligned}
g_n' = {} & \frac{1}{4}\Big\{ e^{2\alpha l} - e^{-2\alpha l} + g_n \left[e^{2\alpha l} + e^{-2\alpha l} + 2\cos 2\beta l \right] \\
& + r_n \left[\left(g_\gamma^2 + b_\gamma^2 \right) \left(e^{2\alpha l} + e^{-2\alpha l} + 2\cos 2\beta l \right) \right. \\
& + 2g_\gamma \left(e^{2\alpha l} - e^{-2\alpha l} \right) + 4b_\gamma \sin 2\beta l \\
& \left. + e^{2\alpha l} + e^{-2\alpha l} - 2\cos 2\beta l \right] \Big\} - r_n' \left(g_\gamma' + b_\gamma' \right)
\end{aligned}
\tag{6.44}
$$

These equations are the normalized versions of Equations (6.29) – (6.32).

Example 6.4 For the purpose of illustrating the effect of a transmission line on noise performance this example has been computed. For an active element with the following

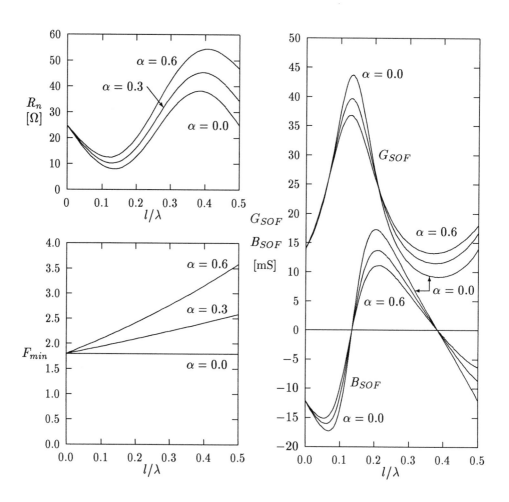

Figure 6.8: Noise parameters from Example 6.4 as functions of reference plane.

noise parameters:

$$R_n = 25\ \Omega, \qquad G_n = 4.8\ \text{mS}, \quad \text{and} \qquad Y_\gamma = 2 + j\,12\ \text{mS}$$

the noise performance of a preceding transmission line of a length up to half a wavelength $(\lambda/2)$ and three different attenuation constants $(\alpha = 0.0, 0.3$ and 0.6 Np$/\lambda)$ has been computed and expressed by Equations $(4.84) - (4.86)$ as

$$F_{e\,min} = 1 + 2\left(R_n G_\gamma + \sqrt{R_n G_n + (R_n G_\gamma)^2}\right)$$

$$Y_{SOF} = \sqrt{G_n/R_n + G_\gamma^2} - j\,B_\gamma$$

and R_n.

The noise parameters as functions of normalized length and with the attenuation constant α as parameter are shown in Figure 6.8(a), (b) and (c). One obvious use of this figure is to choose a length of the transmission line which makes the real part of $Y_{SOF} = 20$ mS. Then the imaginary part can be removed by a stub and the transmission line two-port combination has optimum noise performance for $Y_S = 20$ mS. Another possibility is to choose the imaginary part of $Y_{SOF} = 0$ S and then add a quarter wave transmission line transformer. This also gives a match with optimum noise figure, but, of course, both cases are narrow band matchings.

6.2.1 The equivalent noise two-port of a lossy transmission line

The results above can be used to derive the equivalent noise two-port of a lossy transmission line. Consider two two-ports in cascade. The first is the transmission line and the second is a lossless line of length zero. The noise parameters of the second two-port are all zero. The noise parameters of the transmission line are then derived from the Equations $(6.29) - (6.32)$ letting R_n, G_n and Y_γ be equal to zero. Thus for a lossy transmission line the parameters are

$$R_n = \frac{Z_0}{4}\left(e^{2\alpha l} - e^{-2\alpha l}\right) = \frac{Z_0}{2}\sinh 2\alpha l \qquad (6.45)$$

$$G_\gamma = \frac{1}{Z_0}\left(\frac{e^{2\alpha l} + e^{-2\alpha l} - 2}{e^{2\alpha l} - e^{-2\alpha l}}\right)$$

$$= \frac{1}{Z_0}\left(\coth 2\alpha l - \frac{1}{\sinh 2\alpha l}\right) \qquad (6.46)$$

$$B_\gamma = 0 \qquad (6.47)$$

$$G_n = G_\gamma = \frac{1}{Z_0}\left(\frac{e^{2\alpha l} + e^{-2\alpha l} - 2}{e^{2\alpha l} - e^{-2\alpha l}}\right)$$

$$= \frac{1}{Z_0}\left(\coth 2\alpha l - \frac{1}{\sinh 2\alpha l}\right) \qquad (6.48)$$

From these results it is easy to find F_{min} and Y_{SOF} from Equations (4.84) and (4.86):

$$
\begin{aligned}
F_{min} &= 1 + 2\left[R_n G_\gamma + \sqrt{R_n G_n + (R_n G_\gamma)^2}\right] \\[2mm]
&= 1 + 2\left[\frac{e^{2\alpha l} + e^{-2\alpha l} - 2}{4}\right. \\[2mm]
&\qquad\qquad \left. + \sqrt{\frac{1}{4}\left(e^{2\alpha l} + e^{-2\alpha l} - 2\right) + \frac{1}{16}\left(e^{2\alpha l} + e^{-2\alpha l} - 2\right)^2}\right] \\[2mm]
&= e^{2\alpha l}
\end{aligned}
$$

$$(6.49)$$

$$
\begin{aligned}
Y_{SOF} &= \sqrt{\frac{G_n}{R_n} + G_\gamma^2} - j\,B_\gamma \\[2mm]
&= \sqrt{\frac{4\left(e^{2\alpha l} + e^{-2\alpha l} - 2\right)}{Z_0^2\left(e^{2\alpha l} - e^{-2\alpha l}\right)^2} + \frac{\left(e^{2\alpha l} + e^{-2\alpha l} - 2\right)^2}{Z_0^2\left(e^{2\alpha l} - e^{-2\alpha l}\right)^2}} - j\,0 \\[2mm]
&= \frac{1}{Z_0}
\end{aligned}
$$

$$(6.50)$$

It has been shown above that the minimum noise factor is obtained for a source admittance equal to the characteristic admittance for the transmission line and that the noise factor rises exponentially with the loss factor α and with the length l of the transmission line.

6.3   Noise parameters of interconnected two-ports

A network consisting of linear two-ports each with known small signal and noise parameters can often be replaced by one two-port with small signal and noise parameters derived from the individual parameters. The small signal parameters are computed by general circuit theory and for the noise parameters two methods exist.

One of these methods is to use the definition for the extended noise factor for a two-port, Equation (3.6), or for the extended effective noise temperature, Equation (3.4). In these equations the output noise power density is computed from contributions from the noise parameters, the source and perhaps some one-ports. The same result is then computed from the unknown noise parameters of the total circuit. The two results must be identically equal to each other for all values of the source immittance, and by choosing four values of this source immittance the four resulting equations may be solved for the new noise parameters.

The other method, introduced by Albinsson [6], consists of two parts. The first part aims at transforming the noise parameters of each individual two-port one at a time to the input of the overall two-port, replacing the other individual two-ports with their noise free equivalents. The second part is to combine these individual two-ports to a noise two-port for the resulting network and from this extract the resulting noise parameters.

The first of these two methods is used for the case of two parallel connected two-ports and for two two-ports in cascade. The first part of Albinsson's method is dependent on the circuit configuration, but the second part is very general and will be examined in this section.

6.3.1 Two-ports in parallel

Equations have been derived for two two-ports in parallel. Each two-port is described by the chain or ABCD parameters and by the R_n, G_n and Y_γ noise parameters. If more two-ports are connected in parallel they can be combined by taking two at a time until complete. The parameters from one two-port are all primed and from the other double-primed. The resulting parameters are unprimed.

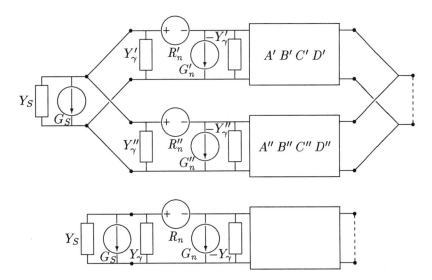

Figure 6.9: Two noisy two-ports in parallel and equivalent circuit.

The two circuits in Figure 6.9 have the same excess noise factor which with parameters from the combined circuit is expressed as

$$F_e - 1 = \frac{1}{G_S}(G_n + R_n|Y_S + Y_\gamma|^2) \qquad (6.51)$$

The excess noise factor can also be expressed as the ratio of the noise power density at the output from noise generators belonging to the two two-ports, to the output noise power density from the source. As it is the ratio of power densities and as they are proportional to the square of the output short circuit current, the excess noise factor can also be expressed as

$$F_e - 1 \quad = \quad \frac{|I_o{}_{G'_n}|^2 + |I_o{}_{G''_n}|^2 + |I_o{}_{R'_n}|^2 + |I_o{}_{R''_n}|^2}{|I_o{}_{G_S}|^2} \tag{6.52}$$

where

$$I_o{}_{G_S} \quad = \quad \frac{B' + B''}{B'D'' + B''D' + Y_S B' B''} I_{G_S}$$

$$I_o{}_{G'_n} \quad = \quad \frac{B' + B''}{B'D'' + B''D' + Y_S B' B''} I_{G'_n}$$

$$I_o{}_{G''_n} \quad = \quad \frac{B' + B''}{B'D'' + B''D' + Y_S B' B''} I_{G''_n}$$

$$I_o{}_{R'_n} \quad = \quad \frac{Y_S B'' + Y'_\gamma(B' + B'') + D'' - D'}{B'D'' + B''D' + Y_S B' B''} E_{R'_n}$$

$$I_o{}_{R''_n} \quad = \quad \frac{Y_S B' + Y''_\gamma(B' + B'') + D' - D''}{B'D'' + B''D' + Y_S B' B''} E_{R''_n}$$

are the output short circuit currents from each noise generator. The identity can now be expressed as

$$\frac{1}{G_S}\left(G_n + R_n |Y_S + Y_\gamma|^2\right) \equiv \frac{1}{G_S}\Bigg(G'_n + G''_n$$

$$+ R'_n \frac{|Y_S B'' + Y'_\gamma(B' + B'') + D'' - D'|^2}{|B' + B''|^2}$$

$$+ R''_n \frac{|Y_S B' + Y''_\gamma(B' + B'') + D' - D''|^2}{|B' + B''|^2}\Bigg)$$

From this identity the results for parallel connection are

$$R_n \quad = \quad \frac{R'_n |B''|^2 + R''_n |B'|^2}{|B' + B''|^2} \tag{6.53}$$

$$G_n \quad = \quad G'_n + G''_n + \frac{R'_n R''_n |B'Y'_\gamma - B''Y''_\gamma + D'' - D'|^2}{R'_n |B''|^2 + R''_n |B'|^2} \tag{6.54}$$

$$Y_\gamma \quad = \quad \frac{R'_n B''^*[Y'_\gamma(B' + B'') + D'' - D'] + R''_n B'^*[Y''_\gamma(B' + B'') + D' - D'']}{R'_n |B''|^2 + R''_n |B'|^2}$$

$$\tag{6.55}$$

In this way it is possible to compute the noise parameters of a known transistor in parallel with a feedback two-port. If this two-port consists of a piece of transmission line in series with a feedback element the noise parameters of the feedback two-port should be computed first. It is therefore necessary to investigate the noise parameters of two-ports in cascade.

6.3.2 Two-ports in cascade

The same procedure as above is used for calculating the noise parameters for the equivalent two-port of two two-ports in cascade. Again, if more two-ports are in cascade, they are taken two at a time. Let the small signal and noise parameters of the first two-port be primed, and let the parameters of the second be double-primed. The resulting parameters are unprimed.

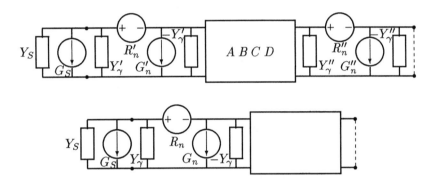

Figure 6.10: Two noisy two-ports in cascade and equivalent circuit.

Computing the noise currents just after the second-stage noise two-port (and thus before the noise free equivalent of the second stage, which does not add more noise power but only amplifies it with the same amplification), gives

$$I_o G_S = \frac{1}{Y_S B' + D'} I_{G_S}$$

$$I_o G'_n = \frac{1}{Y_S B' + D'} I_{G'_n}$$

$$I_o R'_n = \frac{Y_S + Y'_\gamma}{Y_S B' + D'} E_{R'_n}$$

$$I_o G''_n = I_{G''_n}$$

$$I_o R''_n = \left(Y''_\gamma + \frac{Y_S A' + C'}{Y_S B' + D'} \right) E_{R''_n}$$

From these equations the following identity can be derived:

$$\frac{1}{G_S}\left(G_n + R_n|Y_S + Y_\gamma|^2\right) \equiv \frac{1}{G_S}\left(G_n' + R_n'|Y_S + Y_\gamma'|^2 + G_n''|B'Y_S + D'|^2\right.$$

$$\left. + R_n''|(B'Y_s + D')Y_\gamma'' + A'Y_S + C'|^2\right)$$

From this identity the results for cascade connection are

$$R_n = R_n' + G_n''|B'|^2 + R_n''|B'Y_\gamma''| + A'|^2 \tag{6.56}$$

$$Y_\gamma = \frac{1}{R_n}\left[R_n'Y_\gamma' + G_n''B'^*D' + R_n''\left(B'Y_\gamma'' + A'\right)^*\left(D'Y_\gamma'' + C'\right)\right] \tag{6.57}$$

$$G_n = G_n' + R_n'|Y_\gamma'|^2 + G_n''|D'|^2 + R_n''|D'Y_\gamma'' + C'|^2 - R_n|Y_\gamma|^2 \tag{6.58}$$

Equations (6.56) – (6.58) can be used to compute the noise parameters of a feedback element which includes two-ports such as transmission lines used to connect one or two lumped feedback elements. The total feedback element can then be connected in parallel with a transistor and the noise parameters are computed by Equations (6.53) – (6.55).

6.3.3 Albinsson's method of interconnected two-ports

This method is divided into two parts. In the network of two-ports the noise parameters of each two-port in the first part is transformed to the input side of the resulting network as the noise parameters $R_{n,i}$, $G_{n,i}$ and $Y_{\gamma,i}$. This part is dependent of the network configuration and must be performed individually for the network considered. As the network is linear the superposition principle is valid and the second part is to perform this superposition.

In order to investigate this second part it should be kept in mind that the noise parameters used by Albinsson are the II noise parameters from [3] which are based on Figure 4.2. The noise voltage e_n is the summation of the transferred noise voltages $e_{n,i}$ and noise current i_n is the summation of the transferred noise currents $i_{n,i}$, thus

$$e_n = \sum_{i=1}^{I} e_{n,i} \tag{6.59}$$

$$i_n = \sum_{i=1}^{I} i_{n,i} \tag{6.60}$$

As the voltages from the different two-ports are uncorrelated it follows from Equations (2.13) and (2.6) that

$$R_n = \sum_{i=1}^{I} R_{n,i} \tag{6.61}$$

From Equations (4.9) and (4.5) it follows that

$$Y_\gamma = \frac{\langle i_n e_n^* \rangle}{\langle |e_n|^2 \rangle} \qquad (6.62)$$

Thus

$$Y_\gamma = \frac{\sum_{i=1}^I \langle i_{n,i} e_{n,i}^* \rangle}{\sum_{i=1}^I \langle |e_{n,i}|^2 \rangle} = \frac{1}{R_n} \sum_{i=1}^I Y_{\gamma,i} R_{n,i} \qquad (6.63)$$

The uncorrelated part of the current in Figure 4.2, i_{n1}, determines G_n. From

$$i_n = i_{n1} + e_n Y_\gamma \qquad (6.64)$$

it follows that

$$i_{n1} + Y_\gamma \sum_{i=1}^I e_{n,i} = \sum_{i=1}^I (i_{n1,i} + e_{n,i} Y_{\gamma,i}) \qquad (6.65)$$

and thus

$$i_{n1} = \sum_{i=1}^I [i_{n1,i} + e_{n,i}(Y_{\gamma,i} - Y_\gamma)] \qquad (6.66)$$

and, as noise from the different two-ports is uncorrelated,

$$G_n = \sum_{i=1}^I (G_{n,i} + R_{n,i}|Y_{\gamma,i} - Y_\gamma|^2) \qquad (6.67)$$

This is another way to compute the noise parameters R_n (Equation (6.61)), Y_γ (Equation (6.62)) and G_n (Equation (6.67)) of interconnected two-ports.

6.3.4 Matrix formulation

Hillbrand and Russer [7] have made a matrix formulation of parallel, series and cascade connections of linear, noisy two-ports. The noisy two-ports are represented either by an admittance representation with a noise free part and two noise current sources, or by an impedance representation with two noise voltage sources, or by a chain representation with a noise current source and a noise voltage source both at the input side. These representations are shown in Figure 6.11.

The noise free circuit matrices are the Y, Z and chain or $ABCD$ matrices, and the correlation matrices are – in the same order – given by

$$C_Y = \frac{1}{2\Delta f} \begin{bmatrix} \langle i_1 i_1^* \rangle & \langle i_1 i_2^* \rangle \\ \langle i_2 i_1^* \rangle & \langle i_2 i_2^* \rangle \end{bmatrix} \qquad (6.68)$$

$$C_Z = \frac{1}{2\Delta f} \begin{bmatrix} \langle u_1 u_1^* \rangle & \langle u_1 u_2^* \rangle \\ \langle u_2 u_1^* \rangle & \langle u_2 u_2^* \rangle \end{bmatrix} \qquad (6.69)$$

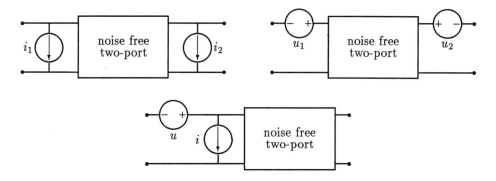

Figure 6.11: Admittance, impedance and chain representations.

$$C_A = \frac{1}{2\,\Delta f} \begin{bmatrix} \langle u\,u^* \rangle & \langle u\,i^* \rangle \\ \langle i\,u^* \rangle & \langle i\,i^* \rangle \end{bmatrix} \qquad (6.70)$$

where Δf is the bandwidth, the factor 2 occurs because the two-sided Fourier transform has been used and $\langle \cdots \rangle$ denotes the ensemble average over processes with identical statistical properties.[2]

Often the correlation matrices can be calculated without knowing the noise sources. If the two-port considered consists of only passive elements the thermal noise from it results in a correlation matrix of either of the two forms:

$$C_Y = 2\,k\,T\,\mathrm{Re}[Y]$$
$$C_Z = 2\,k\,T\,\mathrm{Re}[Z]$$

where $k = 1.38 \times 10^{-23}$ J K^{-1} is Boltzmann's constant and T [K] is the noise temperature of the two-port.

For active two-ports the chain correlation matrix is given by

$$C_A = 2\,k\,T \begin{bmatrix} R_n & \frac{1}{2}(F_{e\,min} - 1) - R_n Y_{SOF}^* \\ \frac{1}{2}(F_{e\,min} - 1) - R_n Y_{SOF} & R_n\,|Y_{SOF}|^2 \end{bmatrix} \qquad (6.71)$$

where R_n is the equivalent noise resistance, $F_{e\,min}$ is the minimum noise factor and Y_{SOF} is the corresponding source admittance.

These matrices can be transformed to any of the three forms by

$$C' = T\,C\,T^\dagger \qquad (6.72)$$

---

[2] See Appendix A.

	From admittance	From impedance	From chain
To admittance	$\begin{bmatrix} 1 & 0 \\ 0 & 1 \end{bmatrix}$	$\begin{bmatrix} Y_{11} & Y_{12} \\ Y_{21} & Y_{22} \end{bmatrix}$	$\begin{bmatrix} -Y_{11} & 1 \\ -Y_{21} & 0 \end{bmatrix}$
To impedance	$\begin{bmatrix} Z_{11} & Z_{12} \\ Z_{21} & Z_{22} \end{bmatrix}$	$\begin{bmatrix} 1 & 0 \\ 0 & 1 \end{bmatrix}$	$\begin{bmatrix} 1 & -Z_{11} \\ 0 & -Z_{21} \end{bmatrix}$
To chain	$\begin{bmatrix} 0 & B \\ 1 & D \end{bmatrix}$	$\begin{bmatrix} 1 & -A \\ 0 & -C \end{bmatrix}$	$\begin{bmatrix} 1 & 0 \\ 0 & 1 \end{bmatrix}$

Table 6.1: Transformation matrices.

where C and C' denote the correlation matrix of the original and resulting representation, respectively. The transformation matrix T is given in Table 6.1 and the dagger ($A^\dagger$) denotes the Hermitian conjugate (of A).

Interconnections of two two-ports in parallel, in series or in cascade result in a correlation matrix given by

$$C_Y = C_{Y_1} + C_{Y_2} \qquad \text{(parallel)} \qquad (6.73)$$

$$C_Z = C_{Z_1} + C_{Z_2} \qquad \text{(series)} \qquad (6.74)$$

$$C_A = C_{A_1} + A_1 \, C_{A_2} \, A_1^\dagger \qquad \text{(cascade)} \qquad (6.75)$$

where the subscripts 1 and 2 refer to the two-ports to be connected.

The noise parameters are obtained from

$$Y_{SOF} = \sqrt{\frac{C_{A,22}}{C_{A,11}} - \left(\frac{\text{Im}[C_{A,12}]}{C_{A,11}}\right)^2} + j\left(\frac{\text{Im}[C_{A,12}]}{C_{A,11}}\right) \qquad (6.76)$$

$$F_{e\,min} = 1 + \frac{C_{A,12} + C_{A,11}Y_{SOF}^*}{kT} \qquad (6.77)$$

$$R_n = C_{A,11} \qquad (6.78)$$

The noise factor as a function of the source impedance Z_S is given by

$$F_e = 1 + \frac{z^\dagger C_A \, z}{2\,kT\,\text{Re}[Z_S]} \qquad (6.79)$$

where

$$z = \begin{bmatrix} 1 \\ Z_S^* \end{bmatrix} \qquad (6.80)$$

This method can be used instead of section 6.1 as demonstrated in [7]. It is also useful when active two-ports are interconnected.

In a more recent paper Dobrowolski [8] has used a wave representation and enlarged the network to be considered. His network consists of interconnected passive multiports which introduce only thermal noise and active linear two-ports. It is also a requirement that the two-ports are interconnected two and two. Dobrowolski's results are applicable to computer aided design of noisy microwave circuits.

6.4 Calculating noise parameters from deembedded data

Pucel et al. [9] have shown a procedure to calculate noise parameters as functions of frequency for field effect transistors, FET's, and high electron mobility transistors, HEMT's. This is especially important at frequencies so high that it is very difficult to measure the noise parameters. If the noise parameters at one frequency and the parasitic elements of a transistor (with its encapsulation) are well known, the intrinsic elements and noise generators with their correlation can be computed by deembedding. As for FET's and HEMT's these noise parameters are approximately independent of frequency and the behaviour of the intrinsic equivalent circuit is well known, so it is possible to compute the S and noise parameters in a large frequency range. Thus computation replaces measurement which is a great advantage as noise parameter measurements are rather difficult at frequencies above $10 - 20$ GHz. The authors claim sufficient accuracy up to at least 40 GHz.

6.4.1 Matrix formulation of the deembedding procedure

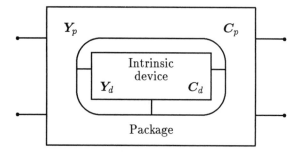

Figure 6.12: Y parameter and noise correlation matrices for a packaged two-port.

The deembedding uses the equivalent circuit shown in Figure 6.12 from which the correlation matrices of the packaged device, C_{pd}, and the intrinsic device, C_d, are related by the linear matrix equation

$$C_{pd} \;=\; P\,C_p\,P^\dagger \,+\, D\,C_d\,D^\dagger \qquad\qquad (6.81)$$

where C_p denotes the admittance correlation matrix of the packaged four-port, and P and D represent package and device transformation matrices expressible in terms of the admittance parameters of the package and device. The dagger denotes the Hermitian (conjugate transpose) of the associated matrix. Equation (6.81) is used when the intrinsic parameters are embedded in parasitic elements of the encapsulation.

Deembedding is performed by solving Equation (6.81) for C_d:

$$C_d \;=\; D^{-1}(C_{pd} - P\,C_p\,P^\dagger)D^{\dagger^{-1}} \qquad\qquad (6.82)$$

Under special conditions the matrix D may not possess an inverse, and the deembedding procedure fails. This may occur when certain non-reciprocal elements, such as isolators, are considered part of the package.

In order to find the matrix transformation matrices P and D in terms of the Y parameters of the package and the active device, some quantities are defined:

Y_p : four-port admittance matrix of the package.

Y_d : two-port admittance matrix of the active device.

Y_{pd} : two-port admittance matrix of the packaged device.

C_p : noise correlation matrix of the package.

C_d : noise correlation matrix of the active device.

C_{pd} : noise correlation matrix of the packaged device.

n_p : vector of the noise current sources of the package.

j_d : vector of the noise current sources of the active device.

I_k : identity matrix of order k.

In Figure 6.13 the active device is shown with noise current generators and terminal currents and voltages. In Figure 6.14 the package is shown with two ports to the outside world and two ports to the active device. It is seen that the active device is a two-port, the package a four-port and the packaged device a two-port.

The terminal voltages and currents of the package can be divided into two groups – one belonging to the external ports (with subscript e) and another belonging to

Figure 6.13: Active two-port with noise current generators.

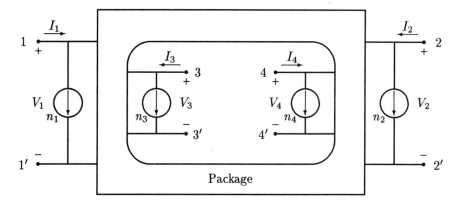

Figure 6.14: Package with noise current generators.

the internal ports (subscript i). In this way the following vectors are defined:

$$
v_e = \begin{bmatrix} V_1 \\ V_2 \end{bmatrix} \qquad v_i = \begin{bmatrix} V_3 \\ V_4 \end{bmatrix}
$$

$$
i_e = \begin{bmatrix} I_1 \\ I_2 \end{bmatrix} \qquad i_i = \begin{bmatrix} I_3 \\ I_4 \end{bmatrix}
$$

and

$$
v = \begin{bmatrix} v_e \\ \cdots \\ v_i \end{bmatrix} = \begin{bmatrix} V_1 \\ V_2 \\ V_3 \\ V_4 \end{bmatrix}
$$

$$
i = \begin{bmatrix} i_e \\ \cdots \\ i_i \end{bmatrix} = \begin{bmatrix} I_1 \\ I_2 \\ I_3 \\ I_4 \end{bmatrix}
$$

In the same manner the package noise source matrices are defined:

$$
\boldsymbol{n}_{pe} = \begin{bmatrix} n_1 \\ n_2 \end{bmatrix}, \qquad \boldsymbol{n}_{pi} = \begin{bmatrix} n_3 \\ n_4 \end{bmatrix}, \qquad \boldsymbol{n}_p = \begin{bmatrix} \boldsymbol{n}_{pe} \\ \cdots \\ \boldsymbol{n}_{pi} \end{bmatrix}
$$

The admittance matrix of the package is partitioned into four 2 × 2 submatrices:

$$
\boldsymbol{Y}_p = \begin{bmatrix} \boldsymbol{Y}_{ee} & \vdots & \boldsymbol{Y}_{ei} \\ \cdots & & \cdots \\ \boldsymbol{Y}_{ie} & \vdots & \boldsymbol{Y}_{ii} \end{bmatrix} \tag{6.83}
$$

and the signal and noise nodal equations for the package can be expressed in terms of the above vectors and matrices either succinctly as

$$
\boldsymbol{i} = \boldsymbol{Y}_p \boldsymbol{v} + \boldsymbol{n}_p \tag{6.84}
$$

or in partitioned form as

$$
\boldsymbol{i}_e = \boldsymbol{Y}_{ee} \boldsymbol{v}_e + \boldsymbol{Y}_{ei} \boldsymbol{v}_i + \boldsymbol{n}_{pe} \tag{6.85}
$$

$$
\boldsymbol{i}_i = \boldsymbol{Y}_{ie} \boldsymbol{v}_e + \boldsymbol{Y}_{ii} \boldsymbol{v}_i + \boldsymbol{n}_{pi} \tag{6.86}
$$

A similar procedure is followed for the active device. With reference to Figure 6.13 the following current and voltage vectors are defined:

$$
\boldsymbol{i}_d = \begin{bmatrix} I_{d3} \\ I_{d4} \end{bmatrix}, \qquad \boldsymbol{v}_d = \begin{bmatrix} V_{d3} \\ V_{d4} \end{bmatrix}, \qquad \boldsymbol{j}_d = \begin{bmatrix} j_3 \\ j_4 \end{bmatrix}
$$

and the nodal equations for the active device can be written as

$$
\boldsymbol{i}_d = \boldsymbol{Y}_d \boldsymbol{v}_d + \boldsymbol{j}_d \tag{6.87}
$$

Applying the boundary conditions

$$
\boldsymbol{i}_d = -\boldsymbol{i}_i, \qquad \boldsymbol{v}_d = \boldsymbol{v}_i
$$

to Equation (6.87), inserting (6.87) into (6.86) and solving for $\boldsymbol{v}_i$ the following equation is obtained:

$$
\boldsymbol{v}_i = -(\boldsymbol{Y}_{ii} + \boldsymbol{Y}_d)^{-1}(\boldsymbol{Y}_{ie} \boldsymbol{v}_e + \boldsymbol{n}_{pi} + \boldsymbol{j}_d) \tag{6.88}
$$

Inserting Equation (6.88) into (6.85) gives

$$
\boldsymbol{i}_e = \boldsymbol{Y}_{ee} \boldsymbol{v}_e - \boldsymbol{Y}_{ei}(\boldsymbol{Y}_{ii} + \boldsymbol{Y}_d)^{-1}(\boldsymbol{Y}_{ie} \boldsymbol{v}_e + \boldsymbol{n}_{pi} + \boldsymbol{j}_d) + \boldsymbol{n}_{pe}
$$

This formula can also be written as

$$i_e = Y_e v_e + i_n \qquad (6.89)$$

where the first term represents the signal component of the current and the second term the noise component. It is evident that Y_e represents the admittance matrix of the two-port packaged device. It is given by the expression

$$Y_e = Y_{ee} + D Y_{ie} \qquad (6.90)$$

where the matrix D is defined as

$$D = -Y_{ei} (Y_{ii} + Y_d)^{-1} \qquad (6.91)$$

The noise matrix term i_n denotes the sum of the contributions of the package and the active device. In matrix form it can be represented as

$$i_n = P n_p + D j_d \qquad (6.92)$$

where

$$P = [I_2 \vdots D] \qquad (6.93)$$

Now the correlation noise matrix of the packaged device

$$C_{pd} = \langle i_n i_n^\dagger \rangle \qquad (6.94)$$

where $\langle \cdots \rangle$ denotes the ensemble average over processes with identical statistical properties.[3] As no correlation exists between the noise sources in the device and the package,

$$C_{pd} = P \langle n_p n_p^\dagger \rangle P^\dagger + D \langle j_p j_p^\dagger \rangle D^\dagger \qquad (6.95)$$

The definitions

$$C_p = \langle n_p n_p^\dagger \rangle \quad \text{and} \quad C_d = \langle j_p j_p^\dagger \rangle \qquad (6.96)$$

lead to

$$C_{pd} = P C_p P^\dagger + D C_d D^\dagger \qquad (6.97)$$

which is Equation (6.81). This equation can be reduced to one involving matrices of order 2 only, when C_p is partitioned in the same way as Y_p (Equation (6.83)).

$$C_{pd} = C_{ee} + D C_{ie} + C_{ei} D^\dagger + D (C_{ii} + C_d) D^\dagger \qquad (6.98)$$

---

[3] Please note that the statistical noise averages, hence correlation matrices, are normalized to $2 k T_0 \Delta f$, where $k = 1.38 \times 10^{-23}$ J K^{-1} is Boltzmann's constant, $T_0 = 290$ K is the standard noise temperature and Δf is the noise bandwidth in Hz. The factor 2 (instead of 4) is chosen because both positive and negative frequencies are included. In any case, this factor cancels out in the final noise parameter expressions.

The deembedding procedure is to find C_d from (6.98) and thus

$$C_d = D^{-1}(C_{pd} - C_{ee})D^{\dagger-1} - C_{ie}D^{\dagger-1} - D^{-1}C_{ei} - C_{ii} \qquad (6.99)$$

This equation is equivalent to (6.82) but it includes only matrices of order 2. It now remains to find the expressions for the correlation matrices C_{pd} and C_p for the packaged device and the package in order to be able to compute the correlation matrix C_d for the device from (6.99).

Twiss [10] has shown that for a linear and passive network, such as the package, which generates only thermal noise, its noise correlation matrix is

$$C_p = \begin{bmatrix} C_{ee} & \vdots & C_{ei} \\ \cdots & & \cdots \\ C_{ie} & \vdots & C_{ii} \end{bmatrix} = \tfrac{1}{2}(Y_p + Y_p^{\dagger}) \qquad (6.100)$$

It should be noted that this equation does not require reciprocity, but in the rest of this section reciprocity is assumed.

Hillbrand and Russer [7] have shown how the noise correlation matrix in ABCD form relates to the noise parameters. Let C_A denote this correlation matrix. From [7] or Equation (6.71) it is seen that

$$C_{A,11} = R_n \qquad (6.101)$$

$$C_{A,12} = \tfrac{1}{2}(F_{e\,min} - 1) - R_n Y_{SOF}^* = R_n Y_\gamma^* \qquad (6.102)$$

$$C_{A,21} = C_{A,12}^* = R_n Y_\gamma \qquad (6.103)$$

$$C_{A,22} = R_n|Y_{SOF}|^2 = G_n + R_n|Y_\gamma|^2 \qquad (6.104)$$

C_{pd} is the correlation noise matrix in admittance form which is obtained from C_A by

$$C_{pd} = V C_A V^{\dagger} \qquad (6.105)$$

where V is expressed in terms of elements of the admittance matrix Y_e (Equation (6.90)) as

$$V = \begin{bmatrix} -Y_{e,11} & 1 \\ -Y_{e,21} & 0 \end{bmatrix} \qquad (6.106)$$

The reverse transformation is given by

$$C_A = V^{-1}C_{pd}V^{\dagger-1} \qquad (6.107)$$

from which

$$F_{e\,min} = 1 + 2(C_{A,12} + C_{A,11}Y_{SOF}^*) \qquad (6.108)$$

$$R_n = C_{A,11} \qquad (6.109)$$

$$Y_{SOF} = \sqrt{\frac{C_{A,22}}{C_{A,11}} - \left(\frac{\mathrm{Im}[C_{A,12}]}{C_{A,11}}\right)^2} + j\left(\frac{\mathrm{Im}[C_{A,12}]}{C_{A,11}}\right) \qquad (6.110)$$

or R_n, G_n and Y_γ are expressed in terms of C_A as

$$R_n = C_{A,11} \qquad (6.111)$$

$$G_n = C_{A,22} - \frac{C_{A,12}\,C_{A,21}}{C_{A,11}} \qquad (6.112)$$

$$Y_\gamma = \frac{C_{A,21}}{C_{A,11}} \qquad (6.113)$$

In many cases this deembedding procedure, which here is performed with matrix algebra, can be performed as shown in Example 6.3 in section 6.1 and in section 6.2, where sections of transmission line in a similar way can be removed from the encapsulation by adding the negative lengths.

6.4.2   Calculating noise parameters at a new frequency

Now the correlation matrix of the active device C_d is known at one frequency. Since the two-port is linear it is possible to write C_d as a superposition of two terms:

$$C_d = T N T^\dagger + S N_t S^\dagger \qquad (6.114)$$

where T and S are transformation matrices which are functions of the equivalent circuit parameters. N_t represents the thermal noise contributions included with the active device and is known from the equivalent circuit parameters. N represents the noise from the noise generators in the intrinsic transistor and from [11] Equation (6.114), N can be expressed as

$$N = T^{-1}(C_d - S N_t S^\dagger)T^{\dagger-1} \qquad (6.115)$$

Now consider a frequency change. The frequency dependence of N is known.[4] N_t represents thermal noise and thus the noise generators are independent of frequency, but the influence of frequency dependent elements shapes the noise spectra of the thermal noise generators in their transformation to the reference planes corresponding to the matrix N_t. The transformation matrices, T and S, should also

---

[4]N is often independent of frequency, e.g. in case of a FET without flicker noise.

be calculated at the new frequency and the noise parameters are calculated in accordance with the method outlined below.

Deembedding procedure:

1. At some frequency, where the noise parameters $F_{e\,min}$, Y_{SOF} and R_n are known, calculate C_A from Equations (6.101) – (6.104).

2. From measurements or simulation calculate the package admittance Y_p and partition it as shown in Equation (6.83).

3. Calculate the device admittance matrix Y_d from its equivalent circuit.

4. Calculate D by Equation (6.91).

5. Calculate V by Equation (6.106).

6. Calculate C_{pd} by Equation (6.105).

7. Calculate and partition C_p by Equation (6.100).

8. Calculate C_d by Equation (6.99).

9. Calculate T, S and N_t for the particular device.

10. Calculate N by Equation (6.115).

Change to a new frequency and at this frequency:

1. Calculate T, S and N_t.

2. Calculate C_d by Equation (6.114) using N from step 10 above.

3. Determine Y_p from measurement or simulation and partition as in Equation (6.83).

4. Calculate Y_d from the equivalent circuit of the device.

5. Calculate D by Equation (6.91).

6. Calculate Y_e by Equation (6.90).

7. Calculate and partition C_p by Equation (6.100).

8. Calculate C_{pd} by Equation (6.98).

9. Calculate V by Equation (6.106).

10. Calculate C_A by Equation (6.107).

11. Calculate the new noise parameters from Equations (6.108) – (6.110).

6.5   Transformer coupled feedback

Consider an active two-port with feedback consisting of two transformers – one
connected to give voltage series feedback and the other to give current parallel
feedback as shown in Figure 6.15. Let the two transformers be identical. This type
of circuit is often used but how does its noise two-port look? Neither of the above
mentioned methods leads to a result.

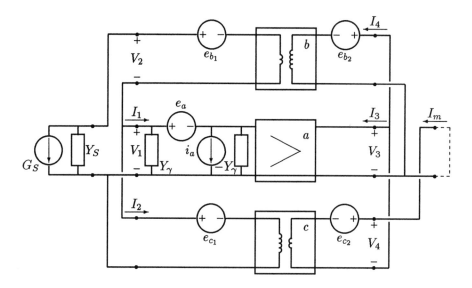

Figure 6.15: Amplifier with double feedback.

The method used here is to compute the noise factor of the circuit in Figure
6.15 and also to compute the noise factor of its equivalent circuit with the unknown
noise two-port. As these two noise factor expressions are identical the new (primed)
noise parameters can be found by setting the two expressions equal for four different
values of source admittances with the real part positive. This method has also been
used to get the results in section 6.1.

In Figure 6.15 the two transformers are characterized by their Z parameters and
the transistor by its Y parameters:

$$\mathbf{Z}_b = \begin{bmatrix} Z_1 & Z_0 \\ Z_0 & Z_2 \end{bmatrix} \tag{6.116}$$

$$\mathbf{Z}_c = \begin{bmatrix} Z_2 & Z_0 \\ Z_0 & Z_1 \end{bmatrix} \tag{6.117}$$

$$\mathbf{Y}_a = \begin{bmatrix} Y_{11} & Y_{12} \\ Y_{21} & Y_{22} \end{bmatrix} \tag{6.118}$$

where $Z_1 = r_1 + j\omega L_1$, $Z_2 = r_2 + j\omega L_2$ and $Z_0 = j\omega M$, corresponding to the impedances of the two inductances and the mutual inductance. The noise currents and voltages are all uncorrelated and the equivalent noise two-port of the amplifier is shown. The noise currents and voltages are determined from the noise parameters by Equations (2.7) and (2.9).

As the noise sources are uncorrelated, the noise factor equals the sum of the output noise powers originating from the noise sources and from the source at standard noise temperature divided by the output noise power from the source at standard noise temperature. There are seven noise sources and each of them gives a contribution to the output noise power. Eight (numbered) equations can be written as follows where the right sides of seven of the equations are either zero or contain a noise source:

$$-V_1 + Z_2 I_2 + Z_0 I_3 + Z_0 I_4 = \begin{cases} 0 \\ -e_{c_1} \end{cases} \tag{6.119}$$

$$-V_2 + Z_1 I_1 + Z_1 I_2 + Z_0 I_4 = \begin{cases} 0 \\ -e_{b_1} \end{cases} \tag{6.120}$$

$$-V_3 + Z_0 I_1 + Z_0 I_2 + Z_2 I_4 = \begin{cases} 0 \\ -e_{b_2} \end{cases} \tag{6.121}$$

$$-V_4 + Z_0 I_2 + Z_1 I_3 + Z_1 I_4 = \begin{cases} 0 \\ -e_{c_2} \end{cases} \tag{6.122}$$

$$\begin{bmatrix} Y_{11} & Y_{12} \\ Y_{21} & Y_{22} \end{bmatrix} \begin{bmatrix} V_1 \\ V_3 \end{bmatrix} + \begin{bmatrix} -1 & 0 \\ 0 & -1 \end{bmatrix} \begin{bmatrix} I_1 \\ I_3 \end{bmatrix} = \begin{bmatrix} 0 \\ 0 \end{bmatrix} \text{ or } \begin{bmatrix} -i_a \\ 0 \end{bmatrix} \tag{6.123}$$

$$\begin{bmatrix} Y_{11} & Y_{12} \\ Y_{21} & Y_{22} \end{bmatrix} \begin{bmatrix} V_1 \\ V_3 \end{bmatrix} + \begin{bmatrix} -1 & 0 \\ 0 & -1 \end{bmatrix} \begin{bmatrix} I_1 \\ I_3 \end{bmatrix} = \begin{bmatrix} 0 \\ 0 \end{bmatrix} \text{ or } \begin{bmatrix} -e_a (Y_\gamma - Y_{11}) \\ e_a Y_{21} \end{bmatrix}$$

$$\tag{6.124}$$

$$Y_S V_1 + Y_S V_2 + I_1 + I_2 = \begin{cases} 0 \\ -i_{G_S} \end{cases} \tag{6.125}$$

$$V_3 + V_4 = 0 \tag{6.126}$$

The noise sources are taken one at a time and in the other equations the zeros are used. These equations can be written in matrix form as shown:

$$
\begin{bmatrix}
-1 & 0 & 0 & 0 & 0 & Z_2 & Z_0 & Z_0 \\
0 & -1 & 0 & 0 & Z_1 & Z_1 & 0 & Z_0 \\
0 & 0 & -1 & 0 & Z_0 & Z_0 & 0 & Z_2 \\
0 & 0 & 0 & -1 & 0 & Z_0 & Z_1 & Z_1 \\
Y_{11} & 0 & Y_{12} & 0 & -1 & 0 & 0 & 0 \\
Y_{21} & 0 & Y_{22} & 0 & 0 & 0 & -1 & 0 \\
Y_S & Y_S & 0 & 0 & 1 & 1 & 0 & 0 \\
0 & 0 & 1 & 1 & 0 & 0 & 0 & 0
\end{bmatrix}
\begin{bmatrix}
V_1 \\ V_2 \\ V_3 \\ V_4 \\ I_1 \\ I_2 \\ I_3 \\ I_4
\end{bmatrix}
= \quad a \qquad (6.127)
$$

Here for each noise contribution a is equal to one of the seven vectors below containing only one noise source each:

$$
\begin{bmatrix} 0 \\ 0 \\ 0 \\ 0 \\ 0 \\ 0 \\ -i_{GS} \\ 0 \end{bmatrix}
\begin{bmatrix} 0 \\ -e_{b_1} \\ 0 \\ 0 \\ 0 \\ 0 \\ 0 \\ 0 \end{bmatrix}
\begin{bmatrix} 0 \\ 0 \\ -e_{b_2} \\ 0 \\ 0 \\ 0 \\ 0 \\ 0 \end{bmatrix}
\begin{bmatrix} -e_{c_1} \\ 0 \\ 0 \\ 0 \\ 0 \\ 0 \\ 0 \\ 0 \end{bmatrix}
\begin{bmatrix} 0 \\ 0 \\ 0 \\ -e_{c_2} \\ 0 \\ 0 \\ 0 \\ 0 \end{bmatrix}
\begin{bmatrix} 0 \\ 0 \\ 0 \\ 0 \\ -e_a(Y_\gamma - Y_{11}) \\ e_a Y_{21} \\ 0 \\ 0 \end{bmatrix}
\begin{bmatrix} 0 \\ 0 \\ 0 \\ 0 \\ -i_a \\ 0 \\ 0 \\ 0 \end{bmatrix}
$$

Solving Equation (6.127) seven times for $I_m = I_3 + I_4$ where $m = 0, 1, \ldots, 6$ it follows from Cramer's rule [13] that I_m is of the form

$$
I_m = \frac{(a_m Y_S + b_m) s_m}{D}
$$

where $s_0 = i_{GS}$, $s_1 = e_{b_1}$, $s_2 = e_{b_2}$, $s_3 = e_{c_1}$, $s_4 = e_{c_2}$, $s_5 = e_a$, $s_6 = i_a$ and D is the determinant. a_m and b_m are dependent on the matrix Y and Z_0, Z_1 and Z_2. It is not necessary to compute D as it is not used in the following, but it is also of the form $a_7 Y_S + b_7$. By use of the algebraic computer program Maple V [12] the quantities $a_0 - a_6$ and $b_0 - b_6$ are computed as:

$$a_0 = 0$$

$$b_0 = Y_{11} Z_0 Z_2 - Y_{12} Z_0^2 - Y_{21} Z_2^2 + Y_{22} Z_0 Z_2 + 2 Z_0$$

$$a_1 = Y_{11} Z_0 Z_2 - Y_{12} Z_0^2 - Y_{21} Z_2^2 + Y_{22} Z_0 Z_2 + 2 Z_0$$

$$b_1 = 0$$

$$a_2 = - Y_{11} Z_1 Z_2 + Y_{12} Z_0 Z_1 + Y_{21} Z_0 Z_2 - Y_{22} Z_0^2 - Z_1 - Z_2$$

$$b_2 = -1 - Y_{11} Z_2 + Y_{12} Z_0$$

$$a_3 = \Delta_Y Z_0 Z_1 Z_2 - \Delta_Y Z_0^3 + Y_{11} Z_0 Z_1 - Y_{12} Z_0^2$$
$$+ Y_{21} Z_1 Z_2 - 2 Y_{21} Z_0^2 + Y_{22} Z_0 Z_2 + 2 Z_0$$

$$b_3 = \Delta_Y Z_0 Z_2 + Y_{11} Z_0 + Y_{21} Z_2$$

$$a_4 = \Delta_Y Z_0^2 Z_2 - \Delta_Y Z_1 Z_2^2 - Y_{11} Z_1 Z_2 + Y_{12} Z_0 Z_2 + Y_{21} Z_0 Z_2$$
$$+ Y_{22} Z_0^2 - Y_{22} Z_1 Z_2 - Y_{22} Z_2^2 - Z_1 - Z_2$$

$$b_4 = -1 - \Delta_Y Z_2^2 - Y_{11} Z_2 - Y_{22} Z_2$$

$$a_5 = \Delta_Y Z_0^3 - \Delta_Y Z_0 Z_1 Z_2 - Y_{11} Z_0 Z_1 + Y_{11} Z_0 Z_2 + 2 Y_{21} Z_0^2$$
$$- Y_{21} Z_1 Z_2 - Y_{21} Z_2^2 + Y_\gamma Z_0 Z_1 - Y_\gamma Z_0 Z_2$$
$$+ Y_\gamma Y_{21} Z_0^2 Z_2 - Y_\gamma Y_{21} Z_1 Z_2^2 - Y_\gamma Y_{22} Z_0^3 + Y_\gamma Y_{22} Z_0 Z_1 Z_2$$

$$b_5 = -\Delta_Y Z_0 Z_2 - Y_{11} Z_0 - Y_{21} Z_2 + Y_\gamma Z_0 - Y_\gamma Y_{21} Z_2^2 + Y_\gamma Y_{22} Z_0 Z_2$$

$$a_6 = Y_{21} Z_0^2 Z_2 - Y_{21} Z_1 Z_2^2 - Y_{22} Z_0^3 + Y_{22} Z_0 Z_1 Z_2 + Z_0 Z_1 - Z_0 Z_2$$

$$b_6 = -Y_{21} Z_2^2 + Y_{22} Z_0 Z_2 + Z_0$$

As $|I_m|^2$ is proportional to the mth contribution to the output noise power it follows that

$$F_e = \frac{\sum_{m=0}^6 |I_m|^2}{|I_0|^2} \equiv 1 + \frac{1}{G_S}\left(G_n' + R_n'|Y_S + Y_\gamma'|^2\right) \tag{6.128}$$

where $G_S = \mathrm{Re}[Y_S]$. As G_S is proportional to $|s_0|^2$ – and $4\,k\,T_0\,\Delta f$ can be cancelled in numerator and denominator – the identity reduces to

$$\frac{\sum_{m=1}^6 |(a_m Y_S + b_m)\,s_m|^2}{|a_0 Y_S + b_0|^2} \equiv G_n' + R_n'|Y_S + Y_\gamma'|^2 \tag{6.129}$$

Introducing the two vectors

$$v_1 = (a_1 s_1,\, a_2 s_2,\, \ldots,\, a_6 s_6) \tag{6.130}$$

$$v_o = (b_1 s_1,\, b_2 s_2,\, \ldots,\, b_6 s_6) \tag{6.131}$$

with the scalar product $(u, v) = \sum_i u_i v_i^*$ and norm $||u|| = \sqrt{(u, u)}$ and as $a_0 = 0$, Equation (6.129) can be written as

$$\frac{||v_1 Y_S + v_o||^2}{|b_0|^2} \equiv G_n' + R_n'|Y_S + Y_\gamma'|^2 \tag{6.132}$$

Equation (6.132) leads to

$$R'_n = \frac{||v_1||^2}{|b_0|^2} \geq 0 \tag{6.133}$$

$$2\,\mathrm{Re}[Y_S(v_1, v_o)] = 2\,\mathrm{Re}[Y_S Y'^*_\gamma ||v_1||^2] \tag{6.134}$$

which gives for $Y_S = 1$:

$$\mathrm{Re}[(v_o, v_1)] = \mathrm{Re}[Y'_\gamma ||v_1||^2]$$

and for $Y_S = j$:

$$\mathrm{Im}[(v_o, v_1)] = \mathrm{Im}[Y'_\gamma ||v_1||^2]$$

or

$$Y'_\gamma = \frac{(v_o, v_1)}{||v_1||^2} \tag{6.135}$$

Finally

$$G'_n |b_0|^2 = ||v_o||^2 - ||v_1||^2 |Y'_\gamma|^2$$

$$= ||v_o||^2 - \frac{||v_1||^2 |(v_o, v_1)|^2}{||v_1||^4}$$

$$G'_n = \frac{||v_o||^2 ||v_1||^2 - |(v_o, v_1)|^2}{|b_0|^2 ||v_1||^2} \geq 0 \tag{6.136}$$

and the expressions for the noise parameters R'_n, G'_n and Y'_γ are found.

6.6   Mixed input

In order to obtain a low noise factor and simultaneously a good input power match, it is sometimes the practice to ground the input transistor somewhere between the base and the emitter and thus apply the input signal to both base and emitter by means of a lossless transformer as shown in Figure 6.16.

For such a circuit the noise parameters can be calculated by expressing the output noise power density with the source at standard noise temperature, and from this an expression can be made for F_e. This expression is identically equal to the standard expression for F_e (Equation (4.11)) for all source admittances and thus the following equations are found:

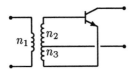

Figure 6.16: Mixed type input circuit.

$$G'_n = \left(\frac{n_2 + n_3}{n_1}\right)^2 G_n \qquad (6.137)$$

$$R'_n = \frac{|n_1^2 Y_{21}|^2}{|n_1 (n_2 + n_3) Y_{21} + n_1 n_3 Y_{22}|^2} R_n \qquad (6.138)$$

$$Y'_\gamma = \left(\frac{n_2 + n_3}{n_1}\right)^2 Y_\gamma + \frac{(n_2 + n_3) n_3}{n_1^2} \frac{Y_{22}}{Y_{21}} Y_\gamma$$

$$- \frac{(n_2 + n_3) n_3}{n_1^2} \frac{\Delta_Y}{Y_{21}} \qquad (6.139)$$

6.7 Transformation to common base and common collector

Equations (6.137) – (6.139) can be used to compute the common base (gate) noise parameters expressed by the common emitter (source) noise and small signal parameters. Let $n_2 = 0$ and $n_3 = n_1$ and one gets:

$$G_n^b = G_n^e \qquad (6.140)$$

$$R_n^b = \frac{|Y_{21\,e}|^2}{|Y_{21\,e} + Y_{22\,e}|^2} R_n^e \qquad (6.141)$$

$$Y_\gamma^b = \left(1 + \frac{Y_{22\,e}}{Y_{21\,e}}\right) Y_\gamma^e - \frac{\Delta_{Y\,e}}{Y_{21\,e}} \qquad (6.142)$$

Transformation from common emitter (source) noise parameters to common collector (drain) noise parameters by use of the Z noise parameters gives the result:

$$r_n^c = r_n^e \qquad (6.143)$$

$$g_n^c = \frac{|Z_{21\,e}|^2}{|Z_{22\,e} - Z_{21\,e}|^2} g_n^e \qquad (6.144)$$

$$Z_\gamma^c = \left(1 - \frac{Z_{22e}}{Z_{21e}}\right) Z_\gamma^e + \frac{\Delta_{Ze}}{Z_{21e}} \qquad (6.145)$$

These results are also given in [14] and [15], but expressed in a different way.

6.8   Noise computations in computer aided design programs

One type of popular optimization program comprises those based on the adjoint network. Without going into detail it is clear that uncorrelated noise contributions should be added as powers or power densities. Every one-port contains one noise generator which is uncorrelated to the other noise sources as they are physically separated. The two-ports contain two partly correlated noise generators which are separated into the equivalent noise two-port as shown in Chapter 4 and thus two uncorrelated sources and a correlation immittance emerge. As equivalent noise multi-ports have not yet been developed to contain uncorrelated sources it is necessary to include correlation matrices.

Dobrowolski's book on computer methods for microwave circuit analysis and design [16] includes linear noise computation of one-, two- and multi-ports.

6.9   References

[1] Engberg, J.: "Simultaneous input power match and noise optimization using feedback", *R69ELS-79*, Electronics Laboratory, General Electric, Syracuse, NY, 1969.

[2] Engberg, J.: "Simultaneous input power match and noise optimization using feedback", *Proc. 4th European Microwave Conference*, Montreux, 1974, pp. 385 – 389.

[3] Rothe, H. & Dahlke, W.: "Theory of noisy fourpoles", *Proc. IRE*, vol. 44, pp. 811 – 818, June 1956.

[4] Lehmann, R. E. & Heston, D. D.: "X-band monolithic series feedback LNA", *IEEE Trans. on Microwave Theory and Techniques*, vol. MTT–33, pp. 1560 – 1566, Dec. 1985.

[5] Engberg, J.: "Noise parameters of embedded noisy two-port networks", *IEE Proc.*, vol. 132, part. H, Feb 1985.

[6] Albinsson, B.: "Noise parameter transformations of interconnected two-port networks", *IEE Proc.*, vol. 134, part H, pp. 125 – 129, April 1987.

[7] Hillbrand, H. & Russer, P. H.: "An efficient method for computer aided noise analysis of linear amplifier networks", *IEEE Trans. on Circuit Systems*, vol. CAS-23, pp. 235 – 238, April 1976.

[8] Dobrowolski, J. A.: "A CAD-oriented method for noise figure computation of two-ports with any internal topology", *IEEE Trans. on Microwave Theory and Techniques*, vol. MTT-37, pp. 15 – 20, January 1989.

[9] Pucel, R. A., Struble, W., Hallgren, R. & Rohde, U. L.: "A general noise de-embedding procedure for packaged two-port linear active devices", *IEEE Trans. on Microwave Theory and Techniques*, vol. MTT-40, pp. 2013 – 2023, November 1992.

[10] Twiss, R. Q.: "Nyquist's and Thevenin's theorems generalized for nonreciprocal linear networks", *Journal of Applied Physics*, vol. 26, pp. 599 – 602, May 1955.

[11] Rohde, U. L.: "New nonlinear noise model for MESFETs including MM-wave application", *Dig. 1990 IEEE Integrated Nonlinear Microwave and Millimeterwave Circuits Workshop*, Duisburg.

[12] Char, B. W., Geddes, K. O., Gonnet, G. H., Leong, B. L., Monagan, M. B. & Watt, S. M.: "Maple V", Springer, 1991.

[13] Hildebrand, F. B.: "Methods of applied mathematics", 2nd edition, Prentice Hall, 1965.

[14] Stangerup, P.: "Electronic noise calculation by computer", *ECR-58*, Danish Research Centre for Applied Electronics, February 1976.

[15] Hagen, J. B.: "Noise parameter transformation for three-terminal amplifiers", *IEEE Trans. on Microwave Theory and Techniques*, vol. MTT-38, pp. 319-321, March 1990.

[16] Dobrowolski, J. A.: "Introduction to computer methods for microwave circuit analysis and design", Artech House, 1991.

Part II

Non-linear systems

7

Noise in non-linear systems: Theory

This chapter deals with the theory of time invariant non-linear noisy multi-port non-autonomous systems. The type of non-linear system considered may contain non-linear one- and multi-port non-linear elements and subsystems, the internal noise sources may be unmodulated (independent) or modulated (dependent), the system may contain dc sources, and the multiple signal input ports may be excited by multiple finite energy[1] signals (also dc) and noise. The method of analysis is based on the use of Volterra series which requires an equivalent circuit description of the non-linear noisy system. The main objective of the present chapter is to derive expressions for the noise and deterministic response at arbitrary response ports in the situation where low level noise is analyzed. Low level noise refers to the situation where the noise is a small perturbation of the deterministic signal regime. This means that the non-linear contributions caused by a non-linear mixing of noise with noise are insignificant.

7.1 Introduction

For non-linear systems the principle of superposition is not valid. This, as a consequence, implies that the (non-linear) transfer function seen from a given noise source in a system to a given output response port does not only depend on the system itself but also on the applied deterministic signals. This makes non-linear noise analysis much more complicated than noise analysis of linear systems. The purpose of the present chapter is to develop a method to analyze low level noise in general time invariant non-linear multi-port non-autonomous systems — general in the sense that the equivalent noise free system may be described by a finite (convergent) multi-port Volterra series. This suggests that the type of system allows: (i) multiple input ports which can be excited by both deterministic signals

---

[1]In an observation time interval of finite length.

(also dc) and noise, (ii) internal dc sources in the system, (iii) multiple unmodulated (independent) internal noise sources, and (iv) multiple modulated (dependent) internal noise sources which may be modulated by multiple arbitrary deterministic signals. The system may have an arbitrary number of responses (or more generally, arbitrary system variables) which may be, e.g., voltages or currents at any node or branch respectively in the underlying network. The low level noise assumption implies that only systems which are small signal linear may be analyzed — this means that contributions caused by non-linear mixing of noise with noise have to be insignificant. However, most of the known devices are small signal linear. Also dc sources in the system are allowed. Traditionally, dc analysis has not been used in relation to Volterra series analysis. This is because (i) it has not previously been possible to analyze multi-port systems, which is required when dc is applied not only to the signal input port, and (ii) Volterra series have traditionally only been used for weakly non-linear systems in which case the influence of the deterministic signals on the dc due to non-linear phenomena is insignificant.

The objective of the present chapter is to determine the ensemble cross-correlation (or autocorrelation) between response Fourier series coefficients at two arbitrary frequencies at arbitrary response ports. This makes it possible to determine average noise powers and average noise power densities which are used extensively in noise analysis. Practical applications are expected to be in the analysis and optimization of noise in mixers with moderate local oscillator levels, interconnected mixers and oscillators, some types of frequency multipliers (e.g. FET types), oscillators (once the oscillation frequency and amplitude are determined, using e.g. [1,2]), and communications systems. The theoretical work may also be used in a combined analysis of intermodulation and noise, and to analyze the noise properties when more complicated deterministic excitations (multiple sinusoidal excitations) are used. Another possible application is in the development of non-linear models of various devices.

The chapter is organized as follows. Section 7.2 presents preliminaries regarding the type of system which is under consideration and discusses the mathematical representation of deterministic signals and noise in the frequency domain. Section 7.3 outlines a method which can be used to represent modulated (dependent) as well as unmodulated (independent) noise sources. Lastly in Section 7.4 expressions are derived for the responses of a non-linear noisy multi-port system.

7.2   Preliminaries

Two major prerequisites for the analysis of noise in non-linear systems are to consider (i) the system description, and (ii) the mathematical representation of noise and deterministic signals. These two problems are treated in the present section.

7.2.1 System description

The type of system under consideration is shown in Figure 7.1. This is a time invariant non-linear non-autonomous multi-port system with multiple internal unmodulated and modulated noise sources, and the system may also contain dc sources. For the analysis it is required that an equivalent circuit description of the non-linear noisy system is available. That is, a detailed non-linear network description of the system must be available. The system is excited by J input signals $\{x_1(f), \ldots, x_J(f)\}$ applied at ports $\{(x, 1), \ldots, (x, J)\}$ — these signals may all include deterministic signals (also dc) and noise. The system has L responses $\{r_1(f), \ldots, r_L(f)\}$ which are present at ports $\{(r, 1), \ldots, (r, L)\}$. The response $r_l(f)$ at port (r, l) where $l \in \{1, 2, \ldots, L\}$ may be any open circuit voltage or short circuit current (or, more generally, an arbitrary system variable) at any node or branch respectively in the underlying non-linear network of the system. Using this system formulation it is possible to determine the noise response, as well as the deterministic signal response (including dc), at any place in the non-linear system. The equivalent noise free non-linear system may contain one- and multi-port non-linear elements and subsystems — e.g. non-linear capacitors, and current generators with a non-linear dependence of two (or more) controlling variables. The primary objective of the present chapter is to determine expressions for the responses $\{r_1(f), \ldots, r_L(f)\}$ and their statistical properties, and to describe the information needed to determine the responses. The statistical properties include the determination of average noise power densities and average noise powers at the response ports as well as the cross-correlation (and autocorrelation) between Fourier series coefficients at arbitrary response ports.

The type of system in Figure 7.1 can not readily be analyzed since it contains internal noise and dc sources. The internal sources, as well as the noise sources applied at the input ports, may be applied at separate external ports provided that the internal topology of the underlying non-linear network is not changed. This is shown in Figure 7.2 where the (unmodulated as well as modulated) noise sources $\{n_1(f), \ldots, n_Q(f)\}$ are applied at ports $\{(n, 1), \ldots, (n, Q)\}$. These noise sources also describe the noise sources at the input ports. Thus the signals $\{s_1(f), \ldots, s_K(f)\}$ applied at ports $\{(s, 1), \ldots, (s, K)\}$ are purely deterministic signals which may also include dc.

The overall system is described up to some maximum order M. This order is the highest order considered, including the possibly non-linear transfer from fundamental (unmodulated) noise source to modulated noise source. For example, if the maximum order of a transfer from unmodulated to modulated noise source is M and the system itself is linear, then the order for the analysis must be M to yield correct results. In this example, of course, the Volterra transfer functions of orders greater than 1 for the linear system are zero. Only the Volterra transfer functions for the modulated noise source of orders greater than 1 are non-zero. Suppose that

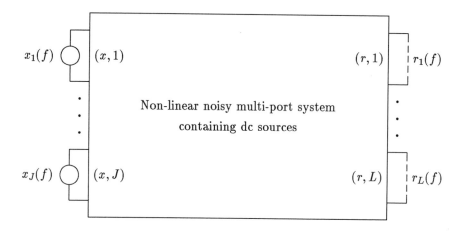

Figure 7.1: Non-linear noisy time invariant multi-port system containing dc sources. Each input signal $x_j(f)$ where $j \in \{1, 2, \ldots, J\}$, which is a voltage or current generator, may contain a deterministic (also dc) signal and noise. Each response $r_l(f)$ where $l \in \{1, 2, \ldots, L\}$ may be either a short circuit current or an open circuit voltage. The noise sources internally in the system may be either unmodulated (independent) or modulated (dependent). The system may also contain dc sources.

the non-linear system in itself is described up to some maximum order M_s, and the maximum order of the transfer from unmodulated to modulated noise source is M_n, then the overall maximum order M *must* be chosen as $M = \max\{M_s, M_n\}$ to yield correct results.

7.2.2   Representation of signals

Fourier series in Volterra series analysis are very attractive since brute force multi-dimensional integration is replaced by addition of multiplicative terms which in a sense is made in one dimension. This makes a Fourier series representation of the signals computationally very efficient. Also the use of Fourier series has advantages of a simple representation of sinusoidal signals, and in the determination of average noise power densities at specified frequencies. Generally the noise and deterministic signals considered are extending over all time, $-\infty < t < \infty$, and have infinite energies. Thus for a real valued noisy signal $x_j(t)$ where $-\infty < t < \infty$ it is given that $\lim_{\tau \to \infty} \int_{-\tau}^{\tau} |x_j(t)|^2 \, dt \ = \ \infty$. This means that the noisy signal $x_j(t)$ is not absolutely integrable and thus it does not generally have a Fourier transform.

   In [3] two standard types of suggestions are given for the frequency domain representation of stationary (random) noise signals. The first suggestion is to represent $x_j(t)$ in a time interval $-\tau < t < \tau$ and to assume $x_j(t) = 0$ for $|t| \geq \tau$. The

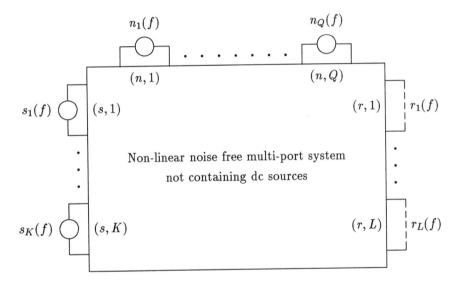

Figure 7.2: Non-linear noise free time invariant multi-port system not containing dc sources with externally applied sources. The signals $\{s_1(f), \ldots, s_K(f)\}$ represent the purely deterministic signals (also dc) applied to the system and internal dc sources. The noise signals $\{n_1(f), \ldots, n_Q(f)\}$ represent noise generated internally in the system and noise entering the system through the signal input ports. The noise sources may be either unmodulated or modulated.

frequency domain representation in this case is the (integral) Fourier transform of $x_j(t)$ given as $x_j(f) = \int_{-\tau}^{\tau} x_j(t)\,\exp[-j2\pi ft]\,dt$. The second suggestion is to assume that the noise signal $x_j(t)$ is periodic with period 2τ such that $x_j(t) = x_j(t + 2i\tau)$ where $i \in \mathcal{Z}$ is an integer. From this, the frequency domain representation is given as a Fourier series. However, the assumption that $x_j(t)$ is periodic has the unfortunate consequence that the autocorrelation function is also periodic such that $\langle x_j(t_1)\,x_j^*(t_2)\rangle = \langle x_j(t_1)\,x_j^*(t_2 + 2i\tau)\rangle$ where $i \in \mathcal{Z}$ is an integer. This also implies that the coefficients of the Fourier series are orthogonal — two Fourier series coefficients $\tilde{x}_{j_1}(p_1\xi)$ and $\tilde{x}_{j_2}(p_2\xi)$ are said to be orthogonal if $\langle \tilde{x}_{j_1}(p_1\xi)\,\tilde{x}_{j_2}^*(p_2\xi)\rangle = 0$ for all integers $p_1, p_2 \in \mathcal{Z}$ except for $p_1 = p_2$ where $\langle \cdot \rangle$ denotes the ensemble average over noise processes with identical statistical properties. Since $\langle x_j(t_1)\,x_j^*(t_2)\rangle$ is generally not periodic for the type of signals considered in the present work this suggestion is not useful. If the systems under consideration are linear (single response), which is the case for the work in [3], then it is not a problem that $\langle x_j(t_1)\,x_j^*(t_2)\rangle$ is periodic, because there is no need for any evaluation between Fourier series coefficients at different frequencies. However, since the systems considered in the present work are non-linear, periodicity of $x_j(t)$ can not be assumed since there may very well be a correlation between Fourier series coefficients at different frequencies. This is, for example, the case for modulated noise sources — even in the case where the fundamental noise source in itself (without modulation) generates white noise (it is very simple to show special cases which illustrate this statement). Since it is most useful to have some kind of a Fourier series representation in the present work, none of the suggestions made in [3] are useful.

One way to achieve the goal of a Fourier series representation without the problems of orthogonal Fourier series coefficients due to periodicity of the signals is described in the following. Assuming that the noise signal $x_j(t)$ has finite energy in the finite time interval $-\tau < t < \tau$ where $\tau > 0$ then $\int_{-\tau}^{\tau} |x_j(t)|^2\,dt < \infty$. In this case $x_j(t)$ can be represented as a Fourier series in the time interval $-\tau < t < \tau$ as

$$x_j(t) \;\; = \;\; \sum_{p=-\infty}^{\infty} \tilde{x}_j(p\xi)\,\exp[j2\pi p\xi t] \tag{7.1}$$

where

$$\tilde{x}_j(p\xi) \;\; = \;\; \xi \int_{-\tau}^{\tau} x_j(t)\,\exp[-j2\pi p\xi t]\,dt \tag{7.2}$$

$$\xi \;\; = \;\; \frac{1}{2\tau} \tag{7.3}$$

In this case $x_j(t)$ is generally non-zero for $t \geq |\tau|$ as opposed to the first method in [3] where $x_j(t) = 0$ for $t \geq |\tau|$. The time interval $[-\tau; \tau]$ is referred to as the observation time interval. In Equation (7.1) the quantity $\tilde{x}_j(p\xi)$ is a complex valued random variable in the random process describing the statistical properties of the noise signal $x_j(t)$. The (integral) Fourier transform of the noise signal $x_j(t)$ in Equation (7.1) is given by

$$x_j(f) \;=\; \mathcal{F}\{x_j(t)\} \tag{7.4}$$

$$=\; \int_{-\infty}^{\infty} x_j(t)\, \exp[-j2\pi ft]\, dt \tag{7.5}$$

$$=\; \sum_{p=-\infty}^{\infty} \tilde{x}_j(p\xi)\, \delta(f - p\xi) \tag{7.6}$$

$$=\; \sum_{p=-\infty}^{\infty} \tilde{x}_j(\xi_p)\, \delta(f - \xi_p), \qquad \xi_p = p\xi \tag{7.7}$$

where $\mathcal{F}\{\cdot\}$ denotes the (integral) Fourier transform, and $\delta(\cdot)$ is the Dirac δ-function [4]. Note from Equation (7.6) that $x_j(f)$ is a two-sided Fourier transform and thus $x_j(f)$ is represented at both positive and negative frequencies. It is very important to maintain both positive and negative frequencies since the mixing of various frequency components is essential to the non-linear noise analysis, and is only correctly analyzed when both positive and negative frequencies are included. It is also seen from Equation (7.6) that the frequency resolution in the spectrum of $x_j(f)$ is given by ξ. This frequency resolution can be made arbitrarily small by choosing τ sufficiently large.

To prove that Equation (7.1) is fulfilled in the time interval $-\tau < t < \tau$ it suffices to show that $\langle |x_j(t) - \sum_{p=-\infty}^{\infty} \tilde{x}_j(p\xi)\, \exp[j2\pi p\xi t]|^2 \rangle \;=\; 0$. This relation may be proved using standard techniques based on the assumption that $x_j(t)$ is real [5]. It can also be shown that if the autocorrelation function $\langle x_j(t_1)\, x_j^*(t_2) \rangle$ is periodic with period 2τ then the coefficients of the Fourier series expansion are orthogonal. However, in general this property is not valid since the autocorrelation function $\langle x_j(t_1)\, x_j^*(t_2) \rangle$ is generally not periodic.

Following the above discussion for both deterministic signals and noise leads to

$$s_k(f) \;=\; \sum_{j_k=1}^{J_k} \tilde{s}_k(\psi_{k,j_k})\, \delta(f - \psi_{k,j_k}) \tag{7.8}$$

where the frequencies $\{\psi_{k,1}, \ldots, \psi_{k,J_k}\}$ are organized such that $\psi_{k,1} \neq \cdots \neq \psi_{k,J_k}$ for all $k \in \{1, 2, \ldots, K\}$, and

$$n_q(f) \;=\; \sum_{p=-\infty}^{\infty} \tilde{n}_q(\xi_p)\, \delta(f - \xi_p) \tag{7.9}$$

where $q \in \{1, 2, \ldots, Q\}$.

As an example of a deterministic time domain signal $s_k(t)$ where $k \in \{1, 2, \ldots, K\}$, consider a sum of a dc term and B_k sinusoidal signals as

$$s_k(t) \quad = \quad \varrho_{k,0} + \sum_{b_k=1}^{B_k} \varrho_{k,b_k} \cos(2\pi \vartheta_{k,b_k} t + \varphi_{k,b_k}) \qquad (7.10)$$

where $\vartheta_{k,b_k} \in \mathcal{R}_+$ for all $k \in \{1, 2, \ldots, K\}$ and $b_k \in \{1, 2, \ldots, B_k\}$, and $\vartheta_{k,1} \neq \cdots \neq \vartheta_{k,B_k}$ for all $k \in \{1, 2, \ldots, K\}$. This leads to $s_k(f) = \mathcal{F}\{s_k(t)\}$ as

$$\begin{aligned} s_k(f) \quad &= \quad \varrho_{k,0}\, \delta(0) \\ &\quad + \frac{1}{2} \sum_{\substack{b_k=-B_k \\ b_k \neq 0}}^{B_k} \varrho_{k,|b_k|}\, \exp[j\,\mathrm{sgn}\{b_k\}\,\varphi_{k,|b_k|}]\, \delta(f - \mathrm{sgn}\{b_k\}\,\vartheta_{k,|b_k|}) \end{aligned}$$

$$(7.11)$$

where $\mathrm{sgn}\{\cdot\}$ is the sign function defined by

$$\mathrm{sgn}\{b\} \quad = \quad \begin{cases} 1 & \text{for } b > 0 \\ 0 & \text{for } b = 0 \\ -1 & \text{for } b < 0 \end{cases} \qquad (7.12)$$

Equation (7.11) can be written according to Equation (7.8) with $J_k = 2B_k + 1$ and

$$\begin{aligned} \psi_{k,2b_k-1} \quad &= \quad \vartheta_{k,b_k}\,, & b_k &\in \{1, 2, \ldots, B_k\} & (7.13) \\ \psi_{k,2b_k} \quad &= \quad -\vartheta_{k,b_k}\,, & b_k &\in \{1, 2, \ldots, B_k\} & (7.14) \\ \psi_{k,2B_k+1} \quad &= \quad 0 & & & (7.15) \end{aligned}$$

and

$$\begin{aligned} \tilde{s}_k(\psi_{k,2b_k-1}) \quad &= \quad \frac{1}{2} \varrho_{k,b_k}\, \exp[j\varphi_{k,b_k}]\,, & b_k &\in \{1, 2, \ldots, B_k\} & (7.16) \\ \tilde{s}_k(\psi_{k,2b_k}) \quad &= \quad \frac{1}{2} \varrho_{k,b_k}\, \exp[-j\varphi_{k,b_k}]\,, & b_k &\in \{1, 2, \ldots, B_k\} & (7.17) \\ \tilde{s}_k(\psi_{k,2B_k+1}) \quad &= \quad \varrho_{k,0} & & & (7.18) \end{aligned}$$

Thus, Equation (7.8) can readily be used.

The notation used in Equations (7.7)–(7.9) implies that sums of pure exponential inputs may be allowed. This type of input may be of significant interest from a theoretical point of view since exponential inputs are frequently used in Volterra series analysis.

7.3   Noise sources

In the analysis of non-linear noisy networks and systems it is sometimes necessary to take into account modulated (dependent) noise sources as well as unmodulated (independent or fundamental) noise sources. That is, the noise generated in a given device or network element may depend on some controlling quantity (or quantities) in the system. For example, shot noise in semiconductor devices is dependent on a deterministic signal applied to it (mainly dc, but also the alternating signals may be of importance if they are not much smaller than the dc signals) [6]. This is because the shot noise is usually proportional to an instantaneous current — this may be viewed as the dc operating point is changing with time due to the excitation. If the deterministic signal is sufficiently large, this causes a modulation of the noise source which in turn leads to correlated noise sidebands — this is even the case if the fundamental noise source (the noise source which is being modulated) generates white noise.

7.3.1 Basic theory

Noise sources are generally included as in Figure 7.3 in which the possibly modulated (dependent) noise source $n_q(f)$ is identical with the response from a non-linear noise free subsystem not containing dc sources with inputs $\{u_{q,1}(f), \ldots, u_{q,I_q}(f)\}$, which are referred to as controlling variables applied at ports $\{(u_q, 1), \ldots, (u_q, I_q)\}$, and a fundamental (unmodulated, independent) noise source $w_q(f)$ where $q \in \{1, 2, \ldots, Q\}$. The non-linear noise free multi-port system in Figure 7.3 *must not* contain dc sources. However, the controlling variables $\{u_{q,1}(f), \ldots, u_{q,I_q}(f)\}$ may have a dc value, and also the fundamental noise source $w_q(t)$ where $q \in \{1, 2, \ldots, Q\}$ may have a non-zero mean value. The controlling variable $u_{q,i_q}(f)$ where $q \in \{1, 2, \ldots, Q\}$ and $i_q \in \{1, 2, \ldots, I_q\}$ may be either an open circuit voltage or a short circuit current (or actually any system variable) at any node or branch respectively in the underlying network of the non-linear system. The non-linear system in Figure 7.3 is described by a multi-port frequency domain Volterra series [7] with inputs $\{w_q(f), u_{q,1}(f), \ldots, u_{q,I_q}(f)\}$ and output $n_q(f)$ where $q \in \{1, 2, \ldots, Q\}$ as

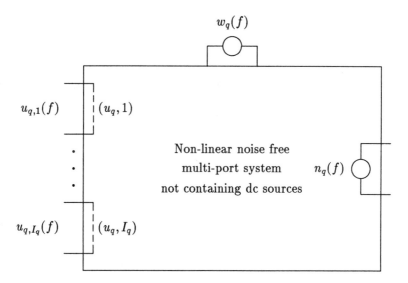

Figure 7.3: Representation of modulated (dependent) and unmodulated (independent) noise sources. The input signals $\{u_{q,1}(f), \ldots, u_{q,I_q}(f)\}$ where $q \in \{1, 2, \ldots, Q\}$ are controlling variables at ports $\{(u_q, 1), \ldots, (u_q, I_q)\}$ — deterministic currents and voltages at any place in the non-linear noisy network. The signal $w_q(f)$ where $q \in \{1, 2, \ldots, Q\}$ is a fundamental (unmodulated) noise voltage or current source, and $n_q(f)$ is the (possibly) modulated noise source.

$$
n_q(f) \;=\; \sum_{m_0=0}^{M} \sum_{m_1=0}^{M} \cdots \sum_{m_{I_q}=0}^{M}
$$

$$
\int_{-\infty}^{\infty} \cdots \int_{-\infty}^{\infty} \int_{-\infty}^{\infty} \cdots \int_{-\infty}^{\infty} \cdots\cdots \int_{-\infty}^{\infty} \cdots \int_{-\infty}^{\infty}
$$

$$
\mathcal{L}_{1,M}(m_0 + m_1 + \cdots + m_{I_q})
$$

$$
(G_q)_{m_0,m_1,\ldots,m_{I_q}}(\Omega_{0,1},\ldots,\Omega_{0,m_0};\Omega_{1,1},\ldots,\Omega_{1,m_1};\cdots
$$

$$
\cdots;\Omega_{I_q,1},\ldots,\Omega_{I_q,m_{I_q}})
$$

$$
w_q(\Omega_{0,1})\cdots w_q(\Omega_{0,m_0})\, u_{q,1}(\Omega_{1,1})\cdots u_{q,1}(\Omega_{1,m_1})\cdots
$$

$$
\cdots u_{q,I_q}(\Omega_{I_q,1})\cdots u_{q,I_q}(\Omega_{I_q,m_{I_q}})
$$

$$
\delta(f - \Omega_{0,1} - \cdots - \Omega_{0,m_0} - \Omega_{1,1} - \cdots - \Omega_{1,m_1} - \cdots
$$

$$
\cdots - \Omega_{I_q,1} - \cdots - \Omega_{I_q,m_{I_q}})
$$

$$
d\Omega_{0,1}\cdots d\Omega_{0,m_0}\, d\Omega_{1,1}\cdots d\Omega_{1,m_1}\cdots
$$

$$
\cdots d\Omega_{I_q,1}\cdots d\Omega_{I_q,m_{I_q}} \tag{7.19}
$$

where

$$
\mathcal{L}_{\alpha,\beta}(\gamma) \;=\;
\begin{cases}
1 & \text{for } \gamma \in \{\alpha, \alpha+1, \ldots, \beta-1, \beta\} \\
0 & \text{otherwise}
\end{cases}
\tag{7.20}
$$

and $(G_q)_{m_0,m_1,\ldots,m_{I_q}}(\cdot)$ is the partly symmetrical multi-port frequency domain Volterra transfer function relating the inputs $\{w_q(f), u_{q,1}(f),\ldots,u_{q,I_q}(f)\}$ to the output $n_q(f)$ of order $m_0 + m_1 + \cdots + m_{I_q}$. In Equation (7.19) the function $\mathcal{L}_{1,M}(\cdot)$ limits the order $m_0 + m_1 + \cdots + m_{I_q}$ to be from 1 to M. Thus, M is the order of truncation of the multi-port Volterra series.

A partly symmetrical multi-port frequency domain Volterra transfer function is defined as follows. Let $\mathcal{P}_{k,l_k}\{\Xi_{k,1},\ldots,\Xi_{k,o_k}\}$ denote permutation $l_k \in \{1,2,\ldots,o_k!\}$ of the $o_k!$ total number of permutations of the frequency variables $\{\Xi_{k,1},\ldots,\Xi_{k,o_k}\}$ where $k \in \{1,2,\ldots,K\}$. Then, a multi-port frequency domain Volterra transfer function $(F_{q,i_q})_{o_1,\ldots,o_K}(\cdot)$ is said to be partly symmetrical if

$$
(F_{q,i_q})_{o_1,\ldots,o_K}(\Xi_{1,1},\ldots,\Xi_{1,o_1};\cdots\cdots;\Xi_{K,1},\ldots,\Xi_{K,o_K})
$$

$$
= (F_{q,i_q})_{o_1,\ldots,o_K}(\mathcal{P}_{1,l_1}\{\Xi_{1,1},\ldots,\Xi_{1,o_1}\};\cdots\cdots;\mathcal{P}_{K,l_K}\{\Xi_{K,1},\ldots,\Xi_{K,o_K}\})
\tag{7.21}
$$

for all the $o_1!\cdots o_K!$ possible permutations of the variables. An unsymmetrical Volterra transfer function $(\mathcal{F}_{q,i_q})_{o_1,\ldots,o_K}(\cdot)$ can be made partly symmetrical as

$$(F_{q,i_q})_{o_1,\ldots,o_K}(\Xi_{1,1},\ldots,\Xi_{1,o_1};\cdots\cdots;\Xi_{K,1},\ldots,\Xi_{K,o_K})$$

$$= \frac{1}{o_1!\cdots o_K!}\sum_{l_1=1}^{o_1!}\cdots\sum_{l_K=1}^{o_K!}$$

$$(\mathcal{F}_{q,i_q})_{o_1,\ldots,o_K}\left(\mathcal{P}_{1,l_1}\{\Xi_{1,1},\ldots,\Xi_{1,o_1}\};\cdots\cdots;\mathcal{P}_{K,l_K}\{\Xi_{K,1},\ldots,\Xi_{K,o_K}\}\right)$$

$$(7.22)$$

All permutations $\{\mathcal{P}_{k,1}\{\cdot\},\ldots,\mathcal{P}_{k,m_k}!\{\cdot\}\}$ are required to be different for all $k \in \{1,2,\ldots,K\}$ to assure (partial) symmetry of $(F_{q,i_q})_{m_1,\ldots,m_K}(\cdot)$.

Note that the assumption of partly symmetrical transfer functions is not a loss of generality, since the response $n_q(f)$ where $q \in \{1,2,\ldots,Q\}$ in Equation (7.19) remains the same if a possibly unsymmetrical frequency domain Volterra transfer function is substituted with the corresponding partly symmetrical transfer function. Note also from Equation (7.19) that there may be several different $(G_q)_{m_0,m_1,\ldots,m_{I_q}}(\cdot)$ multi-port frequency domain Volterra transfer functions of the same order $m_0 + m_1 + \cdots + m_{I_q}$. Observe from Equation (7.19) that even if $(G_q)_{m_0,m_1,\ldots,m_{I_q}}(\cdot)$ where $q \in \{1,2,\ldots,Q\}$ is unsymmetrical then any variables $\{\Omega_{i_q,1},\ldots,\Omega_{i_q,m_{i_q}}\}$ where $k \in \{1,2,\ldots,K\}$ may be permuted and one still obtains the same result for $n_q(f)$. The reason for using partly symmetrical Volterra transfer functions is that the amount of computations required to determine the response may be significantly reduced when the transfer functions are symmetrical compared to when the transfer functions are unsymmetrical.

Full symmetry of a multi-port Volterra transfer function can generally not be utilized. A fully symmetrical Volterra transfer function is defined by

$$(F_{q,i_q})_{o_1,\ldots,o_K}(\Xi_{1,1},\ldots,\Xi_{1,o_1};\cdots\cdots;\Xi_{K,1},\ldots,\Xi_{K,o_K})$$

$$= (F_{q,i_q})_{o_1,\ldots,o_K}(\mathcal{P}_l\{\Xi_{1,1},\ldots,\Xi_{1,o_1};\cdots\cdots;\Xi_{K,1},\ldots,\Xi_{K,o_K}\})\quad (7.23)$$

where $q \in \{1,2,\ldots,Q\}$, and $\mathcal{P}_l\{\cdot\}$ denotes permutation number $l \in \{1,2,\ldots,(o_1 + \cdots + o_K)!\}$ of the frequency variables. The full symmetry in Equation (7.23) is generally not fulfilled since the signals are applied at K possibly different input excitation ports. Thus, two signals from two different input ports may generally not be interchanged to give the same output response. Therefore, multi-port Volterra transfer functions are generally not fully symmetrical, and a fully symmetrical multi-port Volterra transfer function can not be substituted for the possibly unsymmetrical transfer function in Equation (7.19).

The contribution of order 0 for $n_q(f)$ in Equation (7.19), which is identical to the response when no signals (including dc) are applied, is excluded in the systems in Figures 7.2 and 7.3. It is very important to emphasize that the dc sources can not be removed as sources by introducing a zeroth order term in Equation (7.19). This is because the zeroth order contribution, identical to $(G_q)_{0,0,...,0}()$, in this case can not give information on the non-linear interference between dc and other applied deterministic signals. Actually, $(G_q)_{0,0,...,0}()$ is a system contribution which is there but it can not interfere with other signals (it may be called a 'mathematical offset'). It lies in the modelling of non-linear devices that the zeroth order contribution is not of interest (it does not describe any physical behaviour of the circuit). Note that there may very well be interference between dc and other deterministic signals and noise in the system in Figure 7.2. In much work on Volterra series analysis it is stated that the zeroth order contribution $(G_q)_{0,0,...,0}()$ represents the internal dc sources in the system. However, this is generally not correct for the above reasons.

The upper limit for the order $m_0 + m_1 + \cdots + m_{I_q}$ is the maximum order M. Actually, $M \to \infty$ but for practical reasons M must be chosen to have some finite value. Of course, M must be chosen sufficiently large so as not to give incorrect results because of too low a maximum order. For a given M it is sometimes useful to determine limits for the amplitudes of the applied deterministic signals to ensure that the residual components to $n_q(f)$ in Equation (7.19) of orders higher than M are insignificant. This problem is also closely related to the functional modelling of the non-linear elements in the system. The model of a non-linear element is valid for the controlling variable in some finite interval. Therefore, it is a good idea to calculate the deterministic contributions at the controlling variables for the non-linear elements to assure that the elements are not too strongly/weakly excited.

The response for the modulated noise source $n_q(f)$ where $q \in \{1, 2, ..., Q\}$ is limited to exclude zeroth order noise contributions — actually, this is no limitation as in this case the modulated noise source $n_q(f)$ is just a controlled non-linearity, and this type of non-linear element (subsystem) is already included in the non-linear noise free system in Figure 7.2. Using the above assumptions and treating only low level noise, then

$$
\begin{aligned}
n_q(f) = \sum_{m_1=0}^{M-1} \cdots \sum_{m_{I_q}=0}^{M-1} \int_{-\infty}^{\infty} \int_{-\infty}^{\infty} \cdots \int_{-\infty}^{\infty} \cdots \cdots \int_{-\infty}^{\infty} \cdots \int_{-\infty}^{\infty} \\
\mathcal{L}_{0,M-1}(m_1 + \cdots + m_{I_q}) \\
(G_q)_{1,m_1,...,m_{I_q}}(\Omega_{0,1}; \Omega_{1,1}, ..., \Omega_{1,m_1}; \cdots \cdots ; \Omega_{I_q,1}, ..., \Omega_{I_q,m_{I_q}}) \\
w_q(\Omega_{0,1}) \, u_{q,1}(\Omega_{1,1}) \cdots u_{q,1}(\Omega_{1,m_1}) \cdots \cdots u_{q,I_q}(\Omega_{I_q,1}) \cdots u_{q,I_q}(\Omega_{I_q,m_{I_q}}) \\
\delta(f - \Omega_{0,1} - \Omega_{1,1} - \cdots - \Omega_{1,m_1} - \cdots \cdots - \Omega_{I_q,1} - \cdots - \Omega_{I_q,m_{I_q}}) \\
d\Omega_{0,1} \, d\Omega_{1,1} \cdots d\Omega_{1,m_1} \cdots \cdots d\Omega_{I_q,1} \cdots d\Omega_{I_q,m_{I_q}}
\end{aligned}
\tag{7.24}
$$

The multi-port Volterra transfer function $(G_q)_{1,m_1,...,m_{I_q}}(\cdot)$ in Equation (7.24) is referred to as the *modulating function* for the modulated noise source $n_q(t)$ where $q \in \{1, 2, ..., Q\}$. Note from Equation (7.24) that the contribution corresponding to $m_1 = \cdots = m_{I_q} = 0$ is simply a linear filtering of the fundamental (unmodulated) noise source $w_q(f)$, $n_q(f) = (G_q)_{1,0,...,0}(f) \, w_q(f)$, and corresponds to the response when (i) all the controlling variables $u_{q,1}(f) = \cdots = u_{q,I_q}(f) = 0$ where $q \in \{1, 2, \ldots, Q\}$, or (ii) when $M = 1$ meaning that the system is linear. Similarly, if $I_q = 0$ then $n_q(f) = (G_q)_1(f)\, w_q(f)$ in which case $n_q(f)$ is an unmodulated noise source given as a linear filtering of the fundamental (unmodulated) noise source $w_q(f)$ where $q \in \{1, 2, \ldots, Q\}$. Using the formulation in Equation (7.24) it is possible to introduce very complicated modulating functions in a simple way. For example, the modulated noise source $n_q(f)$ where $q \in \{1, 2, \ldots, Q\}$ may depend on several deterministic signals in the non-linear network, and the non-linear network in Figure 7.3. In Table 7.1 some examples of modulated noise sources and the corresponding modulating functions are shown.

7.3.2   Controlling variables

Since low level noise is assumed, which implies that the noise is a small perturbation of the deterministic signal regime, the controlling variables are determined as the contributions to $\{u_{q,1}(f), \ldots, u_{q,I_q}(f)\}$ due to the deterministic signals $\{s_1(f), \ldots, s_K(f)\}$ only. Therefore the modulated noise sources $\{n_1(f), \ldots, n_Q(f)\}$ do not have any feedback impact on the controlling variables, which would be the case for high level modulated noise sources. Thus $u_{q,i_q}(f)$ where $q \in \{1, 2, \ldots, Q\}$ and $i_q \in \{1, 2, \ldots, I_q\}$ can be determined as

$$
\begin{aligned}
u_{q,i_q}(f) \;=\; & \sum_{o_1=0}^{M-1} \cdots \sum_{o_K=0}^{M-1} \int_{-\infty}^{\infty} \cdots \int_{-\infty}^{\infty} \cdots\cdots \int_{-\infty}^{\infty} \cdots \int_{-\infty}^{\infty} \\
& \mathcal{L}_{1,M-1}(o_1 + \cdots + o_K) \\
& (F_{q,i_q})_{o_1,\ldots,o_K}(\Xi_{1,1}, \ldots, \Xi_{1,o_1}; \cdots\cdots; \Xi_{K,1}, \ldots, \Xi_{K,o_K}) \\
& s_1(\Xi_{1,1}) \cdots s_1(\Xi_{1,o_1}) \cdots\cdots s_K(\Xi_{K,1}) \cdots s_K(\Xi_{K,o_K}) \\
& \delta(f - \Xi_{1,1} - \cdots - \Xi_{1,o_1} - \cdots\cdots - \Xi_{K,1} - \cdots - \Xi_{K,o_K}) \\
& d\Xi_{1,1} \cdots d\Xi_{1,o_1} \cdots\cdots d\Xi_{K,1} \cdots d\Xi_{K,o_K}
\end{aligned}
\tag{7.25}
$$

where $(F_{q,i_q})_{o_1,\ldots,o_K}(\cdot)$ is a partly symmetrical multi-port frequency domain Volterra transfer function relating the input signals $\{s_1(f), \ldots, s_K(f)\}$ to the controlling variable $u_{q,i_q}(f)$. The maximum order of the signals $\{s_1(f), \ldots, s_K(f)\}$ is $M - 1$ since the noise signal $w_q(f)$ where $q \in \{1, 2, \ldots, Q\}$ in itself accounts for order 1 in the expression for $n_q(f)$ in Equation (7.24). In Equation (7.25) the signal $s_k(f)$ where $k \in \{1, 2, \ldots, K\}$ is given by the Fourier series in Equation (7.8). Insertion of Equation (7.8) into Equation (7.25) leads to

$n_q(t)$	I_q	$(G_q)_{1,m_1,\ldots,m_{I_q}}(\cdots)$
$C_q w_q(t)$	0	C_q
$C_{q,1} u_{q,1}(t)\, w_q(t)$	1	$\begin{cases} C_{q,1} & \text{for}\quad m_1 = 1 \\ 0 & \text{otherwise} \end{cases}$
$C_{q,1} w_q(t)\, \frac{d}{dt} u_{q,1}(t)$	1	$\begin{cases} j2\pi C_{q,1}\Omega_{1,1} & \text{for}\quad m_1 = 1 \\ 0 & \text{otherwise} \end{cases}$
$C_{q,1} u_{q,1}(t)\, \frac{d}{dt} w_q(t)$	1	$\begin{cases} j2\pi C_{q,1}\Omega_{0,1} & \text{for}\quad m_1 = 1 \\ 0 & \text{otherwise} \end{cases}$
$C_{q,1} \frac{d}{dt}[u_{q,1}(t)\, w_q(t)]$	1	$\begin{cases} j2\pi C_{q,1}(\Omega_{0,1} + \Omega_{1,1}) & \text{for}\quad m_1 = 1 \\ 0 & \text{otherwise} \end{cases}$
$C_{q,1} w_q(t) \int_{-\infty}^{t} u_{q,1}(\tau)\, d\tau$	1	$\begin{cases} C_{q,1}/(j2\pi\Omega_{1,1}) & \text{for}\quad m_1 = 1 \\ 0 & \text{otherwise} \end{cases}$
$C_{q,1} u_{q,1}^2(t)\, w_q(t)$	1	$\begin{cases} C_{q,1} & \text{for}\quad m_1 = 2 \\ 0 & \text{otherwise} \end{cases}$
$C_{q,1} u_{q,1}(t)\, u_{q,2}(t)\, w_q(t)$	2	$\begin{cases} C_{q,1} & \text{for}\quad m_1 = m_2 = 1 \\ 0 & \text{otherwise} \end{cases}$

Table 7.1: Examples of partly symmetrical multi-port frequency domain Volterra transfer functions (modulating functions) versus the time domain modulated noise function $n_q(t)$ and the number of controlling variables I_q where $q \in \{1, 2, \ldots, Q\}$. A partly symmetrical $(G_q)_{1,m_1,\ldots,m_{I_q}}(\cdot)$ multi-port frequency domain Volterra transfer function is derived from the expression for the time domain modulated noise function $n_q(t)$ where $q \in \{1, 2, \ldots, Q\}$ by (integral) Fourier transforming $n_q(t)$, then making a direct parameter identification from Equation (7.24).

$$
\begin{aligned}
u_{q,i_q}(f) \;=\; & \sum_{o_1=0}^{M-1} \cdots \sum_{o_K=0}^{M-1} \sum_{j_{1,1}=1}^{J_1} \cdots \sum_{j_{1,o_1}=1}^{J_1} \cdots\cdots \sum_{j_{K,1}=1}^{J_K} \cdots \sum_{j_{K,o_K}=1}^{J_K} \\
& \mathcal{L}_{1,M-1}(o_1 + \cdots + o_K) \\
& (F_{q,i_q})_{o_1,\ldots,o_K}(\psi_{1,j_{1,1}}, \ldots, \psi_{1,j_{1,o_1}}; \cdots\cdots; \psi_{K,j_{K,1}}, \ldots, \psi_{K,j_{K,o_K}}) \\
& \tilde{s}_1(\psi_{1,j_{1,1}}) \cdots \tilde{s}_1(\psi_{1,j_{1,o_1}}) \cdots\cdots \tilde{s}_K(\psi_{K,j_{K,1}}) \cdots \tilde{s}_K(\psi_{K,j_{K,o_K}}) \\
& \delta(f - \psi_{1,j_{1,1}} - \cdots - \psi_{1,j_{1,o_1}} - \cdots\cdots - \psi_{K,j_{K,1}} - \cdots - \psi_{K,j_{K,o_K}})
\end{aligned}
$$

(7.26)

The computational cost of determining the controlling variable $u_{q,i_q}(f)$ where $q \in \{1,2,\ldots,Q\}$ and $i_q \in \{1,2,\ldots,I_q\}$ can be reduced significantly by using the symmetry properties of the partly symmetrical multi-port frequency domain Volterra transfer function $(F_{q,i_q})_{o_1,\ldots,o_K}(\cdot)$ as

$$
\begin{aligned}
u_{q,i_q}(f) \;=\; & \sum_{o_1=0}^{M-1} \cdots \sum_{o_K=0}^{M-1} \sum_{j_{1,1}=1}^{J_1} \cdots \sum_{j_{1,o_1}=1}^{J_1} \cdots\cdots \sum_{j_{K,1}=1}^{J_K} \cdots \sum_{j_{K,o_K}=1}^{J_K} \\
& \mathcal{L}_{1,M-1}(o_1 + \cdots + o_K) \\
& \mathcal{A}_1(j_{1,1}, \ldots, j_{1,o_1}) \cdots \mathcal{A}_K(j_{K,1}, \ldots, j_{K,o_K}) \\
& \prod_{k=1}^{K} o_k! \; \prod_{k=1}^{K} \prod_{j_k=1}^{J_k} \left\{ \mathcal{N}_{j_k}(j_{k,1}, \ldots, j_{k,o_k})! \right\}^{-1} \\
& (F_{q,i_q})_{o_1,\ldots,o_K}(\psi_{1,j_{1,1}}, \ldots, \psi_{1,j_{1,o_1}}; \cdots\cdots; \psi_{K,j_{K,1}}, \ldots, \psi_{K,j_{K,o_K}}) \\
& \tilde{s}_1(\psi_{1,j_{1,1}}) \cdots \tilde{s}_1(\psi_{1,j_{1,o_1}}) \cdots\cdots \tilde{s}_K(\psi_{K,j_{K,1}}) \cdots \tilde{s}_K(\psi_{K,j_{K,o_K}}) \\
& \delta(f - \psi_{1,j_{1,1}} - \cdots - \psi_{1,j_{1,o_1}} - \cdots\cdots - \psi_{K,j_{K,1}} - \cdots - \psi_{K,j_{K,o_K}})
\end{aligned}
$$

(7.27)

where

$$
\mathcal{A}_k(j_{k,1}, \ldots, j_{k,o_k}) \;=\; \begin{cases} 1 & \text{for } \; j_{k,1} \leq \cdots \leq j_{k,o_k} \\ 0 & \text{otherwise} \end{cases}
$$

(7.28)

and

$$
\mathcal{N}_{j_k}(j_{k,1}, \ldots, j_{k,o_k}) \;=\; \text{number of } \{j_{k,1}, \ldots, j_{k,o_k}\} \text{ which}
$$
$$
\text{are equal to } j_k \in \{1,2,\ldots,J_k\}
$$

(7.29)

The $\mathcal{A}_k(\cdot)$ functions in Equation (7.27) may be used to significantly reduce the number of sum-terms in the expression for the controlling variable $u_{q,i_q}(f)$ where $q \in \{1,2,\ldots,Q\}$ and $i_q \in \{1,2,\ldots,I_q\}$. Note from Equation (7.27) that the only difference between determining $u_{q_1,i_{q_1}}(f)$ and $u_{q_2,i_{q_2}}(f)$ where $q_1, q_2 \in \{1,2,\ldots,Q\}$

with $q_1 \neq q_2$ is that $(F_{q_1,i_{q_1}})_{o_1,\dots,o_K}(\cdot)$ must be replaced with $(F_{q_2,i_{q_2}})_{o_1,\dots,o_K}(\cdot)$ respectively. This may also be used to reduce the number of computations required to determine $u_{q,i_q}(f)$ for all $q \in \{1, 2, \dots, Q\}$ and $i_q \in \{1, 2, \dots, I_q\}$.

To avoid noise contributions to the output response $r_l(f)$ where $l \in \{1, 2, \dots, L\}$ at port (r, l) due to frequencies caused by orders higher than $M - 1$ it is necessary to keep track of the orders of the frequencies of the individual contributions to $u_{q,i_q}(f)$ where $q \in \{1, 2, \dots, Q\}$ and $i_q \in \{1, 2, \dots, I_q\}$ in Equation (7.27). This is done by defining frequency sets $\mathcal{S}_1, \mathcal{S}_2, \dots, \mathcal{S}_{M-1}$ as

$$
\begin{aligned}
\mathcal{S}_o &= \left\{ \psi(\boldsymbol{p}_o) \,\middle|\, \boldsymbol{p}_o \in \{0, 1, 2, \dots, o\}^{(J_1 + \dots + J_K) \times 1} \,\wedge\, \|\boldsymbol{p}_o\| = o \right\} \quad (7.30)\\
&= \{\Psi_{o,1}, \dots, \Psi_{o,E_o}\} \quad (7.31)
\end{aligned}
$$

where $o \in \{1, 2, \dots, M - 1\}$ is the order of the given frequencies, and

$$
\psi(\boldsymbol{p}_o) = \boldsymbol{\psi}^T \boldsymbol{p}_o \quad (7.32)
$$

and

$$
\begin{aligned}
\boldsymbol{p}_o &= \left[p_{1,1}, \dots, p_{1,J_1}, \dots \dots, p_{K,1}, \dots, p_{K,J_K} \right]^T \\
&\in \{0, 1, 2, \dots, o\}^{(J_1 + \dots + J_K) \times 1} \quad (7.33)\\
\boldsymbol{\psi} &= \left[\psi_{1,1}, \dots, \psi_{1,J_1}, \dots \dots, \psi_{K,1}, \dots, \psi_{K,J_K} \right]^T \\
&\in \mathcal{R}^{(J_1 + \dots + J_K) \times 1} \quad (7.34)
\end{aligned}
$$

In Equation (7.30) the quantity $\| \cdot \|$ denotes the sum of all elements in the vector, i.e.

$$
\|\boldsymbol{p}_o\| = p_{1,1} + \dots + p_{1,J_1} + \dots \dots + p_{K,1} + \dots + p_{K,J_K} \quad (7.35)
$$

Equation (7.35) follows since $p_{k,j_k} \in \mathcal{Z}_{0+}$ for all $k \in \{1, 2, \dots, K\}, j_k \in \{1, 2, \dots, J_k\}$. Note from Equation (7.30) that there may be one or more frequencies which belongs to more than one of the sets $\mathcal{S}_1, \mathcal{S}_2, \dots, \mathcal{S}_{M-1}$. For example, if the applied deterministic excitation signals are sinusoids (not pure exponentials) then an even order contribution of order $o_1 \in \{2, 4, 6, \dots\}$ gives some responses at the same frequencies as for lower order contributions $o_2 \in \{2, 4, \dots, o_1 - 2\}$. Note that

$$
\mathcal{S}_1 = \bigcup_{k=1}^{K} \bigcup_{j_k=1}^{J_K} \{\psi_{k,j_k}\} \quad (7.36)
$$

and thus $\mathcal{S}_1$ contains all the applied frequencies regardless of the ports at which they are applied. Also that in a recursive form

$$S_{o+1} = \left\{ \Psi_{o,1} + \psi_{1,1}, \ldots, \Psi_{o,1} + \psi_{1,J_1}, \ldots \right.$$
$$\ldots, \Psi_{o,1} + \psi_{K,1}, \ldots, \Psi_{o,1} + \psi_{K,J_K}, \ldots \ldots$$
$$\ldots \ldots, \Psi_{o,E_o} + \psi_{1,1}, \ldots, \Psi_{o,E_o} + \psi_{1,J_1}, \ldots$$
$$\left. \ldots, \Psi_{o,E_o} + \psi_{K,1}, \ldots, \Psi_{o,E_o} + \psi_{K,J_K} \right\} \tag{7.37}$$
$$= \left\{ \Psi_{o+1,1}, \ldots, \Psi_{o+1,E_{o+1}} \right\} \tag{7.38}$$

and in this case $E_{o+1} \leq E_o(J_1 + \cdots + J_K)$, which means that $E_o \leq (J_1 + \cdots + J_K)^o$. Furthermore

$$\Psi_{o_1,e_{o_1}} + \Psi_{o_2,e_{o_2}} \quad \in \quad S_{o_1+o_2} \tag{7.39}$$

where $o_1, o_2 \in \{1, 2, \ldots, M-1\}$, $e_{o_1} \in \{1, 2, \ldots, E_{o_1}\}$ and $e_{o_2} \in \{1, 2, \ldots, E_{o_2}\}$. Thus, $u_{q,i_q}(f)$ in Equation (7.27) can be rewritten into the following:

$$u_{q,i_q}(f) = \sum_{o=1}^{M-1} \sum_{e=1}^{E_o} {}^o\tilde{u}_{q,i_q}(\Psi_{o,e}) \, \delta(f - \Psi_{o,e}) \tag{7.40}$$
$$= \sum_{o=1}^{M-1} {}^o\tilde{u}_{q,i_q}(S_o) \, \delta(f, S_o) \tag{7.41}$$

where

$${}^o\tilde{u}_{q,i_q}(S_o) = \left[{}^o\tilde{u}_{q,i_q}(\Psi_{o,1}), \ldots, {}^o\tilde{u}_{q,i_q}(\Psi_{o,E_o}) \right]^T \tag{7.42}$$
$$\delta(f, S_o) = \left[\delta(f - \Psi_{o,1}), \ldots, \delta(f - \Psi_{o,E_o}) \right]^T \tag{7.43}$$

and where $q \in \{1, 2, \ldots, Q\}$, $i_q \in \{1, 2, \ldots, I_q\}$ and E_o is the number of different frequencies of order $o \in \{1, 2, \ldots, M-1\}$. Table 7.2 shows the upper limit for E_o versus the order o and the number of input frequencies $J_1 + \cdots + J_K$. E_o is identical to the upper limit listed in Table 7.2 when the input frequencies are incommensurate up to order o. Note from Table 7.2 that the upper limits in many cases are significantly lower than $(J_1 + \cdots + J_K)^o$. This is because the $E_o \leq (J_1 + \cdots + J_K)^o$ condition does not use the fact that the addition of frequencies is associative.

A set of frequencies $\{\omega_1, \ldots, \omega_Q\}$ is defined as incommensurate up to order o if

$$\omega^T p_{1,o} \neq \omega^T p_{2,o} \quad \text{for all} \quad p_{1,o} \neq p_{2,o}$$
$$\text{and} \quad p_{1,o}, p_{2,o} \in \{0, 1, \ldots, o\}^{Q \times 1} \tag{7.44}$$

where $\omega = [\omega_1, \ldots, \omega_Q]^T$, the vector 1-norms $\|p_{1,o}\|, \|p_{2,o}\| \in \{1, 2, \ldots, o\}$, $o \in \{1, 2, \ldots, M\}$ and for

1. Exponential inputs $\exp[j(2\pi\omega_1 t + \phi_1)], \ldots, \exp[j(2\pi\omega_Q t + \phi_Q)]$ where $\omega_1, \ldots,$ $\omega_Q \in \mathcal{R}$ with $|\omega_1| \neq \cdots \neq |\omega_Q|$. Here the frequencies $\{\omega_1, \ldots, \omega_Q\}$ may be negative as well as positive.

2. Sinusoidal inputs $\cos(2\pi\omega_1 t + \phi_1), \ldots, \cos(2\pi\omega_Q t + \phi_Q)$ where $\omega_1, \ldots, \omega_Q \in \mathcal{R}_{0+}$. Here all the frequencies $\{\omega_1, \ldots, \omega_Q\}$ must be positive (zero included).

If a set of frequencies $\{\omega_1, \ldots, \omega_Q\}$ is not incommensurate up to order o, it is commensurate for orders higher than or equal to o. In the literature, e.g. [8,9,10,11], the commensurability concept does not depend on the order, which is not in accordance with the above definition. However, the traditional definition actually assumes that the order is infinite, which of course is not very relevant, since all practical uses of the Volterra series technique are of limited order (traditionally order 2 or 3). For a further discussion on this see [7, footnote 1]. From the definition it is seen that

1. If $0 \in \{\omega_1, \ldots, \omega_Q\}$ then the set $\{\omega_1, \ldots, \omega_Q\}$ is incommensurate up to order $o = 1$ and commensurate for order $o \in \{2, 3, \ldots, \infty\}$.

2. Any set of frequencies $\{\omega_1, \ldots, \omega_Q\}$ where $\omega_1 \neq \cdots \neq \omega_Q$ is incommensurate up to at least order $o = 1$.

The concept of commensurability is very important in the analysis and is frequently used in the following. The definition of commensurability does not depend on the number of input ports for the non-linear system but only on the *overall* frequencies applied. Intuitively, a set of frequencies $\{\omega_1, \ldots, \omega_Q\}$ is incommensurate up to order o if a frequency $\omega_j \in \{\omega_1, \ldots, \omega_Q\}$ *can not* be given as an intermodulation product or harmonics of the other frequencies up to order o.

o	Upper limit for E_o							
	$J_1 + \cdots + J_K$							
	1	2	3	4	5	6	7	8
1	1	2	3	4	5	6	7	8
2	1	3	6	10	15	21	28	36
3	1	4	10	20	35	56	84	120
4	1	5	15	35	70	126	210	330
5	1	6	21	56	126	252	462	792
6	1	7	28	84	210	462	924	1716

Table 7.2: Upper limit for the number of frequencies E_o versus the order o and the number of applied frequencies $J_1 + \cdots + J_K$. Note that the upper limit for E_o applies to the situation of pure exponential inputs.

The coefficient $^o\widetilde{u}_{q,i_q}(\Psi_{o,e})$ where $q \in \{1,2,\ldots,Q\}$, $i_q \in \{1,2,\ldots,I_q\}$, $o \in \{1,2,\ldots,M-1\}$ and $e \in \{1,2,\ldots,E_o\}$ in Equation (7.40) may be determined by comparing Equations (7.27) and (7.40). $^o\widetilde{u}_{q,i_q}(\Psi_{o,e})$ is the Fourier series coefficient of order o at the frequency $\Psi_{o,e}$ for port number (u_q,i_q) where $q \in \{1,2,\ldots,Q\}$ and $i_q \in \{1,2,\ldots,I_q\}$. Note that only the controlling variables

$$\{u_1(f),\ldots,u_I(f)\} \quad = \quad \bigcup_{q=1}^{Q}\{u_{q,1}(f),\ldots,u_{q,I_q}(f)\} \tag{7.45}$$

should be determined since one controlling variable may very well control more than one noise source, and thus $I \in \{0,1,\ldots,I_1+\cdots+I_Q\}$. If none of the noise sources $\{n_1(f),\ldots,n_Q(f)\}$ are modulated then $I = 0$, and if none of the controlling variables are identical, i.e. $\cap_{q=1}^{Q}\{u_{q,1}(f),\ldots,u_{q,I_q}(f)\} = \emptyset$ where $\emptyset$ is the empty set, then $I = I_1+\cdots+I_Q$.

7.3.3 Modulated noise source

The possibly modulated noise source $n_q(f)$ where $q \in \{1,2,\ldots,Q\}$ is given as the response from a non-linear noise free system not containing internal sources with the deterministic modulating input signals $\{u_{q,1}(f),\ldots,u_{q,I_q}(f)\}$ and a fundamental (unmodulated) noise source $w_q(f)$. Insertion of Equation (7.40) into Equation (7.24) leads to

$$n_q(f) \quad = \quad \sum_{m_1=0}^{M-1}\cdots\sum_{m_{I_q}=0}^{M-1}\sum_{p=-\infty}^{\infty}$$

$$\sum_{o_{1,1}=1}^{M-1}\sum_{e_{1,1}=1}^{E_{o_{1,1}}}\cdots\sum_{o_{1,m_1}=1}^{M-1}\sum_{e_{1,m_1}=1}^{E_{o_{1,m_1}}}\cdots\cdots\sum_{o_{I_q,1}=1}^{M-1}\sum_{e_{I_q,1}=1}^{E_{o_{I_q,1}}}\cdots\sum_{o_{I_q,m_{I_q}}=1}^{M-1}\sum_{e_{I_q,m_{I_q}}=1}^{E_{o_{I_q,m_{I_q}}}}$$

$$\mathcal{L}_{1,M-1}(o_{1,1}+\cdots+o_{1,m_1}+\cdots\cdots+o_{I_q,1}+\cdots+o_{I_q,m_{I_q}})$$

$$(G_q)_{1,m_1,\ldots,m_{I_q}}\left(\xi_p;\Psi_{o_{1,1},e_{1,1}},\ldots,\Psi_{o_{1,m_1},e_{1,m_1}};\cdots\right.$$

$$\left.\cdots;\Psi_{o_{I_q,1},e_{I_q,1}},\ldots,\Psi_{o_{I_q,m_{I_q}},e_{I_q,m_{I_q}}}\right)$$

$$\widetilde{w}_q(\xi_p)\,^{o_{1,1}}\widetilde{u}_{q,1}(\Psi_{o_{1,1},e_{1,1}})\cdots^{o_{1,m_1}}\widetilde{u}_{q,1}(\Psi_{o_{1,m_1},e_{1,m_1}})\cdots$$

$$\cdots^{o_{I_q,1}}\widetilde{u}_{q,I_q}(\Psi_{o_{I_q,1},e_{I_q,1}})\cdots^{o_{I_q,m_{I_q}}}\widetilde{u}_{q,I_q}(\Psi_{o_{I_q,m_{I_q}},e_{I_q,m_{I_q}}})$$

$$\delta\left(f-\xi_p-\Psi_{o_{1,1},e_{1,1}}-\cdots-\Psi_{o_{1,m_1},e_{1,m_1}}-\cdots\right.$$

$$\left.\cdots-\Psi_{o_{I_q,1},e_{I_q,1}}-\cdots-\Psi_{o_{I_q,m_{I_q}},e_{I_q,m_{I_q}}}\right) \tag{7.46}$$

Note from Equation (7.46) that there are infinitely many sum terms in the expression for $n_q(f)$. This is due to the fact that the fundamental (unmodulated) noise source $w_q(f)$ where $q \in \{1,2,\ldots,Q\}$ is represented as a Fourier series at an infinite number of frequencies. The effects of the modulating signals $\{u_{q,1}(f), \ldots, u_{q,I_q}(f)\}$ where $q \in \{1, 2, \ldots, Q\}$ are (among other things) to frequency shift the noise component corresponding to a given frequency, and to change (modulate) the amplitude of the fundamental noise source component. The computational cost of determining $n_q(f)$ where $q \in \{1, 2, \ldots, Q\}$ in Equation (7.46) can be reduced significantly by using the symmetry properties of the partly symmetrical multi-port frequency domain Volterra transfer function $(G_q)_{1,m_1,\ldots,m_{I_q}}(\cdot)$ which leads to

$$
\begin{aligned}
n_q(f) \;=\; & \sum_{m_1=0}^{M-1} \cdots \sum_{m_{I_q}=0}^{M-1} \sum_{p=-\infty}^{\infty} \\
& \sum_{o_{1,1}=1}^{M-1} \sum_{e_{1,1}=1}^{E_{o_{1,1}}} \cdots \sum_{o_{1,m_1}=1}^{M-1} \sum_{e_{1,m_1}=1}^{E_{o_{1,m_1}}} \cdots \\
& \cdots \sum_{o_{I_q,1}=1}^{M-1} \sum_{e_{I_q,1}=1}^{E_{o_{I_q,1}}} \cdots \sum_{o_{I_q,m_{I_q}}=1}^{M-1} \sum_{e_{I_q,m_{I_q}}=1}^{E_{o_{I_q,m_{I_q}}}} \\
& \mathcal{L}_{1,M-1}(o_{1,1} + \cdots + o_{1,m_1} + \cdots \cdots + o_{I_q,1} + \cdots + o_{I_q,m_{I_q}}) \\
& \mathcal{A}_1(o_{1,1}, \ldots, o_{1,m_1}) \cdots \mathcal{A}_{I_q}(o_{I_q,1}, \ldots, o_{I_q,m_{I_q}}) \\
& \mathcal{A}_1(e_{1,1}, \ldots, e_{1,m_1}) \cdots \mathcal{A}_{I_q}(e_{I_q,1}, \ldots, e_{I_q,m_{I_q}}) \\
& \prod_{i_q=1}^{I_q} m_{i_q}! \\
& \prod_{i_q=1}^{I_q} \prod_{o_{i_q}=1}^{M-1} \prod_{e_{i_q}=1}^{E_{o_{i_q}}} \left\{ \mathcal{N}_{o_{i_q},e_{i_q}}(o_{i_q,1}, \ldots, o_{i_q,m_{i_q}}; e_{i_q,1}, \ldots, e_{i_q,m_{i_q}})! \right\}^{-1} \\
& (G_q)_{1,m_1,\ldots,m_{I_q}}\Big(\xi_p; \Psi_{o_{1,1},e_{1,1}}, \ldots, \Psi_{o_{1,m_1},e_{1,m_1}}; \cdots \\
& \qquad \cdots ; \Psi_{o_{I_q,1},e_{I_q,1}}, \ldots, \Psi_{o_{I_q,m_{I_q}},e_{I_q,m_{I_q}}}\Big) \\
& \widetilde{w}_q(\xi_p)\,^{o_{1,1}}\widetilde{u}_{q,1}(\Psi_{o_{1,1},e_{1,1}}) \cdots {}^{o_{1,m_1}}\widetilde{u}_{q,1}(\Psi_{o_{1,m_1},e_{1,m_1}}) \cdots \\
& \qquad \cdots {}^{o_{I_q,1}}\widetilde{u}_{q,I_q}(\Psi_{o_{I_q,1},e_{I_q,1}}) \cdots {}^{o_{I_q,m_{I_q}}}\widetilde{u}_{q,I_q}(\Psi_{o_{I_q,m_{I_q}},e_{I_q,m_{I_q}}}) \\
& \delta\Big(f - \xi_p - \Psi_{o_{1,1},e_{1,1}} - \cdots - \Psi_{o_{1,m_1},e_{1,m_1}} - \cdots \\
& \qquad \cdots - \Psi_{o_{I_q,1},e_{I_q,1}} - \cdots - \Psi_{o_{I_q,m_{I_q}},e_{I_q,m_{I_q}}}\Big)
\end{aligned}
\tag{7.47}
$$

where

$$
\begin{aligned}
& \mathcal{N}_{o_{i_q},e_{i_q}}(o_{i_q,1}, \ldots, o_{i_q,m_{i_q}}; e_{i_q,1}, \ldots, e_{i_q,m_{i_q}}) \\
= \;& \text{Number of } \{(o_{i_q,1}, e_{i_q,1}), \ldots, (o_{i_q,m_{i_q}}, e_{i_q,m_{i_q}})\} \quad \text{which} \\
& \text{are equal to } (o_{i_q}, e_{i_q})
\end{aligned}
\tag{7.48}
$$

and $\mathcal{A}_{i_q}(o_{i_q,1}, \ldots, o_{i_q,m_{i_q}})$ and $\mathcal{A}_{i_q}(e_{i_q,1}, \ldots, e_{i_q,m_{i_q}})$ are defined by Equation (7.28). In Equation (7.46), $\mathcal{L}_{0,M-1}(\cdot)$ is used with a different argument compared to Equation (7.24). This is because otherwise there would be contributions of a total order greater than M. This is avoided by using Equation (7.46). Thus, from Equation (7.46) it is seen that the (possibly) modulated noise source $n_q(f)$ where $q \in \{1, 2, \ldots, Q\}$ can be written as the Fourier series in Equation (7.9) with some rather complicated Fourier series coefficients $\tilde{n}_q(\xi_p)$ where $p \in \mathcal{Z}$ is an integer. However, this requires that

$$\exists\, p' \in \mathcal{Z}: \quad \xi_{p'} = \xi_p + \Psi_{o_{1,1},e_{1,1}} + \cdots + \Psi_{o_{1,m_1},e_{1,m_1}} + \cdots$$
$$\cdots + \Psi_{o_{I_q,1},e_{I_q,1}} + \cdots + \Psi_{o_{I_q,m_{I_q}},e_{I_q,m_{I_q}}} \qquad (7.49)$$

for any integer $p \in \mathcal{Z}$, $o_{i_q,l} \in \{1, 2, \ldots, M-1\}$, and $e_{i_q,l} \in \{1, 2, \ldots, E_{o_{i_q,l}}\}$ where $i_q \in \{1, 2, \ldots, I_q\}$ and $l \in \{1, 2, \ldots, m_{i_q}\}$. Equivalently, the requirement in Equation (7.49) can be expressed as

$$\frac{1}{\xi}\Big[\Psi_{o_{1,1},e_{1,1}} + \cdots + \Psi_{o_{1,m_1},e_{1,m_1}} + \cdots$$
$$\cdots + \Psi_{o_{I_q,1},e_{I_q,1}} + \cdots + \Psi_{o_{I_q,m_{I_q}},e_{I_q,m_{I_q}}} \Big] \in \mathcal{Z} \qquad (7.50)$$

for any integer $p \in \mathcal{Z}$, $o_{i_q,l} \in \{1, 2, \ldots, M-1\}$, $e_{i_q,l} \in \{1, 2, \ldots, E_{o_{i_q,l}}\}$ where $i_q \in \{1, 2, \ldots, I_q\}$ and $l \in \{1, 2, \ldots, m_{i_q}\}$. The fulfilment of this requirement is of no concern since the frequency resolution in the Fourier series representation $\xi = 1/(2\tau)$ can be made arbitrarily small by choosing τ sufficiently large (2τ is the time interval of observation). For example, if all applied frequencies are integers then $2\tau = 1$ fulfils the equivalent requirements in Equations (7.49) and (7.50). Assuming that the requirements in Equations (7.49) and (7.50) are fulfilled, then the coefficient $\tilde{n}_q(\xi_p)$ where $q \in \{1, 2, \ldots, Q\}$ in the Fourier series representation of $n_q(f)$ in Equation (7.47) can be determined as

$$\tilde{n}_q(\xi_p) \;=\; \sum_{m_1=0}^{M-1} \cdots \sum_{m_{I_q}=0}^{M-1}$$

$$\sum_{o_{1,1}=1}^{M-1}\sum_{e_{1,1}=1}^{E_{o_{1,1}}} \cdots \sum_{o_{1,m_1}=1}^{M-1}\sum_{e_{1,m_1}=1}^{E_{o_{1,m_1}}} \cdots$$

$$\cdots \sum_{o_{I_q,1}=1}^{M-1}\sum_{e_{I_q,1}=1}^{E_{o_{I_q,1}}} \cdots \sum_{o_{I_q,m_{I_q}}=1}^{M-1}\sum_{e_{I_q,m_{I_q}}=1}^{E_{o_{I_q,m_{I_q}}}}$$

$$\mathcal{L}_{1,M-1}(o_{1,1}+\cdots+o_{1,m_1}+\cdots\cdots+o_{I_q,1}+\cdots+o_{I_q,m_{I_q}})$$

$$\mathcal{A}_1(o_{1,1},\ldots,o_{1,m_1})\cdots\mathcal{A}_{I_q}(o_{I_q,1},\ldots,o_{I_q,m_{I_q}})$$

$$\mathcal{A}_1(e_{1,1},\ldots,e_{1,m_1})\cdots\mathcal{A}_{I_q}(e_{I_q,1},\ldots,e_{I_q,m_{I_q}})$$

$$\prod_{i_q=1}^{I_q} m_{i_q}!$$

$$\prod_{i_q=1}^{I_q}\prod_{o_{i_q}=1}^{M-1}\prod_{e_{i_q}=1}^{E_{o_{i_q}}}\left\{\mathcal{N}_{o_{i_q},e_{i_q}}(o_{i_q,1},\ldots,o_{i_q,m_{i_q}};\,e_{i_q,1},\ldots,e_{i_q,m_{i_q}})!\right\}^{-1}$$

$$(G_q)_{1,m_1,\ldots,m_{I_q}}\Big(\xi_p - \Psi_{o_{1,1},e_{1,1}} - \cdots - \Psi_{o_{1,m_1},e_{1,m_1}} - \cdots$$

$$\cdots - \Psi_{o_{I_q,1},e_{I_q,1}} - \cdots - \Psi_{o_{I_q,m_{I_q}},e_{I_q,m_{I_q}}};$$

$$\Psi_{o_{1,1},e_{1,1}},\ldots,\Psi_{o_{1,m_1},e_{1,m_1}};\cdots$$

$$\cdots;\Psi_{o_{I_q,1},e_{I_q,1}},\ldots,\Psi_{o_{I_q,m_{I_q}},e_{I_q,m_{I_q}}}\Big)$$

$$\tilde{w}_q\Big(\xi_p - \Psi_{o_{1,1},e_{1,1}} - \cdots - \Psi_{o_{1,m_1},e_{1,m_1}} - \cdots$$

$$\cdots - \Psi_{o_{I_q,1},e_{I_q,1}} - \cdots - \Psi_{o_{I_q,m_{I_q}},e_{I_q,m_{I_q}}}\Big)$$

$${}^{o_{1,1}}\tilde{u}_{q,1}(\Psi_{o_{1,1},e_{1,1}})\cdots{}^{o_{1,m_1}}\tilde{u}_{q,1}(\Psi_{o_{1,m_1},e_{1,m_1}})\cdots$$

$$\cdots{}^{o_{I_q,1}}\tilde{u}_{q,I_q}(\Psi_{o_{I_q,1},e_{I_q,1}})\cdots{}^{o_{I_q,m_{I_q}}}\tilde{u}_{q,I_q}(\Psi_{o_{I_q,m_{I_q}},e_{I_q,m_{I_q}}}) \qquad (7.51)$$

For the noise response it is not of interest directly to operate on the expression for $n_q(f)$ but rather on the Fourier series coefficient of $\delta(0)$ in the expression for $n_q(f)$. This is because (i) the noise representation has a (practically — due to a very small ξ) continuos frequency spectrum, and (ii) only low level noise is included in the analysis, in which case it is not necessary to consider contributions caused by non-linear mixing of noise with noise. In Equation (7.51) a part of the argument frequency to $\tilde{w}_q(\cdot)$ is

$$\Psi_{o_{1,1},e_{1,1}} + \cdots + \Psi_{o_{1,m_1},e_{1,m_1}} + \cdots\cdots + \Psi_{o_{I_q,1},e_{I_q,1}} + \cdots + \Psi_{o_{I_q,m_{I_q}},e_{I_q,m_{I_q}}}$$

$$\in \; \mathcal{S}_{o_{1,1}+\cdots+o_{1,m_1}+\cdots\cdots+o_{I_q,1}+\cdots+o_{I_q,m_{I_q}}} \tag{7.52}$$

$$\in \; \bigcup_{o=1}^{M-1} \mathcal{S}_o \tag{7.53}$$

provided that $m_1 + \cdots + m_{I_q} \in \{1, 2, \ldots, M - 1\}$. In Equation (7.51) the factor $\mathcal{L}_{1,M-1}$ implies that there are only (possibly) non-zero contributions to $\tilde{n}_q(\xi_p)$ for $m_1 + \cdots + m_{I_q} \in \{0, 1, \ldots, M-1\}$. The contribution for $m_1 + \cdots + m_{I_q} = 0$ is $\tilde{n}_q(\xi_p) = (G_q)_{1,0,\ldots,0}(\xi_p)\tilde{w}_q(\xi_p)$ which is the linear transfer from unmodulated to modulated noise source (actually, this contribution is unmodulated since it does not depend on the controlling u-signals). This contribution is included in the formulation of frequency sets by defining a frequency set $\mathcal{S}_0$ of order 0 as

$$\mathcal{S}_0 \;=\; \{\Psi_{0,1}\} \tag{7.54}$$

$$\;=\; \{0\} \tag{7.55}$$

which means that $E_0 = 1$. Using the above and collecting terms in Equation (7.51) leads to

$$\tilde{n}_q(\xi_p) \;=\; \sum_{o=0}^{M-1} \sum_{e=1}^{E_o} t_{q,o}(\xi_p, \Psi_{o,e}) \, \tilde{w}_q(\xi_p - \Psi_{o,e}) \tag{7.56}$$

where $q \in \{1, 2, \ldots, Q\}$, and $t_{q,o}(\Psi_{o,e})$ is a $o'th$ order transfer coefficient from unmodulated noise source $w_q(\xi_p - \Psi_{o,e})$ where $o \in \{0, 1, \ldots, M-1\}$, $e \in \{1, 2, \ldots, E_o\}$ to modulated noise source Fourier series coefficient $\tilde{n}_q(\xi_p)$. It is convenient to rewrite $\tilde{n}_q(\xi_p)$ where $q \in \{1, 2, \ldots, Q\}$ in Equation (7.56) into

$$\tilde{n}_q(\xi_p) \;=\; \boldsymbol{t}_q^T(\xi_p) \, \tilde{\boldsymbol{w}}_q(\xi_p) \tag{7.57}$$

where

$$\boldsymbol{t}_q(\xi_p) \;=\; \left[t_q(\xi_p, \Psi_1), \ldots, t_q(\xi_p, \Psi_E) \right]^T \;\in\; \mathcal{C}^{E \times 1} \tag{7.58}$$

$$\tilde{\boldsymbol{w}}_q(\xi_p) \;=\; \left[\tilde{w}_q(\xi_p - \Psi_1), \ldots, \tilde{w}_q(\xi_p - \Psi_E) \right]^T \;\in\; \mathcal{C}^{E \times 1} \tag{7.59}$$

and

$$\{\Psi_1, \ldots, \Psi_E\} \;=\; \bigcup_{o=0}^{M-1} \mathcal{S}_o \tag{7.60}$$

Note from Equations (7.30), (7.31) and (7.60) that $E \in \{1, 2, \ldots, E_0 + E_1 + \cdots + E_{M-1}\}$. If the input frequencies $\{\psi_{1,1}, \ldots, \psi_{1,J_1}, \ldots \ldots, \psi_{K,1}, \ldots, \psi_{K,J_K}\}$ are incommensurate up to order $M - 1$ then $E = E_0 + E_1 + \cdots + E_{M-1}$.

Great care must be taken in determining $t_q(\xi_p)$ where $q \in \{1, 2, \ldots, Q\}$ from Equations (7.56) and (7.57) when frequencies $\{\psi_{1,1}, \ldots, \psi_{1,J_1}, \ldots \ldots, \psi_{K,1}, \ldots, \psi_{K,J_K}\}$ are commensurate. This is because when the frequencies are commensurate a given $t_{q,o}(\xi_p, \Psi_e)$ where $q \in \{1, 2, \ldots, Q\}$ and $e \in \{1, 2, \ldots, E\}$ may e.g. consist of the sum of $t_{q,o_1}(\xi_p, \Psi_e)$ and $t_{q,o_2}(\xi_p, \Psi_e)$ where $o_1, o_2 \in \{0, 1, \ldots, M - 1\}$ and $o_1 \neq o_2$. This is not the case when the frequencies are incommensurate since in this case there is only one $t_{q,o}(\xi_p, \Psi_e)$ of interest where $q \in \{1, 2, \ldots, Q\}$, $o \in \{0, 1, \ldots, M - 1\}$ and $e \in \{1, 2, \ldots, E\}$.

The cross-correlation between two arbitrary Fourier series coefficients of the two modulated noise sources $n_{q_1}(f)$ and $n_{q_2}(f)$ where $q_1, q_2 \in \{1, 2, \ldots, Q\}$ is

$$\langle \tilde{n}_{q_1}(\xi_{p_1}) \, \tilde{n}_{q_2}^*(\xi_{p_2}) \rangle \;=\; \langle t_{q_1}^T(\xi_{p_1}) \, \tilde{w}_{q_1}(\xi_{p_1}) \, t_{q_2}^\dagger(\xi_{p_2}) \, \tilde{w}_{q_2}^*(\xi_{p_2}) \rangle \tag{7.61}$$

where $\langle \cdot \rangle$ indicates the ensemble average over noise processes with identical statistical properties, $[\cdot]^*$ indicates the conjugate of a vector (or matrix), and $[\cdot]^\dagger$ indicates the conjugate transpose (Hermitian conjugate) of a vector (or matrix). Equation (7.61) can be written as

$$\langle \tilde{n}_{q_1}(\xi_{p_1}) \, \tilde{n}_{q_2}^*(\xi_{p_2}) \rangle \;=\; t_{q_1}^T(\xi_{p_1}) \, \widetilde{W}_{q_1,q_2}(\xi_{p_1}, \xi_{p_2}) \, t_{q_2}^*(\xi_{p_2}) \tag{7.62}$$

where $\widetilde{W}_{q_1,q_2}(\xi_{p_1}, \xi_{p_2})$ is a noise cross-correlation matrix for the fundamental (unmodulated) noise sources $w_{q_1}(f)$ and $w_{q_2}(f)$ where $q_1, q_2 \in \{1, 2, \ldots, Q\}$, defined as

$$
\begin{aligned}
&\widetilde{W}_{q_1,q_2}(\xi_{p_1}, \xi_{p_2}) \\
&= \; \langle \tilde{w}_{q_1}(\xi_{p_1}) \, \tilde{w}_{q_2}^\dagger(\xi_{p_2}) \rangle \\
&= \begin{bmatrix} \langle \tilde{w}_{q_1}(\xi_{p_1} - \Psi_1) \, \tilde{w}_{q_2}^*(\xi_{p_2} - \Psi_1) \rangle & \cdots & \langle \tilde{w}_{q_1}(\xi_{p_1} - \Psi_1) \, \tilde{w}_{q_2}^*(\xi_{p_2} - \Psi_E) \rangle \\ \vdots & & \vdots \\ \langle \tilde{w}_{q_1}(\xi_{p_1} - \Psi_E) \, \tilde{w}_{q_2}^*(\xi_{p_2} - \Psi_1) \rangle & \cdots & \langle \tilde{w}_{q_1}(\xi_{p_1} - \Psi_E) \, \tilde{w}_{q_2}^*(\xi_{p_2} - \Psi_E) \rangle \end{bmatrix}
\end{aligned}
$$

$$\tag{7.63}$$
$$\tag{7.64}$$

Note that the matrix $\widetilde{W}_{q_1,q_2}(\xi_{p_1}, \xi_{p_2})$ where $q_1, q_2 \in \{1, 2, \ldots, Q\}$ describes the cross-correlation (and autocorrelation) between Fourier series coefficients $\tilde{w}_{q_1}(f_1)$ and $\tilde{w}_{q_2}(f_2)$ at all frequencies $\xi_{p_1} - f_1, \xi_{p_2} - f_2 \in \{\Psi_1, \ldots, \Psi_E\}$. Note also from Equations (7.63) and (7.64) that

$$\widetilde{W}_{q_2,q_1}(\xi_{p_2}, \xi_{p_1}) \;=\; \widetilde{W}_{q_1,q_2}^\dagger(\xi_{p_1}, \xi_{p_2}) \tag{7.65}$$

This relation may be used to reduce some computations in the determination of the noise (cross-)correlation matrices for the fundamental noise sources.

7.3.4   Fundamental noise sources

A typical type of fundamental time domain noise source is a Gaussian white noise process with specified mean and standard deviation. Traditionally, Gaussian white noise sources are with mean values equal to zero, but in the present work the fundamental noise sources may have non-zero mean values. Due to this the noise is actually not "white" since not all frequencies have the same average power density. However, this type of noise will be referred to as white noise even though it has a non-zero mean. The fundamental time domain Gaussian white noise source with mean μ_q and standard deviation σ_q has a time domain cross-correlation given by

$$\langle w_q(t_1)\, w_q^*(t_2)\rangle \quad = \quad \sigma_q^2\, \delta(t_1 - t_2) + \mu_q^2, \qquad t_1, t_2 \in [-\tau; \tau] \qquad (7.66)$$

for noise source $q \in \{1, 2, \ldots, Q\}$ where $\delta(\cdot)$ is the Dirac delta-function. It can be shown that the corresponding cross-correlation between two arbitrary frequency domain Fourier series coefficients evaluated in the time interval $[-\tau; \tau]$ is given by

$$\langle \tilde{w}_q(\xi_{p_1})\, \tilde{w}_q^*(\xi_{p_2})\rangle \quad = \quad \frac{\sigma_q^2}{2\tau}\frac{\sin[\pi(p_1 - p_2)]}{\pi(p_1 - p_2)} + \mu_q^2\frac{\sin[\pi p_1]}{\pi p_1}\frac{\sin[\pi p_2]}{\pi p_2} \qquad (7.67)$$

$$= \quad \begin{cases} \sigma_q^2/(2\tau) + \mu_q^2 & \text{for } p_1 = p_2 = 0 \\ \sigma_q^2/(2\tau) & \text{for } p_1 = p_2 \neq 0 \\ 0 & \text{otherwise} \end{cases} \qquad (7.68)$$

where $p_1, p_2 \in \mathcal{Z}$. Note that there are only contributions to $\langle \tilde{w}_q(\xi_{p_1})\, \tilde{w}_q^*(\xi_{p_2})\rangle$ when $p_1 = p_2$. When $p_1 = p_2 \neq 0$ only the standard deviation is of importance, and when $p_1 = p_2 = 0$ both the standard deviation and the mean value are of importance.

In some cases it may be useful to make a comparative study by a discrete time domain simulation. In this case the description in time is discretized such that $t \in \{-t_\Lambda, \ldots, -t_1, t_0, t_1, \ldots, t_\Lambda\}$ with $t_\lambda - t_{\lambda-1} = 2\tau/(2\Lambda + 1)$ for all $\lambda \in \{-\Lambda + 1, -\Lambda + 2, \ldots, \Lambda\}$, and

$$t_\lambda \quad = \quad \frac{2\tau}{2\Lambda + 1}\lambda \qquad (7.69)$$

where $\lambda \in \{-\Lambda, \ldots, -1, 0, 1, \ldots, \Lambda\}$. Note from Equation (7.69) that $t_{-\lambda} = -t_\lambda$. In this case it can be shown that the Fourier series coefficient is given by

$$\tilde{w}_q(\xi_p) \quad = \quad \frac{1}{2\Lambda + 1}\sum_{\lambda=-\Lambda}^{\Lambda} w_q(t_\lambda)\, \exp\left[-j2\pi p\frac{\lambda}{2\Lambda + 1}\right] \qquad (7.70)$$

where $p \in \{-\Lambda, \ldots, -1, 0, 1, \ldots, \Lambda\}$. Thus the one time domain stochastic variable is represented as $2\Lambda + 1$ frequency domain stochastic variables. Assuming that $w_q(t)$ where $q \in \{1, 2, \ldots, Q\}$ is a white noise Gaussian stochastic process with mean μ_q and standard deviation σ_q it can be shown that

$$\langle w_q(t_{\lambda_1})\, w_q^*(t_{\lambda_2})\rangle \quad = \quad \begin{cases} \sigma_q^2 + \mu_q^2 & \text{for} \quad \lambda_1 = \lambda_2 \\ \mu_q^2 & \text{for} \quad \lambda_1 \neq \lambda_2 \end{cases} \tag{7.71}$$

where $\lambda_1, \lambda_2 \in \{-\Lambda, \ldots, -1, 0, 1, \ldots, \Lambda\}$. Then the cross-correlation between two arbitrary frequency domain Fourier coefficients is

$$\langle \widetilde{w}_q(\xi_{p_1})\, \widetilde{w}_q^*(\xi_{p_2})\rangle \quad = \quad \begin{cases} \sigma_q^2/(2\Lambda + 1) + \mu_q^2 & \text{for} \quad p_1 = p_2 = 0 \\ \sigma_q^2/(2\Lambda + 1) & \text{for} \quad p_1 = p_2 \neq 0 \\ 0 & \text{for} \quad p_1 \neq p_2 \end{cases} \tag{7.72}$$

Note that $\langle |\widetilde{w}_q(\xi_p)|^2\rangle$ where $q \in \{1, 2, \ldots, Q\}$ and $p \in \{-\Lambda, \ldots, -1, 0, 1, \ldots, \Lambda\}$ is the average signal power at the frequency ξ_p. This means that

$$\sum_{p=-\Lambda}^{\Lambda} \langle |\widetilde{w}_q(\xi_p)|^2\rangle \quad = \quad \sigma_q^2 + \mu_q^2 \tag{7.73}$$

is the total average noise signal power.

7.3.5 Some special cases

The following are special cases related to Equation (7.62):

- If two zero mean noise sources $w_{q_1}(t)$ and $w_{q_2}(t)$ where $q_1, q_2 \in \{1, 2, \ldots, Q\}$ are uncorrelated then

$$\widetilde{W}_{q_1, q_2}(\xi_{p_1}, \xi_{p_2}) \quad = \quad \mathbf{0} \quad \text{for} \quad \begin{cases} q_1 \neq q_2 \\ q_1 = q_2 \ \wedge \ p_1 - p_2 \neq (\Psi_{e_1} - \Psi_{e_2})/\xi \end{cases} \tag{7.74}$$

for all $e_1, e_2 \in \{1, 2, \ldots, E\}$, where $\mathbf{0} \in \{0\}^{E \times E}$ is the zero matrix.

- If the fundamental (unmodulated) noise source $w_q(f)$ where $q \in \{1, 2, \ldots, Q\}$ generates Gaussian zero mean white noise with standard deviation σ then

$$\widetilde{W}_{q,q}(\xi_p, \xi_p) \quad = \quad \sigma^2\, \boldsymbol{I} \tag{7.75}$$

where $\boldsymbol{I} \in \{0, 1\}^{E \times E}$ is the identity matrix. Note that $\widetilde{W}_{q,q}(\xi_{p_1}, \xi_{p_2})$ where $q \in \{1, 2, \ldots, Q\}$ and $p_1, p_2 \in \mathcal{Z}$ are integers can be the zero matrix of dimension $E \times E$, but it can also be a non-zero and non-identity matrix of dimension $E \times E$ depending on $\{\Psi_1, \ldots, \Psi_E\}$, ξ_{p_1} and ξ_{p_2} when the fundamental noise source $w_q(t)$ where $q \in \{1, 2, \ldots, Q\}$ generates white noise.

- If the noise source $n_q(t)$ where $q \in \{1, 2, \ldots, Q\}$ is unmodulated then $I_q = 0$ and

$$t_q(\xi_p) \quad = \quad \begin{bmatrix} 0, \ldots, 0, t_q(\xi_p, 0), 0, \ldots, 0 \end{bmatrix}^T \ \in \ \mathcal{C}^{E \times 1} \tag{7.76}$$

Thus, in this case $\tilde{n}_q(\xi_p) = t_q(\xi_p, 0)\,\tilde{w}_q(\xi_p)$ where $q \in \{1, 2, \ldots, Q\}$ as seen from Equation (7.57).

As an example consider the following situation:

- Determine $\langle \tilde{n}_q(\xi_p + \Psi)\,\tilde{n}_q^*(\xi_p)\rangle$ when $\{\Psi_1, \ldots, \Psi_E\} = \{0, -\Psi, \Psi\}$ where $\Psi \neq 0$ and $w_q(t)$ is a zero mean white noise source with standard deviation σ_q and autocorrelation function

$$\langle \tilde{w}_q(\xi_{p_1})\,\tilde{w}_q^*(\xi_{p_2})\rangle = \begin{cases} \sigma_q^2/(2\tau) & \text{for } p_1 = p_2 \\ 0 & \text{otherwise} \end{cases} \tag{7.77}$$

where $q \in \{1, 2, \ldots, Q\}$. In this case

$$\begin{aligned} \langle \tilde{n}_q(\xi_p + \Psi)\,\tilde{n}_q^*(\xi_p)\rangle = \ & t_q(\xi_p + \Psi, 0)\,t_q^*(\xi_p, -\Psi)\,\sigma_q^2/(2\tau) \\ & + t_q(\xi_p + \Psi, \Psi)\,t_q^*(\xi_p, 0)\,\sigma_q^2/(2\tau) \end{aligned} \tag{7.78}$$

which is generally different from zero. Note that the modulated noise source $n_q(t)$ is correlated at two different frequencies $\xi_p + \Psi$ and ξ_p even when, as in this case, the fundamental (unmodulated) noise source $w_q(t)$ is a zero mean white noise source.

7.3.6   Algorithm

To determine the cross-correlation $\langle \tilde{n}_{q_1}(\xi_{p_1})\,\tilde{n}_{q_2}^*(\xi_{p_2})\rangle$ where $q_1, q_2 \in \{1, 2, \ldots, Q\}$ the following algorithm can be used:

1. Determine the frequency sets $S_0, S_1, S_2, \ldots, S_{M-1}$ from Equation (7.30).

2. Determine the controlling variables

$$\{\tilde{u}_{q_1,1}(\Psi_e), \ldots, \tilde{u}_{q_1,I_{q_1}}(\Psi_e)\} \cup \{\tilde{u}_{q_2,1}(\Psi_e), \ldots, \tilde{u}_{q_2,I_{q_2}}(\Psi_e)\} \tag{7.79}$$

   for all $\Psi_e \in S_1 \cup S_2 \cup \cdots \cup S_{M-1}$ and $q_1, q_2 \in \{1, 2, \ldots, Q\}$ using Equations (7.40), (7.41) and (7.27).

3. Specify the multi-port frequency domain Volterra transfer functions $(G_{q_1})_{1, m_1, \ldots, m_{I_{q_1}}}(\cdot)$ where $m_1 + \cdots + m_{I_{q_1}} \in \{1, 2, \ldots, M-1\}$, $q_1 \in \{1, 2, \ldots, Q\}$, and $(G_{q_2})_{1, m_1, \ldots, m_{I_{q_2}}}(\cdot)$ where $m_1 + \cdots + m_{I_{q_2}} \in \{1, 2, \ldots, M-1\}$ and $q_2 \in \{1, 2, \ldots, Q\}$.

4. Determine the transfer vectors $t_{q_1}(\xi_{p_1})$ and $t_{q_2}(\xi_{p_2})$ where $q_1, q_2 \in \{1, 2, \ldots, Q\}$ and $p_1, p_2 \in \mathcal{Z}$ using Equations (7.51)–(7.59).

5. Specify the noise cross-correlation matrix $\widetilde{W}_{q_1, q_2}(\xi_{p_1}, \xi_{p_2})$ where $q_1, q_2 \in \{1, 2, \ldots, Q\}$ given by Equations (7.63)–(7.64).

6. Determine $\langle \tilde{n}_{q_1}(\xi_{p_1})\,\tilde{n}_{q_2}^*(\xi_{p_2})\rangle$ from Equation (7.62).

7.4 Responses

The response $r_l(f)$ where $l \in \{1, 2, \ldots, L\}$ from port (r, l) Figure 7.2 can be determined as the response from a multi-port Volterra system as

$$
\begin{aligned}
r_l(f) = &\sum_{m_1=0}^{M} \cdots \sum_{m_K=0}^{M} \sum_{o_1=0}^{M} \cdots \sum_{o_Q=0}^{M} \\
&\int_{-\infty}^{\infty} \cdots \int_{-\infty}^{\infty} \int_{-\infty}^{\infty} \cdots \int_{-\infty}^{\infty} \int_{-\infty}^{\infty} \cdots \int_{-\infty}^{\infty} \int_{-\infty}^{\infty} \cdots \int_{-\infty}^{\infty} \\
&\mathcal{L}_{1,M}(m_1 + \cdots + m_K + o_1 + \cdots + o_Q) \\
&(H_l)_{m_1,\ldots,m_K,o_1,\ldots,o_Q}(\Omega_{1,1}, \ldots, \Omega_{1,m_1}; \cdots \cdots; \Omega_{K,1}, \ldots, \Omega_{K,m_K}; \\
&\qquad \Xi_{1,1}, \ldots, \Xi_{1,o_1}; \cdots \cdots; \Xi_{Q,1}, \ldots, \Xi_{Q,o_Q}) \\
&s_1(\Omega_{1,1}) \cdots s_1(\Omega_{1,m_1}) \cdots \cdots s_K(\Omega_{K,1}) \cdots s_K(\Omega_{K,m_K}) \\
&n_1(\Xi_{1,1}) \cdots n_1(\Xi_{1,o_1}) \cdots \cdots n_Q(\Xi_{Q,1}) \cdots n_Q(\Xi_{Q,o_Q}) \\
&\delta(f - \Omega_{1,1} - \cdots - \Omega_{1,m_1} - \cdots \cdots - \Omega_{K,1} - \cdots - \Omega_{K,m_K} \\
&\qquad - \Xi_{1,1} - \cdots - \Xi_{1,o_1} - \cdots \cdots - \Xi_{Q,1} - \cdots - \Xi_{Q,o_Q}) \\
&d\Omega_{1,1} \cdots d\Omega_{1,m_1} \cdots \cdots d\Omega_{K,1} \cdots d\Omega_{K,m_K} \\
&d\Xi_{1,1} \cdots d\Xi_{1,o_1} \cdots \cdots d\Xi_{Q,1} \cdots d\Xi_{Q,o_Q} \qquad (7.80)
\end{aligned}
$$

where $(H_l)_{m_1,\ldots,m_K,o_1,\ldots,o_Q}(\cdot)$ is a partly symmetrical multi-port frequency domain Volterra transfer function relating the inputs $\{s_1(f), \ldots, s_K(f), n_1(f), \ldots, n_Q(f)\}$ to the output $r_l(f)$. The response $r_l(f)$ where $l \in \{1, 2, \ldots, L\}$ may conveniently be separated into two contributions: (i) a purely deterministic contribution $r_{d,l}(f)$, and (ii) a noise contribution $r_{n,l}(f)$ such that

$$
r_l(f) = r_{d,l}(f) + r_{n,l}(f) \qquad (7.81)
$$

where $l \in \{1, 2, \ldots, L\}$. The deterministic response $r_{d,l}(f)$ consists of the linear transfer of input signals to the output port (r, l) where $l \in \{1, 2, \ldots, L\}$ and of all intermodulation contributions of the applied deterministic input signals up to order M. The noise response $r_{n,l}(f)$ consists of the linear transfer of the noise sources to the output port (r, l) where $l \in \{1, 2, \ldots, L\}$ and of all non-linear transfers due to intermodulation between deterministic signals and noise. Contributions due to mixing of noise with noise are not included due to the low-level noise assumption.

7.4.1 Deterministic response

The deterministic response $r_{d,l}(f)$ at port (r, l) where $l \in \{1, 2, \ldots, L\}$ can be determined from Equation (7.80) with $o_1 = \cdots = o_Q = 0$ as

$$
\begin{aligned}
r_{d,l}(f) \;=\; & \sum_{m_1=0}^{M} \cdots \sum_{m_K=0}^{M} \int_{-\infty}^{\infty} \cdots \int_{-\infty}^{\infty} \cdots \cdots \int_{-\infty}^{\infty} \cdots \int_{-\infty}^{\infty} \\
& \mathcal{L}_{1,M}(m_1 + \cdots + m_K) \\
& (H_l)_{m_1,\ldots,m_K,0,\ldots,0}(\Omega_{1,1},\ldots,\Omega_{1,m_1}; \cdots\cdots; \Omega_{K,1},\ldots,\Omega_{K,m_K}) \\
& s_1(\Omega_{1,1}) \cdots s_1(\Omega_{1,m_1}) \cdots\cdots s_K(\Omega_{K,1}) \cdots s_K(\Omega_{K,m_K}) \\
& \delta(f - \Omega_{1,1} - \cdots - \Omega_{1,m_1} - \cdots\cdots - \Omega_{K,1} - \cdots - \Omega_{K,m_K}) \\
& d\Omega_{1,1} \cdots d\Omega_{1,m_1} \cdots\cdots d\Omega_{K,1} \cdots d\Omega_{K,m_K}
\end{aligned}
\tag{7.82}
$$

The deterministic response $r_{d,l}(f)$ where $l \in \{1,2,\ldots,L\}$ is identical with the response $r_l(f)$ when all noise generators are quiet, i.e. $n_1(f) = \cdots = n_Q(f) = 0$. Thus, using Equation (7.8) for $s_k(f)$ in Equation (7.82) leads to the determination of the contribution $r_{d,l}(f)$ where $l \in \{1,2,\ldots,L\}$ as

$$
\begin{aligned}
r_{d,l}(f) \;=\; & \sum_{m_1=0}^{M} \cdots \sum_{m_K=0}^{M} \sum_{j_{1,1}=1}^{J_1} \cdots \sum_{j_{1,m_1}=1}^{J_1} \cdots\cdots \sum_{j_{K,1}=1}^{J_K} \cdots \sum_{j_{K,m_K}=1}^{J_K} \\
& \mathcal{L}_{1,M}(m_1 + \cdots + m_K) \\
& (H_l)_{m_1,\ldots,m_K,0,\ldots,0}(\psi_{1,j_{1,1}},\ldots,\psi_{1,j_{1,m_1}}; \cdots\cdots; \psi_{K,j_{K,1}},\ldots,\psi_{K,j_{K,m_K}}) \\
& \tilde{s}_1(\psi_{1,j_{1,1}}) \cdots \tilde{s}_1(\psi_{1,j_{1,m_1}}) \cdots\cdots \tilde{s}_K(\psi_{K,j_{K,1}}) \cdots \tilde{s}_K(\psi_{K,j_{K,m_K}}) \\
& \delta(f - \psi_{1,j_{1,1}} - \cdots - \psi_{1,j_{1,m_1}} - \cdots\cdots - \psi_{K,j_{K,1}} - \cdots - \psi_{K,j_{K,m_K}})
\end{aligned}
\tag{7.83}
$$

Using the symmetry properties of the partly symmetrical multi-port frequency domain Volterra transfer function $(H_l)_{m_1,\ldots,m_K,0,\ldots,0}(\cdot)$ leads to

$$
\begin{aligned}
r_{d,l}(f) \;=\; & \sum_{m_1=0}^{M} \cdots \sum_{m_K=0}^{M} \sum_{j_{1,1}=1}^{J_1} \cdots \sum_{j_{1,m_1}=1}^{J_1} \cdots\cdots \sum_{j_{K,1}=1}^{J_K} \cdots \sum_{j_{K,m_K}=1}^{J_K} \\
& \mathcal{L}_{1,M}(m_1 + \cdots + m_K) \\
& \mathcal{A}_1(j_{1,1},\ldots,j_{1,m_1}) \cdots \mathcal{A}_K(j_{K,1},\ldots,j_{K,m_K}) \\
& \prod_{k=1}^{K} m_k! \prod_{k=1}^{K} \prod_{j_k=1}^{J_k} \left\{ \mathcal{N}_{j_k}(j_{k,1},\ldots,j_{k,m_k})! \right\}^{-1} \\
& (H_l)_{m_1,\ldots,m_K,0,\ldots,0}(\psi_{1,j_{1,1}},\ldots,\psi_{1,j_{1,m_1}}; \cdots \\
& \qquad \cdots; \psi_{K,j_{K,1}},\ldots,\psi_{K,j_{K,m_K}}) \\
& \tilde{s}_1(\psi_{1,j_{1,1}}) \cdots \tilde{s}_1(\psi_{1,j_{1,m_1}}) \cdots \\
& \qquad \cdots \tilde{s}_K(\psi_{K,j_{K,1}}) \cdots \tilde{s}_K(\psi_{K,j_{K,m_K}}) \\
& \delta(f - \psi_{1,j_{1,1}} - \cdots - \psi_{1,j_{1,m_1}} - \cdots \\
& \qquad \cdots - \psi_{K,j_{K,1}} - \cdots - \psi_{K,j_{K,m_K}})
\end{aligned}
\tag{7.84}
$$

where $\mathcal{A}_k(j_{k,1}, \ldots, j_{k,m_k})$ is defined by Equation (7.28) and $\mathcal{N}_{j_k}(j_{k,1}, \ldots, j_{k,m_k})$ is defined by Equation (7.29). Equation (7.84) can be rewritten as

$$r_{d,l}(f) \quad = \quad \sum_{c=1}^{C} \tilde{r}_{d,l}(\chi_c)\, \delta(f - \chi_c) \tag{7.85}$$

where

$$\mathcal{X} \quad = \quad \left\{\psi(\boldsymbol{p}_o) \middle| \; \boldsymbol{p}_o \in \{0, 1, 2, \ldots, M\}^{(J_1 + \cdots + J_K) \times 1} \right.$$
$$\left. \wedge \; \|\boldsymbol{p}_o\| \in \{1, 2, \ldots, M\}\right\} \tag{7.86}$$
$$= \quad \{\chi_1, \ldots, \chi_C\} \tag{7.87}$$

and $\psi(\boldsymbol{p}_o)$ and $\boldsymbol{p}_o$ are given by Equations (7.32) and (7.33) respectively. The frequencies in the set $\mathcal{X}$ are organized such that $\chi_1 \neq \cdots \neq \chi_C$. The Fourier series coefficients $\tilde{r}_{d,l}(\chi_c)$ where $l \in \{1, 2, \ldots, L\}$ and $c \in \{1, 2, \ldots, C\}$ may be determined from Equation (7.84).

Note from Equations (7.30), (7.60) and (7.86) that $\bigcup_{o=1}^{M-1} \mathcal{S}_o \subseteq \mathcal{X}$. The only case where $\bigcup_{o=1}^{M-1} \mathcal{S}_o = \mathcal{X}$ is if and only if $\psi_{k,j_k} = 0$ for all $k \in \{1, 2, \ldots, K\}$ and $j_k \in \{1, 2, \ldots, J_k\}$. This corresponds to both frequency sets consisting of dc and harmonics of dc (which is also located at dc).

In some cases it is convenient to write the deterministic response at port (r, l) where $l \in \{1, 2, \ldots, L\}$ as

$$r_{d,l}(f) \quad = \quad \sum_{p=-\infty}^{\infty} \tilde{r}_{d,l}(\xi_p)\, \delta(f - \xi_p) \tag{7.88}$$

In this case the deterministic signal is represented at the same frequencies as the noise signals. However, for Equation (7.88) to be correct it *must* be that

$$\exists\, p \in \mathcal{Z}\!: \;\; \xi_p \; = \; \psi_{1,j_{1,1}} + \cdots + \psi_{1,j_{1,m_1}} + \cdots\cdots + \psi_{K,j_{K,1}} + \cdots + \psi_{K,j_{K,m_K}} \tag{7.89}$$

for all $m_1, \ldots, m_K \in \{0, 1, \ldots, M\}$, $m_1 + \cdots + m_K \in \{1, 2, \ldots, M\}$, $j_{k,l} \in \{1, 2, \ldots, J_k\}$ for $k \in \{1, 2, \ldots, K\}$ and $l \in \{1, 2, \ldots, m_k\}$. Equivalently, the requirement is

$$\frac{1}{\xi}\left[\psi_{1,j_{1,1}} + \cdots + \psi_{1,j_{1,m_1}} + \cdots\cdots + \psi_{K,j_{K,1}} + \cdots + \psi_{K,j_{K,m_K}}\right] \quad \in \quad \mathcal{Z} \tag{7.90}$$

Since $\xi = 1/(2\tau)$ can be chosen arbitrarily small it is always possible to choose a ξ such that the requirements in Equations (7.89) and (7.90) are fulfilled. The formulation in Equation (7.88) means that

$$\tilde{r}_{d,l}(\xi_p) \quad = \quad 0 \qquad \text{for} \qquad \xi_p \notin \mathcal{X} \tag{7.91}$$

and $\tilde{r}_{d,l}(\xi_p)$ *may* be different from 0 if $\xi_p \in \mathcal{X}$, where $\mathcal{X}$ is given by Equation (7.86).

7.4.2   Noise response

The contribution to $r_l(f)$ where $l \in \{1, 2, \ldots, L\}$ due to the noise sources, $r_{n,l}(f)$, can be determined from Equation (7.80) as

$$
\begin{aligned}
r_{n,l}(f) \;=\;& \sum_{q=1}^{Q} \sum_{m_1=0}^{M-1} \cdots \sum_{m_K=0}^{M-1} \int_{-\infty}^{\infty} \cdots \int_{-\infty}^{\infty} \cdots\cdots \int_{-\infty}^{\infty} \cdots \int_{-\infty}^{\infty} \int_{-\infty}^{\infty} \\
& \mathcal{L}_{0,M-1}(m_1 + \cdots + m_K) \\
& (H_l)_{m_1,\ldots,m_K,0,\ldots,0,o_q=1,0,\ldots,0}(\Omega_{1,1}, \ldots, \Omega_{1,m_1}; \cdots \\
& \qquad\qquad \cdots ; \Omega_{K,1}, \ldots, \Omega_{K,m_K}; \Xi_{q,1}) \\
& s_1(\Omega_{1,1}) \cdots s_1(\Omega_{1,m_1}) \cdots \\
& \qquad\qquad \cdots s_K(\Omega_{K,1}) \cdots s_K(\Omega_{K,m_K})\, n_q(\Xi_{q,1}) \\
& \delta(f - \Omega_{1,1} - \cdots - \Omega_{1,m_1} - \cdots \\
& \qquad\qquad \cdots - \Omega_{K,1} - \cdots - \Omega_{K,m_K} - \Xi_{q,1}) \\
& d\Omega_{1,1} \cdots d\Omega_{1,m_1} \cdots\cdots d\Omega_{K,1} \cdots d\Omega_{K,m_K}\, d\Xi_{q,1}
\end{aligned}
\tag{7.92}
$$

In Equation (7.92) the sum over the q-variable means that the contributions to $\tilde{r}_{n,l}(f)$ are determined for each noise source successively by adding all the contributions from the individual noise sources. Insertion of Equations (7.8) and (7.19) into Equation (7.92) leads to

$$
\begin{aligned}
r_{n,l}(f) \;=\;& \sum_{q=1}^{Q} \sum_{m_1=0}^{M-1} \cdots \sum_{m_K=0}^{M-1} \sum_{j_{1,1}=1}^{J_1} \cdots \sum_{j_{1,m_1}=1}^{J_1} \cdots\cdots \sum_{j_{K,1}=1}^{J_K} \cdots \sum_{j_{K,m_K}=1}^{J_K} \sum_{p=-\infty}^{\infty} \\
& \mathcal{L}_{0,M-1}(m_1 + \cdots + m_K) \\
& (H_l)_{m_1,\ldots,m_K,0,\ldots,0,o_q=1,0,\ldots,0}(\psi_{1,j_{1,1}}, \ldots, \psi_{1,j_{1,m_1}}; \cdots \\
& \qquad\qquad \cdots ; \psi_{K,j_{K,1}}, \ldots, \psi_{K,j_{K,m_K}}; \xi_p) \\
& \tilde{s}_1(\psi_{1,j_{1,1}}) \cdots \tilde{s}_1(\psi_{1,j_{1,m_1}}) \cdots \\
& \qquad\qquad \cdots \tilde{s}_K(\psi_{K,j_{K,1}}) \cdots \tilde{s}_K(\psi_{K,j_{K,m_K}})\, \tilde{n}_q(\xi_p) \\
& \delta(f - \psi_{1,j_{1,1}} - \cdots - \psi_{1,j_{1,m_1}} - \cdots \\
& \qquad\qquad \cdots - \psi_{K,j_{K,1}} - \cdots - \psi_{K,j_{K,m_K}} - \xi_p)
\end{aligned}
\tag{7.93}
$$

Using the symmetry properties of the partly symmetrical frequency domain multiport Volterra transfer function $(H_l)_{m_1,\ldots,m_K,0,\ldots,0,o_q=1,0,\ldots,0}(\cdot)$ leads to

$$
\begin{aligned}
r_{n,l}(f) = &\sum_{q=1}^{Q}\sum_{m_1=0}^{M-1}\cdots\sum_{m_K=0}^{M-1}\sum_{j_{1,1}=1}^{J_1}\cdots\sum_{j_{1,m_1}=1}^{J_1}\cdots\cdots\sum_{j_{K,1}=1}^{J_K}\cdots\sum_{j_{K,m_K}=1}^{J_K}\sum_{p=-\infty}^{\infty}\\
&\mathcal{L}_{0,M-1}(m_1+\cdots+m_K)\\
&\mathcal{A}_1(j_{1,1},\ldots,j_{1,m_1})\cdots\mathcal{A}_K(j_{K,1},\ldots,j_{K,m_K})\\
&\prod_{k=1}^{K}m_k!\ \prod_{k=1}^{K}\prod_{j_k=1}^{J_K}\left\{\mathcal{N}_{j_k}(j_{k,1},\ldots,j_{k,m_k})!\right\}^{-1}\\
&(H_l)_{m_1,\ldots,m_K,0,\ldots,0,o_q=1,0,\ldots,0}(\psi_{1,j_{1,1}},\ldots,\psi_{1,j_{1,m_1}};\cdots\\
&\qquad\cdots;\psi_{K,j_{K,1}},\ldots,\psi_{K,j_{K,m_K}};\xi_p)\\
&\tilde{s}_1(\psi_{1,j_{1,1}})\cdots\tilde{s}_1(\psi_{1,j_{1,m_1}})\cdots\\
&\qquad\cdots\tilde{s}_K(\psi_{K,j_{K,1}})\cdots\tilde{s}_K(\psi_{K,j_{K,m_K}})\,\tilde{n}_q(\xi_p)\\
&\delta(f-\psi_{1,j_{1,1}}-\cdots-\psi_{1,j_{1,m_1}}-\cdots\\
&\qquad\cdots-\psi_{K,j_{K,1}}-\cdots-\psi_{K,j_{K,m_K}}-\xi_p)
\end{aligned}
\tag{7.94}
$$

where $l \in \{1, 2, \ldots, L\}$. It is seen from Equation (7.94) that $r_{n,l}(f)$ where $l \in \{1, 2, \ldots, L\}$ can be written as a Fourier series as

$$
r_{n,l}(f) = \sum_{p=-\infty}^{\infty}\tilde{r}_{n,l}(\xi_p)\,\delta(f-\xi_p)
\tag{7.95}
$$

where

$$
\begin{aligned}
\tilde{r}_{n,l}(\xi_p) = &\sum_{q=1}^{Q}\sum_{m_1=0}^{M}\cdots\sum_{m_K=0}^{M}\sum_{j_{1,1}=1}^{J_1}\cdots\sum_{j_{1,m_1}=1}^{J_1}\cdots\cdots\sum_{j_{K,1}=1}^{J_K}\cdots\sum_{j_{K,m_K}=1}^{J_K}\\
&\mathcal{L}_{0,M-1}(m_1+\cdots+m_K)\\
&\mathcal{A}_1(j_{1,1},\ldots,j_{1,m_1})\cdots\mathcal{A}_K(j_{K,1},\ldots,j_{K,m_K})\\
&\prod_{k=1}^{K}m_k!\ \prod_{k=1}^{K}\prod_{j_k=1}^{J_K}\left\{\mathcal{N}_{j_k}(j_{k,1},\ldots,j_{k,m_k})!\right\}^{-1}\\
&(H_l)_{m_1,\ldots,m_K,0,\ldots,0,o_q=1,0,\ldots,0}(\psi_{1,j_{1,1}},\ldots,\psi_{1,j_{1,m_1}};\cdots\\
&\qquad\cdots;\psi_{K,j_{K,1}},\ldots,\psi_{K,j_{K,m_K}};\\
&\qquad\xi_p-\psi_{1,j_{1,1}}-\cdots-\psi_{1,j_{1,m_1}}-\cdots\\
&\qquad\cdots-\psi_{K,j_{K,1}}-\cdots-\psi_{K,j_{K,m_K}})\\
&\tilde{s}_1(\psi_{1,j_{1,1}})\cdots\tilde{s}_1(\psi_{1,j_{1,m_1}})\cdots\\
&\qquad\cdots\tilde{s}_K(\psi_{K,j_{K,1}})\cdots\tilde{s}_K(\psi_{K,j_{K,m_K}})\\
&\tilde{n}_q(\xi_p-\psi_{1,j_{1,1}}-\cdots-\psi_{1,j_{1,m_1}}-\cdots\\
&\qquad\cdots-\psi_{K,j_{K,1}}-\cdots-\psi_{K,j_{K,m_K}})
\end{aligned}
\tag{7.96}
$$

Thus, $\tilde{r}_{n,l}(\xi_p)$ where $l \in \{1, 2, \ldots, L\}$ can be written as

$$\tilde{r}_{n,l}(\xi_p) \quad = \quad \sum_{q=1}^{Q} \boldsymbol{\tau}_{l,q}^T(\xi_p)\, \tilde{\boldsymbol{n}}_q(\xi_p) \tag{7.97}$$

where

$$\boldsymbol{\tau}_{l,q}(\xi_p) \quad = \quad \left[\tau_{l,q}(\xi_p, \Psi_1), \ldots, \tau_{l,q}(\xi_p, \Psi_E) \right]^T \quad \in \quad C^{E \times 1} \tag{7.98}$$

$$\tilde{\boldsymbol{n}}_q(\xi_p) \quad = \quad \left[\tilde{n}_q(\xi_p - \Psi_1), \ldots, \tilde{n}_q(\xi_p - \Psi_E) \right]^T \quad \in \quad C^{E \times 1} \tag{7.99}$$

In Equation (7.97) the vector $\boldsymbol{\tau}_{l,q}(\xi_p)$ where $l \in \{1, 2, \ldots, L\}$ and $q \in \{1, 2, \ldots, Q\}$ describes the non-linear conversion from the possibly modulated noise source $\tilde{n}_q(\xi_p)$ to the noise response $\tilde{r}_{n,l}(\xi_p)$ at port (r, l). Thus, the cross-correlation between two Fourier series coefficients $\tilde{r}_{n,l_1}(\xi_{p_1})$ and $\tilde{r}_{n,l_2}(\xi_{p_2})$ where $l_1, l_2 \in \{1, 2, \ldots, L\}$ at the two response ports (r, l_1) and (r, l_2) can be determined as

$$\langle \tilde{r}_{n,l_1}(\xi_{p_1})\, \tilde{r}_{n,l_2}^*(\xi_{p_2}) \rangle \quad = \quad \sum_{q_1=1}^{Q} \sum_{q_2=1}^{Q} \boldsymbol{\tau}_{l_1,q_1}^T(\xi_{p_1})\, \widetilde{\boldsymbol{N}}_{q_1,q_2}(\xi_{p_1}, \xi_{p_2})\, \boldsymbol{\tau}_{l_2,q_2}^*(\xi_{p_2}) \tag{7.100}$$

where

$$\widetilde{\boldsymbol{N}}_{q_1,q_2}(\xi_{p_1}, \xi_{p_2}) \quad = \quad \langle \tilde{\boldsymbol{n}}_{q_1}(\xi_{p_1})\, \tilde{\boldsymbol{n}}_{q_2}^\dagger(\xi_{p_2}) \rangle \tag{7.101}$$

In Equations (7.100) and (7.101), $\widetilde{\boldsymbol{N}}_{q_1,q_2}(\xi_{p_1}, \xi_{p_2})$ is a noise cross-correlation matrix describing the correlation between the possibly modulated noise sources $\{\tilde{n}_{q_1}(\xi_{p_1} - \Psi_1), \ldots, \tilde{n}_{q_1}(\xi_{p_1} - \Psi_E)\}$ and $\{\tilde{n}_{q_2}(\xi_{p_2} - \Psi_1), \ldots, \tilde{n}_{q_2}(\xi_{p_2} - \Psi_E)\}$ applied to ports (n, q_1) and (n, q_2) respectively where $q_1, q_2 \in \{1, 2, \ldots, Q\}$ at the various frequencies given. Using Equation (7.57) it can be shown that

$$\widetilde{\boldsymbol{N}}_{q_1,q_2}(\xi_{p_1}, \xi_{p_2}) \quad = \quad \sum_{e_1=1}^{E} \sum_{e_2=1}^{E} \boldsymbol{d}_{e_1}\, \boldsymbol{t}_{q_1}^T(\xi_{p_1} - \Psi_{e_1})$$
$$\times \widetilde{\boldsymbol{W}}_{q_1,q_2}(\xi_{p_1} - \Psi_{e_1}, \xi_{p_2} - \Psi_{e_2})\, \boldsymbol{t}_{q_2}^*(\xi_{p_2} - \Psi_{e_2})\, \boldsymbol{d}_{e_2}^T \tag{7.102}$$

where

$$\boldsymbol{d}_e \quad = \quad \left[d_1 = 0, \ldots, d_{e-1} = 0, d_e = 1, d_{e+1} = 0, \ldots, d_E = 0 \right]^T \quad \in \quad \{0, 1\}^{E \times 1} \tag{7.103}$$

Thus, using Equations (7.100) and (7.101) gives

$$\langle \tilde{r}_{n,l_1}(\xi_{p_1}) \, \tilde{r}^*_{n,l_2}(\xi_{p_2}) \rangle = \sum_{q_1=1}^{Q} \sum_{q_2=1}^{Q} \sum_{e_1=1}^{E} \sum_{e_2=1}^{E} \tau^T_{l_1,q_1}(\xi_{p_1}) \, \boldsymbol{d}_{e_1} \, \boldsymbol{t}^T_{q_1}(\xi_{p_1} - \boldsymbol{\Psi}_{e_1})$$

$$\times \tilde{W}_{q_1,q_2}(\xi_{p_1} - \boldsymbol{\Psi}_{e_1}, \xi_{p_2} - \boldsymbol{\Psi}_{e_2})$$

$$\times \boldsymbol{t}^*_{q_2}(\xi_{p_2} - \boldsymbol{\Psi}_{e_2}) \, \boldsymbol{d}^T_{e_2} \, \tau^*_{l_2,q_2}(\xi_{p_2}) \tag{7.104}$$

where $l_1, l_2 \in \{1, 2, \ldots, L\}$. Equation (7.104) can be rewritten as

$$\langle \tilde{r}_{n,l_1}(\xi_{p_1}) \, \tilde{r}^*_{n,l_2}(\xi_{p_2}) \rangle = \sum_{q_1=1}^{Q} \sum_{q_2=1}^{Q} \sum_{e_1=1}^{E} \sum_{e_2=1}^{E} \boldsymbol{\alpha}^T_{l_1,q_1,e_1}(\xi_{p_1}, \boldsymbol{\Psi}_{e_1})$$

$$\times \tilde{W}_{q_1,q_2}(\xi_{p_1} - \boldsymbol{\Psi}_{e_1}, \xi_{p_2} - \boldsymbol{\Psi}_{e_2}) \, \boldsymbol{\alpha}^*_{l_2,q_2,e_2}(\xi_{p_2}, \boldsymbol{\Psi}_{e_2}) \tag{7.105}$$

where $l_1, l_2 \in \{1, 2, \ldots, L\}$, and where

$$\boldsymbol{\alpha}_{l,q,e}(\xi_p, \boldsymbol{\Psi}_e) = \boldsymbol{t}_q(\xi_p - \boldsymbol{\Psi}_e) \, \boldsymbol{d}^T_e \, \tau_{l,q}(\xi_p) \tag{7.106}$$

where $l \in \{1, 2, \ldots, L\}$, $q \in \{1, 2, \ldots, Q\}$ and $e \in \{1, 2, \ldots, E\}$. In Equation (7.106), $\boldsymbol{\alpha}_{l,q,e}(\xi_p, \boldsymbol{\Psi}_e)$ is a vector describing the transfer of the fundamental noise source $\{\tilde{w}_q(\xi_p - \boldsymbol{\Psi}_e - \boldsymbol{\Psi}_1), \ldots, \tilde{w}_q(\xi_p - \boldsymbol{\Psi}_e - \boldsymbol{\Psi}_E)\}$ where $q \in \{1, 2, \ldots, Q\}$ at the given specified frequencies to the noise response $\tilde{r}_{n,l}(\xi_p)$ at port (r, l) where $l \in \{1, 2, \ldots, L\}$.

7.4.3 Total response

The Fourier series coefficient for the response at port (r, l) where $l \in \{1, 2, \ldots, L\}$ at frequency ξ_p where $p \in \mathcal{Z}$ is given by

$$\tilde{r}_l(\xi_p) = \tilde{r}_{d,l}(\xi_p) + \tilde{r}_{n,l}(\xi_p) \tag{7.107}$$

where $\tilde{r}_{d,l}(\xi_p)$ is given by Equations (7.88)–(7.91), and $\tilde{r}_{n,l}(\xi_p)$ is given by Equation (7.97). Thus, the cross-correlation between two arbitrary Fourier series coefficients at arbitrary ports and frequencies can be determined as

$$\langle \tilde{r}_{l_1}(\xi_{p_1}) \, \tilde{r}^*_{l_2}(\xi_{p_2}) \rangle = \tilde{r}_{d,l_1}(\xi_{p_1}) \, \tilde{r}^*_{d,l_2}(\xi_{p_2}) + \langle \tilde{r}_{n,l_1}(\xi_{p_1}) \, \tilde{r}^*_{n,l_2}(\xi_{p_2}) \rangle$$

$$+ \tilde{r}_{d,l_1}(\xi_{p_1}) \, \langle \tilde{r}^*_{n,l_2}(\xi_{p_2}) \rangle$$

$$+ \tilde{r}^*_{d,l_2}(\xi_{p_2}) \, \langle \tilde{r}_{n,l_1}(\xi_{p_1}) \rangle \tag{7.108}$$

where $l_1, l_2 \in \{1, 2, \ldots, L\}$ and $p_1, p_2 \in \mathcal{Z}$. In many cases the ensemble average of a Fourier series coefficient is zero in which case the latter two terms in Equation (7.108) vanish. The expression for $\langle \tilde{r}_{n,l_1}(\xi_{p_1}) \tilde{r}^*_{n,l_2}(\xi_{p_2}) \rangle$ is given by Equation (7.105).

It is recalled that $\langle \tilde{r}_{n,l_1}(\xi_{p_1}) \, \tilde{r}^*_{n,l_2}(\xi_{p_2}) \rangle$ where $l_1, l_2 \in \{1, 2, \ldots, L\}$ and $p_1, p_2 \in \mathcal{Z}$ is a two-sided Fourier series coefficient cross-correlation. Thus, if for example the average signal power is determined for the one-sided case with positive frequencies, i.e. $p_1, p_2 \in \mathcal{Z}_{0+}$, then a factor 2 must be multiplied on the right-hand side of Equation (7.108) if the time domain response signals are real.

7.4.4 Some special cases

Consider the following situations:

- If (i) all the noise sources $\{n_1(f), \ldots, n_Q(f)\}$ are unmodulated, and (ii) the system is linear, then Equation (7.104) leads to

$$
\begin{aligned}
\langle \tilde{r}_{n,l_1}(\xi_{p_1}) \, \tilde{r}^*_{n,l_2}(\xi_{p_2}) \rangle &= \sum_{q_1=1}^{Q} \sum_{q_2=1}^{Q} \tau_{l_1,q_1}(\xi_{p_1}) \, t_{q_1}(\xi_{p_1}) \\
&\times \langle \tilde{w}_{q_1}(\xi_{p_1}) \, \tilde{w}^*_{q_2}(\xi_{p_2}) \rangle \, t^*_{q_2}(\xi_{p_2}) \, \tau^*_{l_2,q_2}(\xi_{p_2}) \qquad (7.109) \\
&= \sum_{q_1=1}^{Q} \sum_{q_2=1}^{Q} (H_{l_1})_{m_1=0,\ldots,m_K=0,0,\ldots,0,o_{q_1}=1,0,\ldots,0}(\xi_{p_1}) \\
&\times (G_{q_1})_1(\xi_{p_1}) \, \langle \tilde{w}_{q_1}(\xi_{p_1}) \, \tilde{w}^*_{q_2}(\xi_{p_2}) \rangle \, (G_{q_2})^*_1(\xi_{p_2}) \\
&\times (H_{l_2})^*_{m_1=0,\ldots,m_K=0,0,\ldots,0,o_{q_2}=1,0,\ldots,0}(\xi_{p_2}) \qquad (7.110)
\end{aligned}
$$

In this special case the analysis is very closely related to a nodal noise analysis of a linear network.

- If (i) all the noise sources $\{n_1(f), \ldots, n_Q(f)\}$ are unmodulated, (ii) the system is linear, and (iii) the fundamental (unmodulated) noise sources $\{w_1(f), \ldots, w_Q(f)\}$
are all uncorrelated, then

$$
\begin{aligned}
\langle |\tilde{r}_{n,l}(\xi_p)|^2 \rangle &= \sum_{q=1}^{Q} \left| (H_l)_{m_1=0,\ldots,m_K=0,0,\ldots,0,o_q=1,0,\ldots,0}(\xi_p) \right|^2 \\
&\times \left| (G_q)_1(\xi_p) \right|^2 \langle |\tilde{w}_q(\xi_p)|^2 \rangle \qquad (7.111)
\end{aligned}
$$

7.5 References

[1] Chua, L. O. & Tang, Y.-S.: "Nonlinear oscillation via Volterra series", *IEEE Trans. Circuits and Systems*, vol. 29, no. 3, pp. 150–168, 1982.

[2] Hu, Y., Obregon, J. J. & Mollier, J.-C.: "Nonlinear analysis of microwave FET oscillators using Volterra series", *IEEE Trans. Microwave Theory and Techniques*, vol. 37, no. 11, pp. 1689–1693, 1989.

[3] Haus, H. A. (Chairman of IRE Subcommittee 7.9 on Noise) et al.: "Representation of noise in linear twoports", *Proc. IRE*, vol. 48, no. 1, pp. 69–74, 1960.

[4] Lighthill, M. J.: "Introduction to Fourier analysis and generalized functions", Cambridge University Press, 1958.

[5] Papoulis, A.: "Probability, random variables, and stochastic processes", McGraw–Hill, 1965.

[6] Held, D. N. & Kerr, A. R.: "Conversion loss and noise of microwave and millimeterwave mixers: part I – theory", *IEEE Trans. Microwave Theory and Techniques*, vol. 26, no. 2, pp. 49–55, 1978.

[7] Larsen, T.: "Determination of Volterra transfer functions of non-linear multi-port networks", *Int. J. Circuit Theory and Applications*, vol. 21, no. 2, pp. 107–131, 1993.

[8] Chua, L. O. & Ng, C.-Y.: "Frequency domain analysis of nonlinear systems: formulation of transfer functions", *IEE Journal on Electronic Circuits and Systems*, vol. 3, no. 6, pp. 257–269, 1979.

[9] Chua, L. O. & Ng, C.-Y.: "Frequency domain analysis of nonlinear systems: general theory", *IEE Journal on Electronic Circuits and Systems*, vol. 3, no. 4, pp. 165–185, 1979.

[10] Maas, S. A.: "A general-purpose computer program for the Volterra-series analysis of nonlinear microwave circuits", *IEEE Microwave Theory and Techniques Symposium Digest*, pp. 311–314, 1988.

[11] Steer, M. B., Chang, C.-R. & Rhyne, G. W.: "Computer-aided analysis of nonlinear microwave circuits using frequency-domain nonlinear analysis techniques: the state of the art", *Int. J. Microwave and Millimeter-wave Computer-aided Engineering*, vol. 1, no. 2, pp. 181–200, 1991.

8

Noise in non-linear systems: Examples and Conclusion

This chapter contains three examples for the analysis of low-level noise in non-linear networks and systems. The first example illustrates the properties of a modulated noise source. The second and third examples illustrate the analysis of noise in two types of networks. The examples have been constructed to facilitate comparative numerical simulations of the networks. The numerical simulations turn out to agree very well with the predicted theoretical results.

Generally, the simulations are performed as follows. First a number of $2\Lambda + 1$ fundamental noise samples, which are in accordance with the specified statistical properties, are generated for each noise source by a pseudo random number generator. Then the modulated time domain noise samples are determined from the modulating function, the fundamental noise samples are processed by the network equation(s), and next these time domain samples are Fourier series transformed. The Fourier series coefficients for the frequency points of interest are saved. Then new noise samples are generated for the next iteration and so forth. Thus, the number of iterations is the same as the number of ensembles in the stochastic process for the noise. Denoting the number of iterations by Γ, the value for $\langle \tilde{z}_1(\xi_{p_1}) \tilde{z}_2^*(\xi_{p_2}) \rangle$ where z_1 and z_2 are stochastic variables and $p_1, p_2 \in \{-\Lambda, \ldots, -1, 0, 1, \ldots, \Lambda\}$ is given by

$$\langle \tilde{z}_1(\xi_{p_1}) \tilde{z}_2^*(\xi_{p_2}) \rangle_\Gamma = \frac{1}{\Gamma} \sum_{\gamma=1}^{\Gamma} \tilde{z}_1^{(\gamma)}(\xi_{p_1}) [\tilde{z}_2^{(\gamma)}(\xi_{p_2})]^* \tag{8.1}$$

where $\tilde{z}_1^{(\gamma)}(\xi_{p_1})$ and $\tilde{z}_2^{(\gamma)}(\xi_{p_2})$ are Fourier series coefficients for the ensemble number (iteration number) γ at frequencies ξ_{p_1} and ξ_{p_2}. Equation (8.1) can be used to illustrate the convergence of the result as more and more iterations are processed. If the result of a simulation is to be given as a single number it is most convenient to take the average over a number of iterations. This can be done as

$$\langle \tilde{n}_1(\xi_{p_1})\, \tilde{n}_1^*(\xi_{p_2})\rangle_{\Gamma_1}^{\Gamma_2} \quad = \quad \frac{1}{\Gamma_2 - \Gamma_1 + 1} \sum_{\Gamma=\Gamma_1}^{\Gamma_2} \langle \tilde{n}_1(\xi_{p_1})\, \tilde{n}_1^*(\xi_{p_2})\rangle_{\Gamma} \qquad (8.2)$$

where the average is made over the simulated results for the iteration number $\Gamma \in \{\Gamma_1, \Gamma_1 + 1, \ldots, \Gamma_2\}$.

8.1   Example 1

PROBLEM: First an example of a modulated noise source is considered. The non-linear noisy system under consideration is shown in Figure 8.1. For this system the number of signal input ports $K = 1$ and the number of controlling variables $Q = 1$ (and the number of output ports is $L = 1$ though this is not of interest in this example, since only the noise source is investigated). The modulated noise source current n_1 depends on the controlling current $u_{1,1}$. The objective of the example is to determine $\langle \tilde{n}_1(\xi_{p_1})\, \tilde{n}_1^*(\xi_{p_2})\rangle$ where $p_1, p_2 \in \mathcal{Z}$ are integers, and to compare the theoretical results with numerical experiments. The maximum order is chosen to be $M = 3$.

THEORY: The system in Figure 8.1 is excited by the deterministic signal $s_1(f)$ given by Equation (7.8) as

$$s_1(f) \quad = \quad \sum_{j_1=1}^{4} \tilde{s}_1(\psi_{1,j_1})\, \delta(f - \psi_{1,j_1}) \qquad (8.3)$$

where $\psi_{1,1} = \vartheta_{1,1} > 0$, $\psi_{1,2} = -\vartheta_{1,1}$, $\psi_{1,3} = \vartheta_{1,2} > 0$, and $\psi_{1,4} = -\vartheta_{1,2}$ with $\vartheta_{1,1} \neq \vartheta_{1,2}$ and

$$\tilde{s}_1(\vartheta_{1,1}) \quad = \quad \frac{1}{2}\, \varrho_{1,1}\, \exp[j\, \varphi_{1,1}] \qquad (8.4)$$

$$\tilde{s}_1(-\vartheta_{1,1}) \quad = \quad \frac{1}{2}\, \varrho_{1,1}\, \exp[-j\, \varphi_{1,1}] \qquad (8.5)$$

$$\tilde{s}_1(\vartheta_{1,2}) \quad = \quad \frac{1}{2}\, \varrho_{1,2}\, \exp[j\, \varphi_{1,2}] \qquad (8.6)$$

$$\tilde{s}_1(-\vartheta_{1,2}) \quad = \quad \frac{1}{2}\, \varrho_{1,2}\, \exp[-j\, \varphi_{1,2}] \qquad (8.7)$$

and thus $J_1 = 4$. The time domain noise signal $n_1(t)$ is given by

$$n_1(t) \quad = \quad C_2\, u_{1,1}(t)\, w_1(t) + C_3\, u_{1,1}^2(t)\, w_1(t) \qquad (8.8)$$

where the fundamental (unmodulated) noise source $w_1(t)$ is a white Gaussian noise source with mean μ_1 and standard deviation σ_1. Equations (7.24) and (8.8) lead to $I_1 = 1$ and

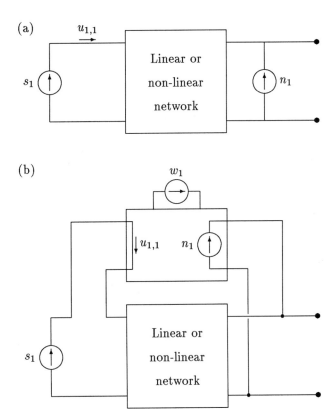

Figure 8.1: Example 1: Non-linear noisy system with a modulated noise source. (a) Basic system where the noise source n_1 is modulated by the controlling variable $u_{1,1}$. (b) Equivalent representation of the modulated noise source in the same form as in Figure 7.2 where w_1 is the fundamental (unmodulated) noise source.

$$(G_1)_{1,m_1}(\Omega_{0,1}; \Omega_{1,1}, \ldots, \Omega_{1,m_1}) \quad = \quad \begin{cases} C_2 & \text{for} \quad m_1 = 1 \\ C_3 & \text{for} \quad m_1 = 2 \\ 0 & \text{otherwise} \end{cases} \tag{8.9}$$

The frequency sets S_0, S_1 and S_2 can be determined as

$$
\begin{aligned}
S_0 &= \{\Psi_{0,1}\}, \qquad E_0 = 1 & (8.10) \\
&= \{0\} & (8.11)
\end{aligned}
$$

$$
\begin{aligned}
S_1 &= \{\Psi_{1,1}, \Psi_{1,2}, \Psi_{1,3}, \Psi_{1,4}\}, \qquad E_1 = 4 & (8.12) \\
&= \{\vartheta_{1,1}, -\vartheta_{1,1}, \vartheta_{1,2}, -\vartheta_{1,2}\} & (8.13)
\end{aligned}
$$

$$
\begin{aligned}
S_2 &= \{\Psi_{2,1}, \Psi_{2,2}, \Psi_{2,3}, \Psi_{2,4}, \Psi_{2,5}, \Psi_{2,6}, \Psi_{2,7}, \Psi_{2,8}, \Psi_{2,9}\}, \qquad E_2 = 9 \\
& \tag{8.14}
\end{aligned}
$$

$$
\begin{aligned}
&= \{2\vartheta_{1,1}, -2\vartheta_{1,1}, 2\vartheta_{1,2}, -2\vartheta_{1,2}, 0, \vartheta_{1,1} + \vartheta_{1,2}, \vartheta_{1,1} - \vartheta_{1,2}, \\
&\qquad \vartheta_{1,1} + \vartheta_{1,2}, -\vartheta_{1,1} - \vartheta_{1,2}\} \tag{8.15}
\end{aligned}
$$

and thus $E = 13$, and

$$
\begin{aligned}
\{\Psi_1, \ldots, \Psi_{13}\} &= \{\vartheta_{1,1}, -\vartheta_{1,1}, \vartheta_{1,2}, -\vartheta_{1,2}, 2\vartheta_{1,1}, -2\vartheta_{1,1}, 2\vartheta_{1,2}, -2\vartheta_{1,2}, \\
&\qquad \vartheta_{1,1} + \vartheta_{1,2}, \vartheta_{1,1} - \vartheta_{1,2}, -\vartheta_{1,1} + \vartheta_{1,2}, -\vartheta_{1,1} - \vartheta_{1,2}, 0\} \tag{8.16}
\end{aligned}
$$

To determine the noise conversion vector $t_1(\xi_p)$ it is observed that $u_{1,1}(f) = s_1(f)$ and thus

$$
\begin{aligned}
u_{1,1}(f) &= {}^1\tilde{u}_{1,1}(\Psi_{1,1})\,\delta(f - \Psi_{1,1}) + {}^1\tilde{u}_{1,1}(\Psi_{1,2})\,\delta(f - \Psi_{1,2}) \\
&\quad + {}^1\tilde{u}_{1,1}(\Psi_{1,3})\,\delta(f - \Psi_{1,3}) + {}^1\tilde{u}_{1,1}(\Psi_{1,4})\,\delta(f - \Psi_{1,4}) \tag{8.17}
\end{aligned}
$$

where

$$
\begin{aligned}
{}^1\tilde{u}_{1,1}(\Psi_{1,1}) &= \tilde{s}_1(\vartheta_{1,1}) & (8.18) \\
{}^1\tilde{u}_{1,1}(\Psi_{1,2}) &= \tilde{s}_1(-\vartheta_{1,1}) & (8.19) \\
{}^1\tilde{u}_{1,1}(\Psi_{1,3}) &= \tilde{s}_1(\vartheta_{1,2}) & (8.20) \\
{}^1\tilde{u}_{1,1}(\Psi_{1,4}) &= \tilde{s}_1(-\vartheta_{1,2}) & (8.21)
\end{aligned}
$$

using

$$(F_{1,1})_{o_1}(\Xi_{1,1},\ldots,\Xi_{1,o_1}) \;=\; \begin{cases} 1 & \text{for } o_1 = 1 \\ 0 & \text{otherwise} \end{cases} \tag{8.22}$$

From Equations (7.57)–(7.59) and (8.8) it is seen that

$$\tilde{n}_1(\xi_p) \;=\; t_1^T(\xi_p)\,\tilde{w}_1(\xi_p) \tag{8.23}$$

where

$$\begin{aligned}
t_1(\xi_p) \;=\; \Big[&C_2\tilde{s}_1(\vartheta_{1,1}),\, C_2\tilde{s}_1(-\vartheta_{1,1}),\, C_2\tilde{s}_1(\vartheta_{1,2}),\, C_2\tilde{s}_1(-\vartheta_{1,2}), \\
&C_3\tilde{s}_1(\vartheta_{1,1})\tilde{s}_1(\vartheta_{1,1}),\, C_3\tilde{s}_1(-\vartheta_{1,1})\tilde{s}_1(-\vartheta_{1,1}), \\
&C_3\tilde{s}_1(\vartheta_{1,2})\tilde{s}_1(\vartheta_{1,2}),\, C_3\tilde{s}_1(-\vartheta_{1,2})\tilde{s}_1(-\vartheta_{1,2}), \\
&2C_3\tilde{s}_1(\vartheta_{1,1})\tilde{s}_1(\vartheta_{1,2}),\, 2C_3\tilde{s}_1(\vartheta_{1,1})\tilde{s}_1(-\vartheta_{1,2}), \\
&2C_3\tilde{s}_1(-\vartheta_{1,1})\tilde{s}_1(\vartheta_{1,2}),\, 2C_3\tilde{s}_1(-\vartheta_{1,1})\tilde{s}_1(-\vartheta_{1,2}), \\
&2C_3\tilde{s}_1(\vartheta_{1,1})\tilde{s}_1(-\vartheta_{1,1}) + 2C_3\tilde{s}_1(\vartheta_{1,2})\tilde{s}_1(-\vartheta_{1,2}) \Big]^T \tag{8.24}
\end{aligned}$$

$$\begin{aligned}
\tilde{w}_1(\xi_p) \;=\; \Big[&\tilde{w}_1(\xi_p - \vartheta_{1,1}),\, \tilde{w}_1(\xi_p + \vartheta_{1,1}),\, \tilde{w}_1(\xi_p - \vartheta_{1,2}), \\
&\tilde{w}_1(\xi_p + \vartheta_{1,2}),\, \tilde{w}_1(\xi_p - 2\vartheta_{1,1}),\, \tilde{w}_1(\xi_p + 2\vartheta_{1,1}), \\
&\tilde{w}_1(\xi_p - 2\vartheta_{1,2}),\, \tilde{w}_1(\xi_p + 2\vartheta_{1,2}),\, \tilde{w}_1(\xi_p - \vartheta_{1,1} - \vartheta_{1,2}), \\
&\tilde{w}_1(\xi_p - \vartheta_{1,1} + \vartheta_{1,2}),\, \tilde{w}_1(\xi_p + \vartheta_{1,1} - \vartheta_{1,2}), \\
&\tilde{w}_1(\xi_p + \vartheta_{1,1} + \vartheta_{1,2}),\, \tilde{w}_1(\xi_p) \Big]^T \tag{8.25}
\end{aligned}$$

Thus the cross-correlation between two arbitrary frequency domain Fourier series coefficients $\tilde{n}_1(\xi_{p_1})$ and $\tilde{n}_1(\xi_{p_2})$ where $p_1, p_2 \in \mathcal{Z}$ is given by

$$\langle \tilde{n}_1(\xi_{p_1})\,\tilde{n}_1^*(\xi_{p_2})\rangle \;=\; t_1^T(\xi_{p_1})\,\langle \tilde{w}_1(\xi_{p_1})\,\tilde{w}_1^\dagger(\xi_{p_2})\rangle\, t_1^*(\xi_{p_2}) \tag{8.26}$$

The result for $\langle \tilde{n}_1(\xi_{p_1})\,\tilde{n}_1^*(\xi_{p_2})\rangle$ remains unchanged for any choice of maximum order as long as $M \geq 3$. The vectors $t_1(\xi_p)$ and $\tilde{w}_1(\xi_p)$ increase in dimension with increasing $M \geq 3$ but merely introduce new zero elements. The cross-correlation $\langle \tilde{n}_1(\xi_{p_1})\,\tilde{n}_1^*(\xi_{p_2})\rangle$ where $p_1, p_2 \in \mathcal{Z}$ for the modulated noise source is generally different from 0 if

$$\begin{aligned}
|\xi_{p_1} - \xi_{p_2}| \;\in\; \Big\{ &0,\, \vartheta_{1,2},\, 2\vartheta_{1,2},\, 3\vartheta_{1,2},\, 4\vartheta_{1,2},\, |\vartheta_{1,1} - 3\vartheta_{1,2}|,\, |\vartheta_{1,1} - 2\vartheta_{1,2}|, \\
&|\vartheta_{1,1} - \vartheta_{1,2}|,\, \vartheta_{1,1},\, \vartheta_{1,1} + \vartheta_{1,2},\, \vartheta_{1,1} + 2\vartheta_{1,2}, \\
&\vartheta_{1,1} + 3\vartheta_{1,2},\, |2\vartheta_{1,1} - 2\vartheta_{1,2}|,\, |2\vartheta_{1,1} - \vartheta_{1,2}|,\, 2\vartheta_{1,1}, \\
&2\vartheta_{1,1} + \vartheta_{1,2},\, 2\vartheta_{1,1} + 2\vartheta_{1,2},\, |3\vartheta_{1,1} - \vartheta_{1,2}|, \\
&3\vartheta_{1,1},\, 3\vartheta_{1,1} + \vartheta_{1,2},\, 4\vartheta_{1,1} \Big\} \tag{8.27}
\end{aligned}$$

The cross-correlation $\langle \tilde{n}_1(\xi_{p_1})\,\tilde{n}_1^*(\xi_{p_2})\rangle$ *may be* 0 even if $|\xi_{p_1}-\xi_{p_2}|$ fulfils Equation (8.27) but then it depends on the conversion vector $t_1(\xi_p)$ given by Equation (8.24). Note that $\langle \tilde{n}_1(\xi_{p_1})\,\tilde{n}_1^*(\xi_{p_2})\rangle$ may be different from 0 if $p_1 \neq p_2$. This is due to the modulation of the fundamental white noise source. When there are $p_1 \neq p_2$ which fulfils Equation (8.27) then $\langle \tilde{w}_1(\xi_{p_1})\,\tilde{w}_1^\dagger(\xi_{p_2})\rangle$ is a non-zero matrix with non-zero off-diagonal elements, e.g. for $\vartheta_{1,1} = 3$ and $\vartheta_{1,2} = 7$ it can be shown that

$$
\langle \tilde{w}_1(\vartheta_{1,1})\,\tilde{w}_1^\dagger(\vartheta_{1,2})\rangle =
\begin{bmatrix}
0 & 0 & \eta_1^{ct} & 0 & 0 & 0 & 0 & 0 & 0 & 0 & 0 & 0 & 0 \\
0 & 0 & 0 & 0 & 0 & 0 & 0 & 0 & 0 & 0 & 0 & 0 & 0 \\
0 & 0 & 0 & 0 & 0 & 0 & 0 & 0 & 0 & 0 & 0 & 0 & 0 \\
0 & \eta_0^{ct} & 0 & 0 & 0 & 0 & 0 & 0 & 0 & 0 & 0 & 0 & 0 \\
0 & 0 & 0 & 0 & 0 & 0 & 0 & 0 & \eta_0^{ct} & 0 & 0 & 0 & 0 \\
0 & 0 & 0 & 0 & 0 & 0 & 0 & 0 & 0 & 0 & 0 & 0 & 0 \\
0 & 0 & 0 & 0 & 0 & 0 & 0 & 0 & 0 & 0 & 0 & 0 & 0 \\
0 & 0 & 0 & 0 & 0 & 0 & 0 & 0 & 0 & 0 & 0 & \eta_0^{ct} & 0 \\
0 & 0 & 0 & 0 & 0 & 0 & \eta_0^{ct} & 0 & 0 & 0 & 0 & 0 & 0 \\
0 & 0 & 0 & 0 & 0 & 0 & 0 & 0 & 0 & 0 & 0 & 0 & \eta_0^{ct} \\
0 & 0 & 0 & 0 & 0 & 0 & 0 & 0 & 0 & 0 & 0 & 0 & 0 \\
0 & 0 & 0 & 0 & 0 & \eta_0^{ct} & 0 & 0 & 0 & 0 & 0 & 0 & 0 \\
0 & 0 & 0 & 0 & 0 & 0 & 0 & 0 & 0 & 0 & 0 & 0 & 0 \\
0 & 0 & 0 & 0 & 0 & 0 & 0 & 0 & 0 & 0 & \eta_0^{ct} & 0 & 0
\end{bmatrix}
\tag{8.28}
$$

where

$$
\eta_a^{ct} \;=\; \frac{\sigma_1^2}{2\tau} + a\,\mu_1^2, \qquad a \in \{0,1\}
\tag{8.29}
$$

The superscript ct for η_a^{ct} denotes that the coefficient is used for the continuous time case.

NUMERICAL EXPERIMENT: The frequency domain autocorrelation function for the Fourier series coefficient of the modulated noise source $\tilde{n}_1(\xi_p)$ at an arbitrary frequency ξ_p where $p \in \mathcal{Z}$ is given by Equation (7.66). To investigate the correctness of Equation (8.26) numerical simulations are performed according to the description in Section 7.3. In this case it can be shown that the autocorrelation is given by

$$
\langle |\tilde{n}_1(\xi_p)|^2 \rangle \;=\;
\begin{cases}
\sigma_1^2\,\|t_1(\xi_p)\|_2^2/(2\Lambda+1) + \mu_1^2\,|t_1(\xi_p,\Psi_{13})|^2 & \text{for } p = 0 \\
\sigma_1^2\,\|t_1(\xi_p)\|_2^2/(2\Lambda+1) & \text{for } p \neq 0
\end{cases}
\tag{8.30}
$$

The following numerical simulations have been performed to illustrate the properties of the modulated noise source:

Simulation 1 The autocorrelation $\langle|\tilde{n}_1(\xi_p)|^2\rangle_\Gamma$ with $p = 17$ is simulated versus the number of iterations Γ. The data and result of the simulation are shown in Figure 8.2. For this situation only the standard deviation and not the mean is of importance as predicted from the theoretical result in Equation (8.30). The simulated result $\langle|\tilde{n}_1(\xi_{17})|^2\rangle_{29901}^{30000} = 15.48$ agrees with the theoretical result $\langle|\tilde{n}_1(\xi_{17})|^2\rangle = 15.42$ with a deviation of 0.37 %.

Simulation 2 The autocorrelation $\langle|\tilde{n}_1(\xi_p)|^2\rangle_\Gamma$ with $p = 0$ is simulated versus the number of iterations Γ. The data and result of the simulation are shown in Figure 8.3. For this situation both the standard deviation and the mean are of importance as predicted from the theoretical result in Equation (8.30). The simulated result $\langle|\tilde{n}_1(\xi_0)|^2\rangle_{29901}^{30000} = 23.74$ agrees with the theoretical result $\langle|\tilde{n}_1(\xi_0)|^2\rangle = 23.76$ with a deviation of -0.09 %.

Simulation 3 The autocorrelation $\langle|\tilde{n}_1(\xi_p)|^2\rangle_{29901}^{30000}$ with $p = 7$ is simulated versus the amplitudes $\varrho_{1,1} = \varrho_{1,2}$ of the input sinusoidals. The data and result of the simulation are shown in Figure 8.4. All simulated values $\langle|\tilde{n}_1(\xi_7)|^2\rangle_{29901}^{30000}$ agree with the theoretical result $\langle|\tilde{n}_1(\xi_7)|^2\rangle$ in Equation (8.30) with a deviation less than ± 1.25 %. As seen from Figure 8.4 the noise level $\langle|\tilde{n}_1(\xi_7)|^2\rangle$ increases with the amplitudes $\varrho_{1,1} = \varrho_{1,2}$ of the deterministic excitation signals.

Simulation 4 The cross-correlation $\langle\tilde{n}_1(\xi_{p_1})\,\tilde{n}_1^*(\xi_{p_2})\rangle_\Gamma$ with $p_1 = 4$ and $p_2 = 6$ for $\tau = 1.0$, $\vartheta_{1,1} = 2.0$ and $\vartheta_{1,2} = 3.0$ is simulated versus the number of iterations Γ (thus, $\xi_{p_1} = \xi_4 = \vartheta_{1,1}$ and $\xi_{p_2} = \xi_6 = \vartheta_{1,2}$). The data and results of the simulations for $\mathrm{Re}[\langle\tilde{n}_1(\xi_{p_1})\,\tilde{n}_1^*(\xi_{p_2})\rangle_\Gamma]$ and $\mathrm{Im}[\langle\tilde{n}_1(\xi_{p_1})\,\tilde{n}_1^*(\xi_{p_2})\rangle_\Gamma]$ are shown in Figures 8.5 and 8.6. The theoretical result for this situation can be predicted from Equation (8.26) observing that

$$\langle\tilde{w}_1(\xi_4)\,\tilde{w}_1^\dagger(\xi_6)\rangle =$$

$$
\begin{bmatrix}
0 & 0 & \eta_1^{dt} & 0 & 0 & 0 & 0 & 0 & 0 & 0 & 0 & 0 & 0 \\
0 & 0 & 0 & 0 & 0 & 0 & 0 & 0 & 0 & \eta_0^{dt} & 0 & 0 & 0 \\
0 & 0 & 0 & 0 & \eta_0^{dt} & 0 & 0 & 0 & 0 & 0 & 0 & 0 & 0 \\
0 & \eta_0^{dt} & 0 & 0 & 0 & 0 & 0 & 0 & 0 & 0 & 0 & 0 & 0 \\
0 & 0 & 0 & 0 & 0 & 0 & 0 & 0 & \eta_0^{dt} & 0 & 0 & 0 & 0 \\
0 & 0 & 0 & \eta_0^{dt} & 0 & 0 & 0 & 0 & 0 & 0 & 0 & 0 & 0 \\
0 & 0 & 0 & 0 & 0 & 0 & 0 & 0 & 0 & 0 & 0 & 0 & 0 \\
0 & 0 & 0 & 0 & 0 & 0 & 0 & 0 & 0 & 0 & 0 & \eta_0^{dt} & 0 \\
0 & 0 & 0 & 0 & 0 & 0 & \eta_0^{dt} & 0 & 0 & 0 & 0 & 0 & 0 \\
0 & 0 & 0 & 0 & 0 & 0 & 0 & 0 & 0 & 0 & 0 & 0 & \eta_0^{dt} \\
\eta_0^{dt} & 0 & 0 & 0 & 0 & 0 & 0 & 0 & 0 & 0 & 0 & 0 & 0 \\
0 & 0 & 0 & 0 & 0 & \eta_0^{dt} & 0 & 0 & 0 & 0 & 0 & 0 & 0 \\
0 & 0 & 0 & 0 & 0 & 0 & 0 & 0 & 0 & 0 & \eta_0^{dt} & 0 & 0 \\
\end{bmatrix}
$$

$$(8.31)$$

where

$$\eta_a^{dt} = \frac{\sigma_1^2}{2\Lambda + 1} + a\,\mu_1^2, \qquad a \in \{0,1\} \tag{8.32}$$

The superscript dt for η_a^{dt} denotes that the coefficient is used for the discrete time case. The simulated value $\mathrm{Re}[\langle \tilde{n}_1(\xi_4)\,\tilde{n}_1^*(\xi_6)\rangle_{29901]}^{30000]} = 14.37$ agrees with the theoretical result $\mathrm{Re}[\langle \tilde{n}_1(\xi_4)\,\tilde{n}_1^*(\xi_6)\rangle] = 14.26$ with a deviation of 0.78 %. The simulated value $\mathrm{Im}[\langle \tilde{n}_1(\xi_{p_1})\,\tilde{n}_1^*(\xi_{p_2})\rangle_{29901]}^{30000]} = 0.00095$ and the theoretical result is $\mathrm{Im}[\langle \tilde{n}_1(\xi_{p_1})\,\tilde{n}_1^*(\xi_{p_2})\rangle] = 0$.

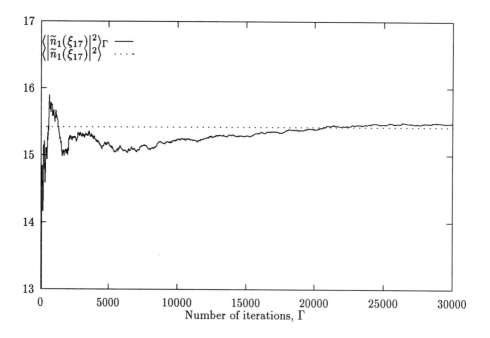

Figure 8.2: Example 1: A numerical experiment to illustrate the properties of the modulated noise source. The full line gives the simulated values $\langle |\tilde{n}_1(\xi_p)|^2\rangle_\Gamma$ versus number of iterations, and the dotted line is the theoretical result $\langle |\tilde{n}_1(\xi_p)|^2\rangle$ predicted from Equation (8.30). The data for the experiment are: $\Lambda = 32$, $\mu_1 = 0.8$, $\sigma_1 = 5.2$, $p = 17$, $\tau = 1.0$, $\varrho_{1,1} = 2.1$, $\vartheta_{1,1} = 2.0$, $\varphi_{1,1} = 0.0$, $\varrho_{1,2} = 1.9$, $\vartheta_{1,2} = 3.0$, $\varphi_{1,2} = 0.0$, $C_2 = 1.4$, and $C_3 = 0.9$. The simulated $\langle |\tilde{n}_1(\xi_p)|^2\rangle_{29901}^{30000} = 15.48$ with a standard deviation of 0.0015. The theoretical result is $\langle |\tilde{n}_1(\xi_{17})|^2\rangle = 15.42$. This corresponds to a deviation of the simulated result from the theoretical result of 0.37 %.

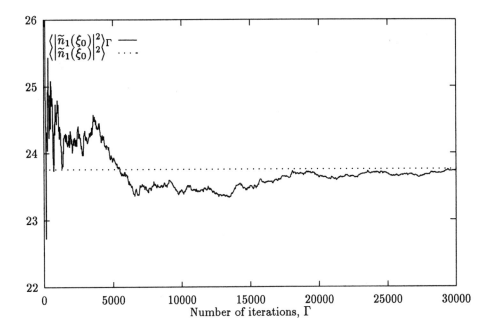

Figure 8.3: Example 1: A numerical experiment to illustrate the properties of the modulated noise source. The full line gives the simulated values $\langle|\tilde{n}_1(\xi_p)|^2\rangle_\Gamma$ versus number of iterations, and the dotted line is the theoretical result $\langle|\tilde{n}_1(\xi_p)|^2\rangle$ predicted from Equation (8.30). The data for the experiment are: $\Lambda = 32$, $\mu_1 = 0.8$, $\sigma_1 = 5.2$, $p = 0$, $\tau = 1.0$, $\varrho_{1,1} = 2.1$, $\vartheta_{1,1} = 2.0$, $\varphi_{1,1} = 0.0$, $\varrho_{1,2} = 1.9$, $\vartheta_{1,2} = 3.0$, $\varphi_{1,2} = 0.0$, $C_2 = 1.4$, and $C_3 = 0.9$. The simulated $\langle|\tilde{n}_1(\xi_0)|^2\rangle_{29901}^{30000} = 23.74$ with a standard deviation of 0.0039. The theoretical result is $\langle|\tilde{n}_1(\xi_0)|^2\rangle = 23.76$. This corresponds to a deviation from the theoretical result of -0.09%.

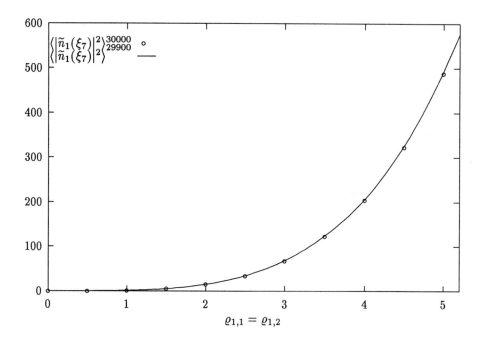

Figure 8.4: Example 1: A numerical experiment to illustrate the properties of the modulated noise source. The full line gives the theoretical values for $\langle |\tilde{n}_1(\xi_7)|^2 \rangle$ versus the amplitudes $\varrho_{1,1} = \varrho_{1,2}$ for the modulating sinusoidals predicted from Equation (8.30), and circles give the simulated numerical values. The simulated values are the average of the last 100 iterations of a total of 30000 iterations (ensembles). All the simulated values $\langle |\tilde{n}_1(\xi_7)|^2 \rangle_{29901}^{30000}$ agree with the theoretical predictions with a deviation less than ± 1.25 %. The data for the experiment are: $\Lambda = 32$, $\mu_1 = 0.8$, $\sigma_1 = 5.2$, $p = 7$, $\tau = 1.0$, $\vartheta_{1,1} = 2.0$, $\varphi_{1,1} = 0.0$, $\vartheta_{1,2} = 3.0$, $\varphi_{1,2} = 0.0$, $C_2 = 1.4$, and $C_3 = 0.9$.

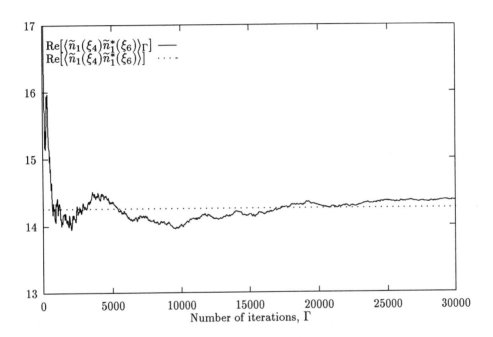

Figure 8.5: Example 1: A numerical experiment to illustrate the properties of the modulated noise source. The full line gives the simulated values $\mathrm{Re}[\langle \tilde{n}_1(\xi_{p_1})\,\tilde{n}_1^*(\xi_{p_2})\rangle_{29901}^{30000}]$ with $p_1 = 4$ and $p_2 = 6$ versus number of iterations, and the dotted line is the theoretical result predicted from Equations (8.26) and (8.31). The data for the experiment are: $\Lambda = 32$, $\mu_1 = 0.8$, $\sigma_1 = 5.2$, $p_1 = 4$, $p_2 = 6$, $\tau = 1.0$, $\varrho_{1,1} = 2.1$, $\vartheta_{1,1} = 2.0$, $\varphi_{1,1} = 0.0$, $\varrho_{1,2} = 1.9$, $\vartheta_{1,2} = 3.0$, $\varphi_{1,2} = 0.0$, $C_2 = 1.4$, and $C_3 = 0.9$. The simulated $\mathrm{Re}[\langle \tilde{n}_1(\xi_4)\,\tilde{n}_1^*(\xi_6)\rangle_{29901}^{30000}] = 14.37$ with a standard deviation of 0.0016. The theoretical result is $\mathrm{Re}[\langle \tilde{n}_1(\xi_4)\,\tilde{n}_1^*(\xi_6)\rangle] = 14.26$. This corresponds to a deviation from the theoretical result of 0.78 %.

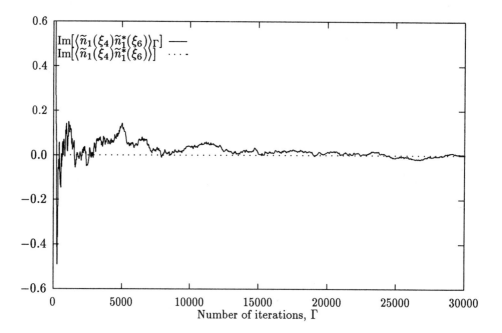

Figure 8.6: Example 1: A numerical experiment to illustrate the properties of the modulated noise source. The full line gives the simulated values $\text{Im}[\langle \tilde{n}_1(\xi_{p_1}) \, \tilde{n}_1^*(\xi_{p_2}) \rangle_\Gamma]$ with $p_1 = 4$ and $p_2 = 6$ versus number of iterations, and the dotted line is the theoretical result predicted from Equations (8.26) and (8.31). The data for the experiment are: $\Lambda = 32$, $\mu_1 = 0.8$, $\sigma_1 = 5.2$, $p_1 = 4$, $p_2 = 6$, $\tau = 1.0$, $\varrho_{1,1} = 2.1$, $\vartheta_{1,1} = 2.0$, $\varphi_{1,1} = 0.0$, $\varrho_{1,2} = 1.9$, $\vartheta_{1,2} = 3.0$, $\varphi_{1,2} = 0.0$, $C_2 = 1.4$, and $C_3 = 0.9$. The simulated value $\text{Im}[\langle \tilde{n}_1(\xi_4) \, \tilde{n}_1^*(\xi_6) \rangle_{29901}^{30000}] = 0.00095$ with a standard deviation of 0.00086. The theoretical result is $\text{Im}[\langle \tilde{n}_1(\xi_4) \, \tilde{n}_1^*(\xi_6) \rangle] = 0$.

8.2 Example 2

PROBLEM: As a simple example of the theory, consider the non-linear time invariant noisy system in Figure 8.7. For this system the number of signal input ports is $K = 1$, the number of controlling variables is $Q = 1$, and the number of output ports is $L = 1$. The deterministic input signal is a single sinusoidal signal, and the single noise source is unmodulated. The objective of the example is to determine the cross-correlation $\langle \tilde{r}_{n,1}(\xi_{p_1}) \tilde{r}^*_{n,1}(\xi_{p_2}) \rangle$ where $p_1, p_2 \in \mathcal{Z}$ are integers.

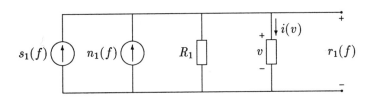

Figure 8.7: Example 2: Non-linear noisy Van der Pool system with one noise sources and one non-linear element.

THEORY: The system in Figure 8.7 is excited by the deterministic signal $s_1(f)$ given by Equation (7.8) as

$$s_1(f) = \sum_{j_1=1}^{2} \tilde{s}_1(\psi_{1,j_1}) \, \delta(f - \psi_{1,j_1}) \tag{8.33}$$

where $\psi_{1,1} = \vartheta_{1,1} > 0$ and $\psi_{1,2} = -\vartheta_{1,1}$, and

$$\tilde{s}_1(\vartheta_{1,1}) = \frac{1}{2} \varrho_{1,1} \, \exp[j \, \varphi_{1,1}] \tag{8.34}$$

$$\tilde{s}_1(-\vartheta_{1,1}) = \frac{1}{2} \varrho_{1,1} \, \exp[-j \, \varphi_{1,1}] \tag{8.35}$$

and thus $J_1 = 2$. The time domain noise signal $n_1(t)$ is given by

$$n_1(t) = C_1 \, w_1(t) \tag{8.36}$$

where $w_1(t)$ is a coloured noise source. Equations (7.24) and (8.36) lead to $I_1 = 0$ and

$$(G_1)_1(\Omega_{0,1}) = C_1 \tag{8.37}$$

The non-linear element is a current/voltage (non-linear conductance) element given by

$$i(v) \quad = \quad i(v(t), t) \quad = \quad g_2\, v^2(t) \tag{8.38}$$

where $g_2, g_3 \in \mathcal{R}$ are real constants. For the system in Figure 8.7 the maximum order considered is chosen to be $M = 8$. Note from Figure 8.7 that the Volterra transfer functions relating inputs $s_1(f)$ and $n_1(f)$ to the output $r_1(f)$ of arbitrary orders are non-zero even though the non-linear element described by Equation (8.38) is of second order. The autocorrelation for the fundamental noise source $\tilde{w}_1(\xi_p)$ is chosen to

$$\langle \tilde{w}_1(\xi_{p_1})\, \tilde{w}_1^*(\xi_{p_2}) \rangle \quad = \quad \begin{cases} \kappa_1/(\kappa_2 + |\xi_{p_1}|^2) & \text{for} \quad p_1 = p_2 \\ 0 & \text{otherwise} \end{cases} \tag{8.39}$$

where $\kappa_1, \kappa_2 \in \mathcal{R}_+$ are positive real constants. The frequency sets $\mathcal{S}_0$, $\mathcal{S}_1$, $\mathcal{S}_2$, $\mathcal{S}_3$, $\mathcal{S}_4$, $\mathcal{S}_5$, $\mathcal{S}_6$, and $\mathcal{S}_7$ can be determined as

$$
\begin{aligned}
\mathcal{S}_0 \;&=\; \{\Psi_{0,1}\}, &\quad E_0 = 1 & \tag{8.40}\\
&=\; \{0\} & \tag{8.41}\\
\mathcal{S}_1 \;&=\; \{\Psi_{1,1}, \Psi_{1,2}\}, &\quad E_1 = 2 & \tag{8.42}\\
&=\; \{\vartheta_{1,1}, -\vartheta_{1,1}\} & \tag{8.43}\\[4pt]
\mathcal{S}_2 \;&=\; \{\Psi_{2,1}, \Psi_{2,2}, \Psi_{2,3}\}, &\quad E_2 = 3 & \tag{8.44}\\
&=\; \{2\vartheta_{1,1}, -2\vartheta_{1,1}, 0\} & \tag{8.45}\\[4pt]
\mathcal{S}_3 \;&=\; \{\Psi_{3,1}, \Psi_{3,2}, \Psi_{3,3}, \Psi_{3,4}\}, &\quad E_3 = 4 & \tag{8.46}\\
&=\; \{3\vartheta_{1,1}, -3\vartheta_{1,1}, \vartheta_{1,1}, -\vartheta_{1,1}\} & \tag{8.47}\\[4pt]
\mathcal{S}_4 \;&=\; \{\Psi_{4,1}, \Psi_{4,2}, \Psi_{4,3}, \Psi_{4,4}, \Psi_{4,5}\}, &\quad E_4 = 5 & \tag{8.48}\\
&=\; \{4\vartheta_{1,1}, -4\vartheta_{1,1}, 2\vartheta_{1,1}, -2\vartheta_{1,1}, 0\} & \tag{8.49}\\[4pt]
\mathcal{S}_5 \;&=\; \{\Psi_{5,1}, \Psi_{5,2}, \Psi_{5,3}, \Psi_{5,4}, \Psi_{5,5}, \Psi_{5,6}\}, &\quad E_5 = 6 & \tag{8.50}\\
&=\; \{5\vartheta_{1,1}, -5\vartheta_{1,1}, 3\vartheta_{1,1}, -3\vartheta_{1,1}, \vartheta_{1,1}, -\vartheta_{1,1}\} & \tag{8.51}\\[4pt]
\mathcal{S}_6 \;&=\; \{\Psi_{6,1}, \Psi_{6,2}, \Psi_{6,3}, \Psi_{6,4}, \Psi_{6,5}, \Psi_{6,6}, \Psi_{6,7}\}, &\quad E_6 = 7 & \tag{8.52}\\
&=\; \{6\vartheta_{1,1}, -6\vartheta_{1,1}, 4\vartheta_{1,1}, -4\vartheta_{1,1}, 2\vartheta_{1,1}, -2\vartheta_{1,1}, 0\} & \tag{8.53}\\[4pt]
\mathcal{S}_7 \;&=\; \{\Psi_{7,1}, \Psi_{7,2}, \Psi_{7,3}, \Psi_{7,4}, \Psi_{7,5}, \Psi_{7,6}, \Psi_{7,7}, \Psi_{7,8}\}, &\quad E_7 = 8 & \tag{8.54}\\
&=\; \{7\vartheta_{1,1}, -7\vartheta_{1,1}, 5\vartheta_{1,1}, -5\vartheta_{1,1}, 3\vartheta_{1,1}, -3\vartheta_{1,1}, \vartheta_{1,1}, -\vartheta_{1,1}\} & \tag{8.55}
\end{aligned}
$$

and thus $E = 15$, and

$$\{\Psi_1, \ldots, \Psi_{15}\} = \{0, \vartheta_{1,1}, -\vartheta_{1,1}, 2\vartheta_{1,1}, -2\vartheta_{1,1}, 3\vartheta_{1,1}, -3\vartheta_{1,1}, 4\vartheta_{1,1}, -4\vartheta_{1,1},$$
$$5\vartheta_{1,1}, -5\vartheta_{1,1}, 6\vartheta_{1,1}, -6\vartheta_{1,1}, 7\vartheta_{1,1}, -7\vartheta_{1,1}\} \qquad (8.56)$$

Using Equations (7.57)–(7.59) and (8.36) leads to

$$\boldsymbol{t}_1(\xi_p) = [C_1, 0, 0, 0, 0, 0, 0, 0, 0, 0, 0, 0, 0, 0, 0, 0]^T \qquad (8.57)$$

The conversion vector $\boldsymbol{\tau}_{1,1}(\xi_p)$ can be determined using Equations (7.98) and (7.96) as

$$\boldsymbol{\tau}_{1,1}(\xi_p) = \left[\tau_{1,1}(\xi_p, \Psi_1), \ldots, \tau_{1,1}(\xi_p, \Psi_{15})\right]^T \qquad (8.58)$$

where

$$
\begin{aligned}
\tau_{1,1}(\xi_p, \Psi_1) = \ & (H_1)_{0,1}(\xi_p) \\
& + 2\,(H_1)_{2,1}(\vartheta_{1,1}, -\vartheta_{1,1}; \xi_p)\,\tilde{s}_1(\vartheta_{1,1})\,\tilde{s}_1(-\vartheta_{1,1}) \\
& + 6\,(H_1)_{4,1}(\vartheta_{1,1}, \vartheta_{1,1}, -\vartheta_{1,1}, -\vartheta_{1,1}; \xi_p) \\
& \quad \tilde{s}_1(\vartheta_{1,1})\,\tilde{s}_1(\vartheta_{1,1})\,\tilde{s}_1(-\vartheta_{1,1})\,\tilde{s}_1(-\vartheta_{1,1}) \\
& + 20\,(H_1)_{6,1}(\vartheta_{1,1}, \vartheta_{1,1}, \vartheta_{1,1}, -\vartheta_{1,1}, -\vartheta_{1,1}, -\vartheta_{1,1}; \xi_p) \\
& \quad \tilde{s}_1(\vartheta_{1,1})\,\tilde{s}_1(\vartheta_{1,1})\,\tilde{s}_1(\vartheta_{1,1})\,\tilde{s}_1(-\vartheta_{1,1})\,\tilde{s}_1(-\vartheta_{1,1})\,\tilde{s}_1(-\vartheta_{1,1})
\end{aligned}
$$
$$(8.59)$$

$$
\begin{aligned}
\tau_{1,1}(\xi_p, \Psi_2) = \ & (H_1)_{1,1}(\vartheta_{1,1}; \xi_p - \vartheta_{1,1})\,\tilde{s}_1(\vartheta_{1,1}) \\
& + 3\,(H_1)_{3,1}(\vartheta_{1,1}, \vartheta_{1,1}, -\vartheta_{1,1}; \xi_p - \vartheta_{1,1}) \\
& \quad \tilde{s}_1(\vartheta_{1,1})\,\tilde{s}_1(\vartheta_{1,1})\,\tilde{s}_1(-\vartheta_{1,1}) \\
& + 10\,(H_1)_{5,1}(\vartheta_{1,1}, \vartheta_{1,1}, \vartheta_{1,1}, -\vartheta_{1,1}, -\vartheta_{1,1}; \xi_p - \vartheta_{1,1}) \\
& \quad \tilde{s}_1(\vartheta_{1,1})\,\tilde{s}_1(\vartheta_{1,1})\,\tilde{s}_1(\vartheta_{1,1})\,\tilde{s}_1(-\vartheta_{1,1})\,\tilde{s}_1(-\vartheta_{1,1}) \\
& + 35\,(H_1)_{7,1}(\vartheta_{1,1}, \vartheta_{1,1}, \vartheta_{1,1}, \vartheta_{1,1}, -\vartheta_{1,1}, -\vartheta_{1,1}, -\vartheta_{1,1}; \xi_p - \vartheta_{1,1}) \\
& \quad \tilde{s}_1(\vartheta_{1,1})\,\tilde{s}_1(\vartheta_{1,1})\,\tilde{s}_1(\vartheta_{1,1})\,\tilde{s}_1(\vartheta_{1,1}) \\
& \quad \tilde{s}_1(-\vartheta_{1,1})\,\tilde{s}_1(-\vartheta_{1,1})\,\tilde{s}_1(-\vartheta_{1,1})
\end{aligned}
$$
$$(8.60)$$

$$
\begin{aligned}
\tau_{1,1}(\xi_p, \Psi_3) \;=\; & (H_1)_{1,1}(-\vartheta_{1,1}; \xi_p + \vartheta_{1,1}) \, \tilde{s}_1(-\vartheta_{1,1}) \\
& + 3\,(H_1)_{3,1}(\vartheta_{1,1}, -\vartheta_{1,1}, -\vartheta_{1,1}; \xi_p + \vartheta_{1,1}) \\
& \quad \tilde{s}_1(\vartheta_{1,1}) \, \tilde{s}_1(-\vartheta_{1,1}) \, \tilde{s}_1(-\vartheta_{1,1}) \\
& + 10\,(H_1)_{5,1}(\vartheta_{1,1}, \vartheta_{1,1}, -\vartheta_{1,1}, -\vartheta_{1,1}, -\vartheta_{1,1}; \xi_p + \vartheta_{1,1}) \\
& \quad \tilde{s}_1(\vartheta_{1,1}) \, \tilde{s}_1(\vartheta_{1,1}) \, \tilde{s}_1(-\vartheta_{1,1}) \, \tilde{s}_1(-\vartheta_{1,1}) \, \tilde{s}_1(-\vartheta_{1,1}) \\
& + 35\,(H_1)_{7,1}(\vartheta_{1,1}, \vartheta_{1,1}, \vartheta_{1,1}, -\vartheta_{1,1}, -\vartheta_{1,1}, -\vartheta_{1,1}, -\vartheta_{1,1}; \xi_p + \vartheta_{1,1}) \\
& \quad \tilde{s}_1(\vartheta_{1,1}) \, \tilde{s}_1(\vartheta_{1,1}) \, \tilde{s}_1(\vartheta_{1,1}) \\
& \quad \tilde{s}_1(-\vartheta_{1,1}) \, \tilde{s}_1(-\vartheta_{1,1}) \, \tilde{s}_1(-\vartheta_{1,1}) \, \tilde{s}_1(-\vartheta_{1,1})
\end{aligned}
$$

(8.61)

$$
\begin{aligned}
\tau_{1,1}(\xi_p, \Psi_4) \;=\; & (H_1)_{2,1}(\vartheta_{1,1}, \vartheta_{1,1}; \xi_p - 2\vartheta_{1,1}) \, \tilde{s}_1(\vartheta_{1,1}) \, \tilde{s}_1(\vartheta_{1,1}) \\
& + 4\,(H_1)_{4,1}(\vartheta_{1,1}, \vartheta_{1,1}, \vartheta_{1,1}, -\vartheta_{1,1}; \xi_p - 2\vartheta_{1,1}) \\
& \quad \tilde{s}_1(\vartheta_{1,1}) \, \tilde{s}_1(\vartheta_{1,1}) \, \tilde{s}_1(\vartheta_{1,1}) \, \tilde{s}_1(-\vartheta_{1,1}) \\
& + 15\,(H_1)_{6,1}(\vartheta_{1,1}, \vartheta_{1,1}, \vartheta_{1,1}, \vartheta_{1,1}, -\vartheta_{1,1}, -\vartheta_{1,1}; \xi_p - 2\vartheta_{1,1}) \\
& \quad \tilde{s}_1(\vartheta_{1,1}) \, \tilde{s}_1(\vartheta_{1,1}) \, \tilde{s}_1(\vartheta_{1,1}) \, \tilde{s}_1(\vartheta_{1,1}) \, \tilde{s}_1(-\vartheta_{1,1}) \, \tilde{s}_1(-\vartheta_{1,1})
\end{aligned}
$$

(8.62)

$$
\begin{aligned}
\tau_{1,1}(\xi_p, \Psi_5) \;=\; & (H_1)_{2,1}(-\vartheta_{1,1}, -\vartheta_{1,1}; \xi_p + 2\vartheta_{1,1}) \, \tilde{s}_1(-\vartheta_{1,1}) \, \tilde{s}_1(-\vartheta_{1,1}) \\
& + 4\,(H_1)_{4,1}(\vartheta_{1,1}, -\vartheta_{1,1}, -\vartheta_{1,1}, -\vartheta_{1,1}; \xi_p + 2\vartheta_{1,1}) \\
& \quad \tilde{s}_1(\vartheta_{1,1}) \, \tilde{s}_1(-\vartheta_{1,1}) \, \tilde{s}_1(-\vartheta_{1,1}) \, \tilde{s}_1(-\vartheta_{1,1}) \\
& + 15\,(H_1)_{6,1}(\vartheta_{1,1}, \vartheta_{1,1}, -\vartheta_{1,1}, -\vartheta_{1,1}, -\vartheta_{1,1}, -\vartheta_{1,1}; \xi_p + 2\vartheta_{1,1}) \\
& \quad \tilde{s}_1(\vartheta_{1,1}) \, \tilde{s}_1(\vartheta_{1,1}) \, \tilde{s}_1(-\vartheta_{1,1}) \, \tilde{s}_1(-\vartheta_{1,1}) \, \tilde{s}_1(-\vartheta_{1,1}) \, \tilde{s}_1(-\vartheta_{1,1})
\end{aligned}
$$

(8.63)

$$
\begin{aligned}
\tau_{1,1}(\xi_p, \Psi_6) \;=\; & (H_1)_{3,1}(\vartheta_{1,1}, \vartheta_{1,1}, \vartheta_{1,1}; \xi_p - 3\vartheta_{1,1}) \, \tilde{s}_1(\vartheta_{1,1}) \, \tilde{s}_1(\vartheta_{1,1}) \, \tilde{s}_1(\vartheta_{1,1}) \\
& + 5\,(H_1)_{5,1}(\vartheta_{1,1}, \vartheta_{1,1}, \vartheta_{1,1}, \vartheta_{1,1}, -\vartheta_{1,1}; \xi_p - 3\vartheta_{1,1}) \\
& \quad \tilde{s}_1(\vartheta_{1,1}) \, \tilde{s}_1(\vartheta_{1,1}) \, \tilde{s}_1(\vartheta_{1,1}) \, \tilde{s}_1(\vartheta_{1,1}) \, \tilde{s}_1(-\vartheta_{1,1}) \\
& + 21\,(H_1)_{7,1}(\vartheta_{1,1}, \vartheta_{1,1}, \vartheta_{1,1}, \vartheta_{1,1}, \vartheta_{1,1}, -\vartheta_{1,1}, -\vartheta_{1,1}; \xi_p - 3\vartheta_{1,1}) \\
& \quad \tilde{s}_1(\vartheta_{1,1}) \, \tilde{s}_1(\vartheta_{1,1}) \, \tilde{s}_1(\vartheta_{1,1}) \, \tilde{s}_1(\vartheta_{1,1}) \, \tilde{s}_1(\vartheta_{1,1}) \\
& \quad \tilde{s}_1(-\vartheta_{1,1}) \, \tilde{s}_1(-\vartheta_{1,1})
\end{aligned}
$$

(8.64)

$$\tau_{1,1}(\xi_p, \Psi_7) = (H_1)_{3,1}(-\vartheta_{1,1}, -\vartheta_{1,1}, -\vartheta_{1,1}; \xi_p + 3\vartheta_{1,1})$$
$$\tilde{s}_1(-\vartheta_{1,1})\,\tilde{s}_1(-\vartheta_{1,1})\,\tilde{s}_1(-\vartheta_{1,1})$$
$$+ 5\,(H_1)_{5,1}(\vartheta_{1,1}, -\vartheta_{1,1}, -\vartheta_{1,1}, -\vartheta_{1,1}, -\vartheta_{1,1}; \xi_p + 3\vartheta_{1,1})$$
$$\tilde{s}_1(\vartheta_{1,1})\,\tilde{s}_1(-\vartheta_{1,1})\,\tilde{s}_1(-\vartheta_{1,1})\,\tilde{s}_1(-\vartheta_{1,1})\,\tilde{s}_1(-\vartheta_{1,1})$$
$$+ 21\,(H_1)_{7,1}(\vartheta_{1,1}, \vartheta_{1,1}, -\vartheta_{1,1}, -\vartheta_{1,1}, -\vartheta_{1,1}, -\vartheta_{1,1}, -\vartheta_{1,1}; \xi_p + 3\vartheta_{1,1})$$
$$\tilde{s}_1(\vartheta_{1,1})\,\tilde{s}_1(\vartheta_{1,1})\,\tilde{s}_1(-\vartheta_{1,1})\,\tilde{s}_1(-\vartheta_{1,1})$$
$$\tilde{s}_1(-\vartheta_{1,1})\,\tilde{s}_1(-\vartheta_{1,1})\,\tilde{s}_1(-\vartheta_{1,1})$$

$$(8.65)$$

$$\tau_{1,1}(\xi_p, \Psi_8) = (H_1)_{4,1}(\vartheta_{1,1}, \vartheta_{1,1}, \vartheta_{1,1}, \vartheta_{1,1}; \xi_p - 4\vartheta_{1,1})$$
$$\tilde{s}_1(\vartheta_{1,1})\,\tilde{s}_1(\vartheta_{1,1})\,\tilde{s}_1(\vartheta_{1,1})\,\tilde{s}_1(\vartheta_{1,1})$$
$$+ 6\,(H_1)_{6,1}(\vartheta_{1,1}, \vartheta_{1,1}, \vartheta_{1,1}, \vartheta_{1,1}, \vartheta_{1,1}, -\vartheta_{1,1}; \xi_p - 4\vartheta_{1,1})$$
$$\tilde{s}_1(\vartheta_{1,1})\,\tilde{s}_1(\vartheta_{1,1})\,\tilde{s}_1(\vartheta_{1,1})\,\tilde{s}_1(\vartheta_{1,1})\,\tilde{s}_1(\vartheta_{1,1})\,\tilde{s}_1(-\vartheta_{1,1})$$

$$(8.66)$$

$$\tau_{1,1}(\xi_p, \Psi_9) = (H_1)_{4,1}(-\vartheta_{1,1}, -\vartheta_{1,1}, -\vartheta_{1,1}, -\vartheta_{1,1}; \xi_p + 4\vartheta_{1,1})$$
$$\tilde{s}_1(-\vartheta_{1,1})\,\tilde{s}_1(-\vartheta_{1,1})\,\tilde{s}_1(-\vartheta_{1,1})\,\tilde{s}_1(-\vartheta_{1,1})$$
$$+ 6\,(H_1)_{6,1}(\vartheta_{1,1}, -\vartheta_{1,1}, -\vartheta_{1,1}, -\vartheta_{1,1}, -\vartheta_{1,1}, -\vartheta_{1,1}; \xi_p + 4\vartheta_{1,1})$$
$$\tilde{s}_1(\vartheta_{1,1})\,\tilde{s}_1(-\vartheta_{1,1})\,\tilde{s}_1(-\vartheta_{1,1})\,\tilde{s}_1(-\vartheta_{1,1})\,\tilde{s}_1(-\vartheta_{1,1})\,\tilde{s}_1(-\vartheta_{1,1})$$

$$(8.67)$$

$$\tau_{1,1}(\xi_p, \Psi_{10}) = (H_1)_{5,1}(\vartheta_{1,1}, \vartheta_{1,1}, \vartheta_{1,1}, \vartheta_{1,1}, \vartheta_{1,1}; \xi_p - 5\vartheta_{1,1})$$
$$\tilde{s}_1(\vartheta_{1,1})\,\tilde{s}_1(\vartheta_{1,1})\,\tilde{s}_1(\vartheta_{1,1})\,\tilde{s}_1(\vartheta_{1,1})\,\tilde{s}_1(\vartheta_{1,1})$$
$$+ 7\,(H_1)_{7,1}(\vartheta_{1,1}, \vartheta_{1,1}, \vartheta_{1,1}, \vartheta_{1,1}, \vartheta_{1,1}, \vartheta_{1,1}, -\vartheta_{1,1}; \xi_p - 5\vartheta_{1,1})$$
$$\tilde{s}_1(\vartheta_{1,1})\,\tilde{s}_1(\vartheta_{1,1})\,\tilde{s}_1(\vartheta_{1,1})\,\tilde{s}_1(\vartheta_{1,1})\,\tilde{s}_1(\vartheta_{1,1})\,\tilde{s}_1(\vartheta_{1,1})\,\tilde{s}_1(-\vartheta_{1,1})$$

$$(8.68)$$

$$\tau_{1,1}(\xi_p, \Psi_{11}) = (H_1)_{5,1}(-\vartheta_{1,1}, -\vartheta_{1,1}, -\vartheta_{1,1}, -\vartheta_{1,1}, -\vartheta_{1,1}; \xi_p + 5\vartheta_{1,1})$$
$$\tilde{s}_1(-\vartheta_{1,1})\,\tilde{s}_1(-\vartheta_{1,1})\,\tilde{s}_1(-\vartheta_{1,1})\,\tilde{s}_1(-\vartheta_{1,1})\,\tilde{s}_1(-\vartheta_{1,1})$$
$$+ 7\,(H_1)_{7,1}(\vartheta_{1,1}, -\vartheta_{1,1}, -\vartheta_{1,1}, -\vartheta_{1,1}, -\vartheta_{1,1}, -\vartheta_{1,1}, -\vartheta_{1,1};$$
$$\xi_p + 5\vartheta_{1,1})$$
$$\tilde{s}_1(\vartheta_{1,1})\,\tilde{s}_1(-\vartheta_{1,1})\,\tilde{s}_1(-\vartheta_{1,1})\,\tilde{s}_1(-\vartheta_{1,1})\,\tilde{s}_1(-\vartheta_{1,1})$$
$$\tilde{s}_1(-\vartheta_{1,1})\,\tilde{s}_1(-\vartheta_{1,1})$$

$$(8.69)$$

$$\tau_{1,1}(\xi_p, \Psi_{12}) = (H_1)_{6,1}(\vartheta_{1,1}, \vartheta_{1,1}, \vartheta_{1,1}, \vartheta_{1,1}, \vartheta_{1,1}, \vartheta_{1,1}; \xi_p - 6\vartheta_{1,1})$$
$$\tilde{s}_1(\vartheta_{1,1})\,\tilde{s}_1(\vartheta_{1,1})\,\tilde{s}_1(\vartheta_{1,1})\,\tilde{s}_1(\vartheta_{1,1})\,\tilde{s}_1(\vartheta_{1,1})\,\tilde{s}_1(\vartheta_{1,1}) \qquad (8.70)$$

$$
\begin{aligned}
\tau_{1,1}(\xi_p, \Psi_{13}) \;=\; & (H_1)_{6,1}(-\vartheta_{1,1}, -\vartheta_{1,1}, -\vartheta_{1,1}, -\vartheta_{1,1}, -\vartheta_{1,1}, -\vartheta_{1,1}; \xi_p + 6\vartheta_{1,1}) \\
& \tilde{s}_1(-\vartheta_{1,1})\,\tilde{s}_1(-\vartheta_{1,1})\,\tilde{s}_1(-\vartheta_{1,1})\,\tilde{s}_1(-\vartheta_{1,1})\,\tilde{s}_1(-\vartheta_{1,1}) \\
& \tilde{s}_1(-\vartheta_{1,1})
\end{aligned}
\tag{8.71}
$$

$$
\begin{aligned}
\tau_{1,1}(\xi_p, \Psi_{14}) \;=\; & (H_1)_{7,1}(\vartheta_{1,1}, \vartheta_{1,1}, \vartheta_{1,1}, \vartheta_{1,1}, \vartheta_{1,1}, \vartheta_{1,1}, \vartheta_{1,1}; \xi_p - 7\vartheta_{1,1}) \\
& \tilde{s}_1(\vartheta_{1,1})\,\tilde{s}_1(\vartheta_{1,1})\,\tilde{s}_1(\vartheta_{1,1})\,\tilde{s}_1(\vartheta_{1,1})\,\tilde{s}_1(\vartheta_{1,1}) \\
& \tilde{s}_1(\vartheta_{1,1})\,\tilde{s}_1(\vartheta_{1,1})
\end{aligned}
\tag{8.72}
$$

$$
\begin{aligned}
\tau_{1,1}(\xi_p, \Psi_{15}) \;=\; & (H_1)_{7,1}(-\vartheta_{1,1}, -\vartheta_{1,1}, -\vartheta_{1,1}, -\vartheta_{1,1}, -\vartheta_{1,1}, -\vartheta_{1,1}, -\vartheta_{1,1}; \\
& \xi_p + 7\vartheta_{1,1})\,\tilde{s}_1(-\vartheta_{1,1})\,\tilde{s}_1(-\vartheta_{1,1})\,\tilde{s}_1(-\vartheta_{1,1}) \\
& \tilde{s}_1(-\vartheta_{1,1})\,\tilde{s}_1(-\vartheta_{1,1})\,\tilde{s}_1(-\vartheta_{1,1})\,\tilde{s}_1(-\vartheta_{1,1})
\end{aligned}
\tag{8.73}
$$

Thus the autocorrelation $\langle \tilde{r}_{n,1}(\xi_{p_1})\tilde{r}_{n,1}^*(\xi_{p_2})\rangle$ where $p_1, p_2 \in \mathcal{Z}$ can be determined using Equation (7.104). The multi-port frequency domain Volterra transfer functions in Equations (8.59)–(8.73) can be determined as

$$
(H_1)_{0,1}(\Xi_{1,1}) \;=\; R_1 \tag{8.74}
$$

$$
(H_1)_{1,1}(\Omega_{1,1}; \Xi_{1,1}) \;=\; -2\,R_1^3 g_2 \tag{8.75}
$$

$$
(H_1)_{2,1}(\Omega_{1,1}, \Omega_{1,2}; \Xi_{1,1}) \;=\; 6\,R_1^5 g_2^2 \tag{8.76}
$$

$$
(H_1)_{3,1}(\Omega_{1,1}, \Omega_{1,2}, \Omega_{1,3}; \Xi_{1,1}) \;=\; -20\,R_1^7 g_2^3 \tag{8.77}
$$

$$
(H_1)_{4,1}(\Omega_{1,1}, \Omega_{1,2}, \Omega_{1,3}, \Omega_{1,4}; \Xi_{1,1}) \;=\; 70\,R_1^9 g_2^4 \tag{8.78}
$$

$$
(H_1)_{5,1}(\Omega_{1,1}, \Omega_{1,2}, \Omega_{1,3}, \Omega_{1,4}, \Omega_{1,5}; \Xi_{1,1}) \;=\; -252\,R_1^{11} g_2^5 \tag{8.79}
$$

$$
(H_1)_{6,1}(\Omega_{1,1}, \Omega_{1,2}, \Omega_{1,3}, \Omega_{1,4}, \Omega_{1,5}, \Omega_{1,6}; \Xi_{1,1}) \;=\; 924\,R_1^{13} g_2^6 \tag{8.80}
$$

$$
(H_1)_{7,1}(\Omega_{1,1}, \Omega_{1,2}, \Omega_{1,3}, \Omega_{1,4}, \Omega_{1,5}, \Omega_{1,6}, \Omega_{1,7}; \Xi_{1,1}) \;=\; -3432\,R_1^{15} g_2^7 \tag{8.81}
$$

Thus, $\langle \tilde{r}_{n,1}(\xi_{p_1})\,\tilde{r}_{n,1}^*(\xi_{p_2})\rangle$ can be determined as

$$
\begin{aligned}
\langle \tilde{r}_{n,1}(\xi_{p_1})\,\tilde{r}_{n,1}^*(\xi_{p_2})\rangle \;=\; & \sum_{e_1=1}^{15}\sum_{e_2=1}^{15} \tau_{1,1}^T(\xi_{p_1})\,\boldsymbol{d}_{e_1}\,\boldsymbol{t}_1^T(\xi_{p_1} - \Psi_{e_1}) \\
& \times \widetilde{W}_{1,1}(\xi_{p_1} - \Psi_{e_1}, \xi_{p_2} - \Psi_{e_2}) \\
& \times \boldsymbol{t}_1^*(\xi_{p_2} - \Psi_{e_2})\,\boldsymbol{d}_{e_2}^T\,\tau_{1,1}^*(\xi_{p_2})
\end{aligned}
\tag{8.82}
$$

by use of Equations (7.105) and (7.106). It can be shown that the autocorrelation $\langle|\tilde{r}_{n,1}(\xi_p)|^2\rangle$ is given by

$$\langle|\tilde{r}_{n,1}(\xi_p)|^2\rangle \quad = \quad C_1^2 \frac{\kappa_1}{\kappa_2 + |\xi_p|^2} \|\tau_{1,1}(\xi_p)\|^2 \tag{8.83}$$

Note that the output autocorrelation $\langle|\tilde{r}_{n,1}(\xi_p)|^2\rangle$ has the same frequency domain shape as the autocorrelation for the fundamental noise source $\langle|\tilde{w}_1(\xi_p)|^2\rangle$. The only difference is the multiplication factor $C_1^2 \|\tau_{1,1}(\xi_p)\|^2$ which depends on the amplitude of the input sinusoidal signal, and on the linear and non-linear network parameters.

NUMERICAL EXPERIMENT: To investigate the correctness of the theoretical results presented in this example, numerical simulations are performed in accordance with Section 7.3. To obtain the desired correlation properties for the fundamental noise source $w_1(t)$ as given by Equation (8.39), $w_1(t)$ is given by a linear filtering of a noise signal $\varepsilon_1(t)$ which is a zero mean white Gaussian noise process with standard deviation σ_1. Choosing the frequency domain transfer function of the linear filter $H_{lin}(f)$ as

$$H_{lin}(f) \quad = \quad \frac{1}{\sigma_1} \sqrt{\kappa_1(2\Lambda + 1)} \frac{1}{\sqrt{\kappa_2 + jf}} \tag{8.84}$$

gives the desired correlation properties for $\tilde{w}_1(\xi_p)$. Thus the fundamental noise source is given by

$$\tilde{w}_1(\xi_{p_1}) \quad = \quad H_{lin}(\xi_{p_1}) \tilde{\varepsilon}_1(\xi_{p_1}) \tag{8.85}$$

which leads to the derived correlation properties for $\tilde{w}_1(\xi_p)$ as given by Equation (8.39). It is necessary to determine fundamental noise samples $w_1(t_{-\Lambda}), \ldots, w_1(t_\Lambda)$ from the noise samples $\varepsilon_1(t_{-\Lambda}), \ldots, \varepsilon_1(t_\Lambda)$ taking into account the linear filtering. One way to do this is first to generate $\varepsilon_1(t_{-\Lambda}), \ldots, \varepsilon_1(t_\Lambda)$ as a zero mean white Gaussian noise process with standard deviation σ_1, then determine all the Fourier series coefficients $\tilde{\varepsilon}_1(\xi_{-\Lambda}), \ldots, \tilde{\varepsilon}_1(\xi_\Lambda)$, then multiply $H_{lin}(\xi_{-\Lambda}), \ldots, H_{lin}(\xi_\Lambda)$ on the corresponding ε-Fourier series coefficients to obtain $\tilde{w}_1(\xi_{-\Lambda}), \ldots, \tilde{w}_1(\xi_\Lambda)$, and then finally translate the w-Fourier series coefficients to the time domain samples $w_1(t_{-\Lambda}), \ldots, w_1(t_\Lambda)$ by inverse Fourier series transformations. Thus $w_1(t_\lambda)$ where $\lambda \in \{-\Lambda, \ldots, -1, 0, 1, \ldots, \Lambda\}$ can be determined from $\varepsilon_1(t_{-\Lambda}), \ldots, \varepsilon_1(t_\Lambda)$ as

$$w_1(t_\lambda) \quad = \quad \frac{1}{\sigma_1} \sqrt{\frac{\kappa_1}{2\Lambda + 1}} \sum_{\lambda_1 = -\Lambda}^{\Lambda} \frac{1}{\sqrt{\kappa_2 + j\lambda_1 \frac{1}{2\tau}}} \exp\left[j2\pi \frac{\lambda\lambda_1}{2\Lambda + 1}\right]$$

$$\times \sum_{\lambda_2 = -\Lambda}^{\Lambda} \varepsilon_1(t_{\lambda_2}) \exp\left[-j2\pi \frac{\lambda_1\lambda_2}{2\Lambda + 1}\right] \tag{8.86}$$

To increase the speed of computation observe that

$$w_1(t_\lambda) = \frac{1}{\sigma_1} \sqrt{\frac{\kappa_1}{2\Lambda+1}} \frac{1}{\sqrt{\kappa_2}} S_2(0)$$

$$+ \frac{2}{\sigma_1} \sqrt{\frac{\kappa_1}{2\Lambda+1}} \sum_{\lambda_1=1}^{\Lambda} \mathrm{Re}\left[\frac{1}{\sqrt{\kappa_2}+j\,\lambda_1\frac{1}{2\tau}} \exp\left[j2\pi\frac{\lambda\lambda_1}{2\Lambda+1}\right] S_2(\lambda_1)\right]$$

(8.87)

where

$$S_2(\lambda_1) = \sum_{\lambda_2=-\Lambda}^{\Lambda} \varepsilon_1(t_{\lambda_2}) \exp\left[-j2\pi\frac{\lambda_1\lambda_2}{2\Lambda+1}\right]$$

(8.88)

$$= \varepsilon_1(t_0)$$

$$+ \sum_{\lambda_2=1}^{\Lambda}\left\{\varepsilon_1(t_{-\lambda_2}) \exp\left[j2\pi\frac{\lambda_1\lambda_2}{2\Lambda+1}\right] + \varepsilon_1(t_{\lambda_2}) \exp\left[-j2\pi\frac{\lambda_1\lambda_2}{2\Lambda+1}\right]\right\}$$

(8.89)

with $S_2(-\lambda_1) = S_2^*(\lambda_1)$ assuming that $\varepsilon_1(t)$ is real valued for all $t \in [-\tau; \tau]$. Note from Equation (8.89) that

$$S_2(0) = \sum_{\lambda_2=-\Lambda}^{\Lambda} \varepsilon_1(t_{\lambda_2})$$

(8.90)

It is also important to note that for arbitrary finite $p_1 \in \mathcal{Z}$ and $\tau \in \mathcal{R}_+$ then

$$\lim_{\kappa_1=\kappa_2\to\infty} H_{lin}(\xi_{p_1}) = 1 \quad \text{for} \quad \sigma_1 = \sqrt{2\Lambda+1}$$

(8.91)

In this case $\tilde{w}_1(\xi_{p_1}) = \tilde{\varepsilon}_1(\xi_{p_1})$ for all $p_1 \in \{-\Lambda,\ldots,-1,0,1,\ldots,\Lambda\}$ which means that $w_1(t_\lambda) = \varepsilon_1(t_\lambda)$ for all $\lambda \in \{-\Lambda,\ldots,-1,0,1,\ldots,\Lambda\}$. This gives a simple (but of course not complete) way to test the correctness of the implemented numerical simulation.

The network equation for the non-linear noisy system in Figure 8.7 is given by

$$g_2\,r_1^2(t) = s_1(t) + n_1(t) - \frac{1}{R_1}r_1(t)$$

(8.92)

Thus the system is described by a second-degree equation as

$$g_2\,R_1\,r_1^2(t) + r_1(t) - R_1[s_1(t) + n_1(t)] = 0$$

(8.93)

The correct solution to this equation is given by

$$r_1(t) = \frac{1}{2\,g_2\,R_1}\left\{-1 + \sqrt{1 + 4\,g_2\,R_1^2[s_1(t) + n_1(t)]}\right\}$$

(8.94)

by which the solution to the linearized system may be found as $r_1(t) = R_1[s_1(t) + n_1(t)]$ as g_2 approaches 0 (this is not the case for the other possible solution to Equation (8.92)).

Simulation 1 The autocorrelation for the noise source $\langle|\widetilde{w}_1(\xi_p)|^2\rangle_{39901}^{40000}$ is simulated versus the frequency point $p \in \{-\Lambda, \ldots, -1, 0, 1, \ldots, \Lambda\}$. The data and results of the simulation are shown in Figure 8.8. All simulated values $\langle|\widetilde{w}_1(\xi_p)|^2\rangle_{39901}^{40000}$ where $|p| \in \{1, 3, 5, 7, 9, 11, 13, 15\}$ agree with the theoretical results $\langle|\widetilde{w}_1(\xi_p)|^2\rangle$ with deviations less than ±1.3 %. To avoid making assumptions on the simulation it is only possible to make direct noise comparisons for $|p| \in \{1, 3, 5, 7, 9, 11, 13, 15\}$ since the simulated results for $|p| \in \{0, 2, 4, 6, 8, 10, 12, 14, 16\}$ contain both a noise contribution and a deterministic distribution which can not be separated without making some assumptions.

Simulation 2 The autocorrelation for the output response $\langle|\widetilde{r}_1(\xi_p)|^2\rangle_{39901}^{40000}$ with $p = 5$ is simulated versus the amplitude $\varrho_{1,1}$ of the input sinusoidal for various maximum orders $M \in \{2, 4, 6, 8\}$. The data and results of the simulation are shown in Figure 8.9. As seen from Figure 8.9 the noise level increases with amplitude $\varrho_{1,1}$. Quite as expected from the theory, the agreement between theory and experiments improves when the maximum order M for the theoretical prediction increases. When $M = 8$ the experimental results $\langle|\widetilde{r}_1(\xi_{11})|^2\rangle_{39901}^{40000}$ and the theoretical results $\langle|\widetilde{r}_1(\xi_{11})|^2\rangle$ agree with deviations less than ±0.7 % for $\varrho_{1,1} \in \{0.0, 0.2, 0.4, \ldots, 1.6\}$ as suggested from Figure 8.9.

Simulation 3 The autocorrelation for the output response $\langle|\widetilde{r}_1(\xi_p)|^2\rangle_{39901}^{40000}$ is simulated versus the frequency point $p \in \{-\Lambda, \ldots, -1, 0, 1, \ldots, \Lambda\}$. The data and results of the simulation are shown in Figure 8.10. All simulated values $\langle|\widetilde{r}_1(\xi_p)|^2\rangle_{39901}^{40000}$ where $|p| \in \{1, 3, 5, 7, 9, 11, 13, 15\}$ agree with the theoretical results $\langle|\widetilde{r}_{n,1}(\xi_p)|^2\rangle$ with deviations less than ±1.6 %. To avoid making assumptions on the simulation it is only possible to make direct noise comparisons for $|p| \in \{1, 3, 5, 7, 9, 11, 13, 15\}$ since the simulated result for $|p| \in \{0, 2, 4, 6, 8, 10, 12, 14\}$ contains both a noise contribution and a deterministic contribution which can not be separated without making some assumptions.

8.3 Example 3

<u>PROBLEM:</u> As an example of the theory consider the non-linear time invariant noisy system in Figure 8.11. For this system $K = 1$, $Q = 2$ and $L = 1$. The objective of the example is to determine the autocorrelation $\langle|\widetilde{r}_{n,1}(\xi_p)|^2\rangle$ where $p \in \mathcal{Z}$ is an

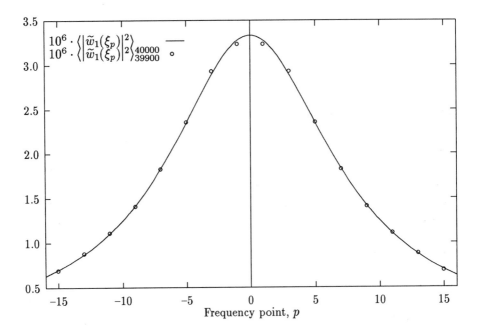

Figure 8.8: Example 2: A numerical experiment to illustrate the properties of the fundamental noise source. The full line gives the theoretical values for $\langle|\tilde{w}_1(\xi_p)|^2\rangle$ versus the frequency point p predicted from Equation (8.39), and circles give the simulated numerical values. The simulated values are the average of the last 100 iterations of a total of 40000 iterations (ensembles). The simulated values agree with the theoretical predictions with deviations less than ± 1.3 %. The data for the experiment are: $\Lambda = 16$, $\sigma_1 = 1.0$, $\kappa_1 = 5 \times 10^{-5}$, $\kappa_2 = 15$, and $\tau = 1.0$.

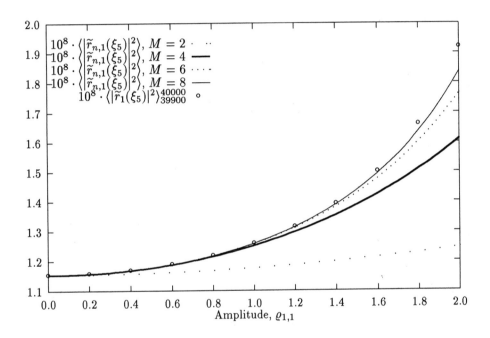

Figure 8.9: Example 2: A numerical experiment to illustrate the properties of the fundamental noise source. The full and dotted lines give the theoretical values for $\langle |\tilde{w}_1(\xi_p)|^2 \rangle$ with $p = 5$ versus the amplitude $\varrho_{1,1}$ of the input signal for various choices of the maximum order M, and the circles give the results for the experiments. The simulated values are the average of the last 100 iterations of a total of 40000 iterations (ensembles). The simulated values agree with the theoretical predictions for $M = 8$ with a deviation less than ± 0.7 % for amplitudes $\varrho_{1,1} \in \{0.0, 0.2, 0.4, \ldots, 1.6\}$. The data for the experiment are: $\Lambda = 16$, $\sigma_1 = 1.0$, $C_1 = 0.07$, $\kappa_1 = 5 \times 10^{-5}$, $\kappa_2 = 15.0$, $\tau = 1.0$, $\vartheta_{1,1} = 1.0$, $\varphi_{1,1} = 0.0$, $g_2 = 0.1$, and $R_1 = 1.0$.

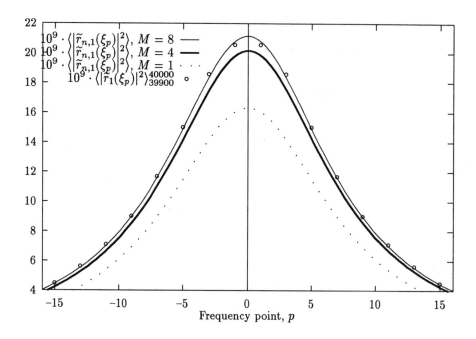

Figure 8.10: Example 2; A numerical experiment to illustrate the properties of the output noise frequency spectrum. The full and dotted lines give the theoretical values for $\langle |\tilde{r}_{n,1}(\xi_p)|^2 \rangle$ with $\varrho_{1,1} = 1.6$ versus the frequency point p and for various maximum orders $M \in \{1, 4, 8\}$ predicted from Equation (8.39), and circles give the simulated numerical values. The simulated values are the average of the last 100 iterations of a total of 40000 iterations (ensembles). The simulated values agree with the theoretical predictions with deviations less than $\pm 1.6\%$. The data for the experiment are: $\Lambda = 16$, $\sigma_1 = 1.0$, $C_1 = 0.07$, $\kappa_1 = 5 \times 10^{-5}$, $\kappa_2 = 15.0$, $\tau = 1.0$, $\vartheta_{1,1} = 1.0$, $\varphi_{1,1} = 0.0$, $g_2 = 0.1$, and $R_1 = 1.0$.

integer, and to compare the result with a direct analytical time domain simulation.

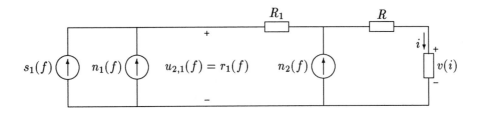

Figure 8.11: Example 3: Non-linear noisy system with two noise sources and one non-linear element.

THEORY: The system in Figure 8.11 is excited by the deterministic signal $s_1(f)$ given by Equation (7.8) as

$$s_1(f) \quad = \quad \sum_{j_1=1}^{2} \tilde{s}_1(\psi_{1,j_1})\, \delta(f - \psi_{1,j_1}) \tag{8.95}$$

where $\psi_{1,1} = \vartheta_{1,1} > 0$ and $\psi_{1,2} = -\vartheta_{1,1}$, and

$$\tilde{s}_1(\vartheta_{1,1}) \quad = \quad \frac{1}{2}\, \varrho_{1,1}\, \exp[j\, \varphi_{1,1}] \tag{8.96}$$

$$\tilde{s}_1(-\vartheta_{1,1}) \quad = \quad \frac{1}{2}\, \varrho_{1,1}\, \exp[-j\, \varphi_{1,1}] \tag{8.97}$$

and thus $J_1 = 2$. The time domain noise signal $n_1(t)$ is given by

$$n_1(t) \quad = \quad C_1\, w_1(t) \tag{8.98}$$

where $w_1(t)$ is a white noise source. Thus the noise source $n_1(t)$ is unmodulated. Equation (8.98) leads to $I_1 = 0$ and

$$(G_1)_1(\Omega_{0,1}) \quad = \quad C_1 \tag{8.99}$$

The time domain noise signal $n_2(t)$ in Figure 8.11 is a modulated noise source given by

$$n_2(t) \quad = \quad C_2\, w_2(t)\, u_{2,1}(t) \tag{8.100}$$

which leads to $I_2 = 1$ and

$$(G_2)_{1,m_1}(\Omega_{0,1}; \Omega_{1,1}, \ldots, \Omega_{1,m_1}) \quad = \quad \begin{cases} C_2 & \text{for} \quad m_1 = 1 \\ 0 & \text{otherwise} \end{cases} \tag{8.101}$$

Note that Equation (8.100) corresponds to the noise source $n_2(f)$ in a sense being amplitude modulated by the controlling variable $u_{2,1}(f)$. The non-linear element is a voltage/current (non-linear resistance) element given by

$$v(i) \quad = \quad v(i(t), t) \quad = \quad a_2\, i^2(t) + a_3\, i^3(t) \tag{8.102}$$

For the system in Figure 8.11 the maximum order considered is chosen to be $M = 4$. All Volterra transfer functions relating $s_1(f)$, $n_1(f)$ and $n_2(f)$ to the response $r_1(f)$ of orders higher than 3 are zero, even though, due to the modulation of noise source 2, there are non-zero noise contributions to the response caused by a total order of 4. The two noise sources w_1 and w_2 are zero mean white noise Gaussian processes with standard deviations σ_1 and σ_2. The continuous time autocorrelations for the two fundamental noise sources $\tilde{w}_1(\xi_p)$ and $\tilde{w}_2(\xi_p)$ are given by

$$\langle \tilde{w}_q(\xi_{p_1})\, \tilde{w}_q^*(\xi_{p_2}) \rangle \quad = \quad \begin{cases} \sigma_q^2/(2\tau) & \text{for} \quad p_1 = p_2 \\ 0 & \text{otherwise} \end{cases} \tag{8.103}$$

where $q \in \{1,2\}$. The cross-correlation between two Fourier series coefficients for the two fundamental noise sources can be shown to be $\langle \tilde{w}_1(\xi_{p_1})\, \tilde{w}_2^*(\xi_{p_2}) \rangle = 0$ for all $p_1, p_2 \in \mathcal{Z}$. The frequency sets $\mathcal{S}_0$, $\mathcal{S}_1$, $\mathcal{S}_2$ and $\mathcal{S}_3$ can be determined as

$$\begin{aligned} \mathcal{S}_0 \quad &= \quad \{\Psi_{0,1}\}, \qquad E_0 = 1 \tag{8.104} \\ &= \quad \{0\} \tag{8.105} \end{aligned}$$

$$\begin{aligned} \mathcal{S}_1 \quad &= \quad \{\Psi_{1,1}, \Psi_{1,2}\}, \qquad E_1 = 2 \tag{8.106} \\ &= \quad \{\vartheta_{1,1}, -\vartheta_{1,1}\} \tag{8.107} \end{aligned}$$

$$\begin{aligned} \mathcal{S}_2 \quad &= \quad \{\Psi_{2,1}, \Psi_{2,2}, \Psi_{2,3}\}, \qquad E_2 = 3 \tag{8.108} \\ &= \quad \{2\vartheta_{1,1}, -2\vartheta_{1,1}, 0\} \tag{8.109} \end{aligned}$$

$$\begin{aligned} \mathcal{S}_3 \quad &= \quad \{\Psi_{3,1}, \Psi_{3,2}, \Psi_{3,3}, \Psi_{3,4}\}, \qquad E_3 = 4 \tag{8.110} \\ &= \quad \{3\vartheta_{1,1}, -3\vartheta_{1,1}, \vartheta_{1,1}, -\vartheta_{1,1}\} \tag{8.111} \end{aligned}$$

and thus $E = 7$, and

$$\{\Psi_1, \ldots, \Psi_7\} \quad = \quad \{0, \vartheta_{1,1}, -\vartheta_{1,1}, 2\vartheta_{1,1}, -2\vartheta_{1,1}, 3\vartheta_{1,1}, -3\vartheta_{1,1}\} \tag{8.112}$$

Observe from Equations (7.57)–(7.59) and (8.98) that this leads to $t_1(\xi_p)$ as

$$t_1(\xi_p) \quad = \quad [C_1, 0, 0, 0, 0, 0, 0]^T \tag{8.113}$$

To determine $t_2(\xi_p)$ observe that

$$^{1}\tilde{u}_{2,1}(\Psi_{1,1}) = (R_1 + R)\,\tilde{s}_1(\vartheta_{1,1}) \qquad (8.114)$$

$$^{1}\tilde{u}_{2,1}(\Psi_{1,2}) = (R_1 + R)\,\tilde{s}_1(-\vartheta_{1,1}) \qquad (8.115)$$

$$^{2}\tilde{u}_{2,1}(\Psi_{2,1}) = a_2\,\tilde{s}_1(\vartheta_{1,1})\tilde{s}_1(\vartheta_{1,1}) \qquad (8.116)$$

$$^{2}\tilde{u}_{2,1}(\Psi_{2,2}) = a_2\,\tilde{s}_1(-\vartheta_{1,1})\tilde{s}_1(-\vartheta_{1,1}) \qquad (8.117)$$

$$^{2}\tilde{u}_{2,1}(\Psi_{2,3}) = 2\,a_2\,\tilde{s}_1(\vartheta_{1,1})\tilde{s}_1(-\vartheta_{1,1}) \qquad (8.118)$$

$$^{3}\tilde{u}_{2,1}(\Psi_{3,1}) = a_3\,\tilde{s}_1(\vartheta_{1,1})\tilde{s}_1(\vartheta_{1,1})\tilde{s}_1(\vartheta_{1,1}) \qquad (8.119)$$

$$^{3}\tilde{u}_{2,1}(\Psi_{3,2}) = a_3\,\tilde{s}_1(-\vartheta_{1,1})\tilde{s}_1(-\vartheta_{1,1})\tilde{s}_1(-\vartheta_{1,1}) \qquad (8.120)$$

$$^{3}\tilde{u}_{2,1}(\Psi_{3,3}) = 3\,a_3\,\tilde{s}_1(\vartheta_{1,1})\tilde{s}_1(\vartheta_{1,1})\tilde{s}_1(-\vartheta_{1,1}) \qquad (8.121)$$

$$^{3}\tilde{u}_{2,1}(\Psi_{3,4}) = 3\,a_3\,\tilde{s}_1(\vartheta_{1,1})\tilde{s}_1(-\vartheta_{1,1})\tilde{s}_1(-\vartheta_{1,1}) \qquad (8.122)$$

using the fact that

$$(F_{2,1})_{o_1}(\Xi_{1,1},\dots,\Xi_{1,o_1}) = \begin{cases} R_1 + R & \text{for} \quad o_1 = 1 \\ a_2 & \text{for} \quad o_1 = 2 \\ a_3 & \text{for} \quad o_1 = 3 \\ 0 & \text{otherwise} \end{cases} \qquad (8.123)$$

This means that the controlling variable $u_{2,1}(f)$ includes up to third-order contributions due to $s_1(f)$. Using the above controlled variable $u_{2,1}(f)$ and Equations (8.100) and (8.101) leads to

$$
\begin{aligned}
t_2(\xi_p) = \big[&2\,C_2 a_2\tilde{s}_1(\vartheta_{1,1})\tilde{s}_1(-\vartheta_{1,1}), \\
&C_2(R_1 + R)\,\tilde{s}_1(\vartheta_{1,1}) + 3\,C_2 a_3\tilde{s}_1(\vartheta_{1,1})\tilde{s}_1(\vartheta_{1,1})\tilde{s}_1(-\vartheta_{1,1}), \\
&C_2(R_1 + R)\,\tilde{s}_1(-\vartheta_{1,1}) + 3\,C_2 a_3\tilde{s}_1(\vartheta_{1,1})\tilde{s}_1(-\vartheta_{1,1})\tilde{s}_1(-\vartheta_{1,1}), \\
&C_2 a_2\tilde{s}_1(\vartheta_{1,1})\tilde{s}_1(\vartheta_{1,1}), \\
&C_2 a_2\tilde{s}_1(-\vartheta_{1,1})\tilde{s}_1(-\vartheta_{1,1}),\; C_2 a_3\tilde{s}_1(\vartheta_{1,1})\tilde{s}_1(\vartheta_{1,1})\tilde{s}_1(\vartheta_{1,1}), \\
&C_2 a_3\tilde{s}_1(-\vartheta_{1,1})\tilde{s}_1(-\vartheta_{1,1})\tilde{s}_1(-\vartheta_{1,1})\big]^{T}
\end{aligned} \qquad (8.124)
$$

Next, the conversion vectors $\tau_{1,1}(\xi_p)$ and $\tau_{1,2}(\xi_p)$ can be determined using Equations (7.98) and (7.96) as

$$\tau_{1,q}(\xi_p) = \big[\tau_{1,q}(\xi_p,\Psi_1),\dots,\tau_{1,q}(\xi_p,\Psi_7)\big]^{T} \qquad (8.125)$$

where

$$
\begin{aligned}
\tau_{1,q}(\xi_p,\Psi_1) = &(H_1)_{0,2-q,q-1}(\xi_p) \\
&+ 2\,(H_1)_{2,2-q,q-1}(\vartheta_{1,1},-\vartheta_{1,1};\xi_p)\,\tilde{s}_1(\vartheta_{1,1})\tilde{s}_1(-\vartheta_{1,1}) \qquad (8.126)
\end{aligned}
$$

$$\tau_{1,q}(\xi_p, \Psi_2) = (H_1)_{1,2-q,q-1}(\vartheta_{1,1}; \xi_p - \vartheta_{1,1})\,\tilde{s}_1(\vartheta_{1,1})$$
$$+ 3\,(H_1)_{3,2-q,q-1}(\vartheta_{1,1}, \vartheta_{1,1}, -\vartheta_{1,1}; \xi_p - \vartheta_{1,1})$$
$$\tilde{s}_1(\vartheta_{1,1})\,\tilde{s}_1(\vartheta_{1,1})\,\tilde{s}_1(-\vartheta_{1,1}) \tag{8.127}$$

$$\tau_{1,q}(\xi_p, \Psi_3) = (H_1)_{1,2-q,q-1}(-\vartheta_{1,1}; \xi_p + \vartheta_{1,1})\,\tilde{s}_1(-\vartheta_{1,1})$$
$$+ 3\,(H_1)_{3,2-q,q-1}(\vartheta_{1,1}, -\vartheta_{1,1}, -\vartheta_{1,1}; \xi_p + \vartheta_{1,1})$$
$$\tilde{s}_1(\vartheta_{1,1})\,\tilde{s}_1(-\vartheta_{1,1})\,\tilde{s}_1(-\vartheta_{1,1}) \tag{8.128}$$

$$\tau_{1,q}(\xi_p, \Psi_4) = (H_1)_{2,2-q,q-1}(\vartheta_{1,1}, \vartheta_{1,1}; \xi_p - 2\vartheta_{1,1})\,\tilde{s}_1(\vartheta_{1,1})\,\tilde{s}_1(\vartheta_{1,1}) \tag{8.129}$$

$$\tau_{1,q}(\xi_p, \Psi_5) = (H_1)_{2,2-q,q-1}(-\vartheta_{1,1}, -\vartheta_{1,1}; \xi_p + 2\vartheta_{1,1})\,\tilde{s}_1(-\vartheta_{1,1})\,\tilde{s}_1(-\vartheta_{1,1}) \tag{8.130}$$

$$\tau_{1,q}(\xi_p, \Psi_6) = (H_1)_{3,2-q,q-1}(\vartheta_{1,1}, \vartheta_{1,1}, \vartheta_{1,1}; \xi_p - 3\vartheta_{1,1})$$
$$\tilde{s}_1(\vartheta_{1,1})\,\tilde{s}_1(\vartheta_{1,1})\,\tilde{s}_1(\vartheta_{1,1}) \tag{8.131}$$

$$\tau_{1,q}(\xi_p, \Psi_7) = (H_1)_{3,2-q,q-1}(-\vartheta_{1,1}, -\vartheta_{1,1}, -\vartheta_{1,1}; \xi_p + 3\vartheta_{1,1})$$
$$\tilde{s}_1(-\vartheta_{1,1})\,\tilde{s}_1(-\vartheta_{1,1})\,\tilde{s}_1(-\vartheta_{1,1}) \tag{8.132}$$

for $q \in \{1, 2\}$. Thus the autocorrelation $\langle |\tilde{r}_{n,1}(\xi_p)|^2 \rangle$ where $p \in \mathcal{Z}$ can be determined using Equation (7.104). The multi-port frequency domain Volterra transfer functions in Equations (8.126)–(8.132) can be determined as

$$(H_1)_{0,1,0}(\Xi_{1,1}) = R_1 + R \tag{8.133}$$

$$(H_1)_{0,0,1}(\Xi_{2,1}) = R \tag{8.134}$$

$$(H_1)_{1,1,0}(\Omega_{1,1}; \Xi_{1,1}) = 2\,a_2 \tag{8.135}$$

$$(H_1)_{1,0,1}(\Omega_{1,1}; \Xi_{2,1}) = 2\,a_2 \tag{8.136}$$

$$(H_1)_{2,1,0}(\Omega_{1,1}, \Omega_{1,2}; \Xi_{1,1}) = 3\,a_3 \tag{8.137}$$

$$(H_1)_{2,0,1}(\Omega_{1,1}, \Omega_{1,2}; \Xi_{2,1}) = 3\,a_3 \tag{8.138}$$

$$(H_1)_{3,1,0}(\Omega_{1,1}, \Omega_{1,2}, \Omega_{1,3}; \Xi_{1,1}) = 0 \tag{8.139}$$

$$(H_1)_{3,0,1}(\Omega_{1,1}, \Omega_{1,2}, \Omega_{1,3}; \Xi_{2,1}) = 0 \tag{8.140}$$

Thus the vectors $\tau_{1,1}(\xi_p)$ and $\tau_{1,2}(\xi_p)$ are given by

$$
\tau_{1,1}(\xi_p) = \Big[R_1 + R + 6a_3\tilde{s}_1(\vartheta_{1,1})\tilde{s}_1(-\vartheta_{1,1}),\, 2a_2\tilde{s}_1(\vartheta_{1,1}),\, 2a_2\tilde{s}_1(-\vartheta_{1,1}),
$$
$$
3a_3\tilde{s}_1(\vartheta_{1,1})\tilde{s}_1(\vartheta_{1,1}),\, 3a_3\tilde{s}_1(-\vartheta_{1,1})\tilde{s}_1(-\vartheta_{1,1}),\, 0,\, 0 \Big]^T \tag{8.141}
$$

and

$$
\tau_{1,2}(\xi_p) = \Big[R + 6a_3\tilde{s}_1(\vartheta_{1,1})\tilde{s}_1(-\vartheta_{1,1}),\, 2a_2\tilde{s}_1(\vartheta_{1,1}),\, 2a_2\tilde{s}_1(-\vartheta_{1,1}),
$$
$$
3a_3\tilde{s}_1(\vartheta_{1,1})\tilde{s}_1(\vartheta_{1,1}),\, 3a_3\tilde{s}_1(-\vartheta_{1,1})\tilde{s}_1(-\vartheta_{1,1}),\, 0,\, 0 \Big]^T \tag{8.142}
$$

The autocorrelation $\langle |\tilde{r}_{n,1}(\xi_p)|^2 \rangle$ is given by

$$
\langle |\tilde{r}_{n,1}(\xi_p)|^2 \rangle = \sum_{q_1=1}^{2} \sum_{q_2=1}^{2} \sum_{e_1=1}^{7} \sum_{e_2=1}^{7} \tau_{1,q_1}^T(\xi_p)\, d_{e_1}\, t_{q_1}^T(\xi_p - \Psi_{e_1})
$$
$$
\times \widetilde{W}_{q_1,q_2}(\xi_p - \Psi_{e_1}, \xi_p - \Psi_{e_2})\, t_{q_2}^*(\xi_p - \Psi_{e_2})\, d_{e_2}^T\, \tau_{1,q_2}^*(\xi_p)
$$
$$
\tag{8.143}
$$

This leads to

$$
\langle |\tilde{r}_{n,1}(\xi_p)|^2 \rangle = \sum_{q_1=1}^{2} \sum_{q_2=1}^{2} \sum_{e_1=1}^{7} \sum_{e_2=1}^{7} \sum_{\gamma_1=1}^{7} \sum_{\gamma_2=1}^{7} \tau_{1,q_1}(\xi_p, \Psi_{e_1})\, \tau_{1,q_2}^*(\xi_p, \Psi_{e_2})
$$
$$
\times t_{q_1}(\xi_p - \Psi_{e_1}, \Psi_{\gamma_1})\, t_{q_2}^*(\xi_p - \Psi_{e_2}, \Psi_{\gamma_2})
$$
$$
\times \langle \tilde{w}_{q_1}(\xi_p - \Psi_{e_1} - \Psi_{\gamma_1})\, \tilde{w}_{q_2}^*(\xi_p - \Psi_{e_2} - \Psi_{\gamma_2}) \rangle \tag{8.144}
$$

where the autocorrelation of the fundamental noise sources is given by Equation (8.103).

NUMERICAL EXPERIMENT: To investigate the correctness of the theoretical results presented in this example, numerical simulations are performed in accordance with Section 7.3. In the simulation it is first necessary to determine the controlling variable $u_{2,1}(t)$. This can be determined as

$$
u_{2,1}(t) = (R_1 + R)\, s_1(t) + a_2\, s_1^2(t) + a_3\, s_1^3(t) \tag{8.145}
$$

where the contribution to the controlling variable due to the noise sources has been ignored (low level noise is assumed). Once the controlling variable $u_{2,1}(t)$ has been determined the response $r_1(t)$ can be determined as

$$
\begin{aligned}
r_1(t) = \; & [s_1(t) + n_1(t)]\,R_1 \; + \; [s_1(t) + n_1(t) + n_2(t)]\,R \\
& + a_2\,[s_1(t) + n_1(t) + n_2(t)]^2 \; + \; a_3\,[s_1(t) + n_1(t) + n_2(t)]^3
\end{aligned}
\tag{8.146}
$$

As a special case it is quite interesting to note that the autocorrelation in the situation where the system is *linear* — but including the (non-linear) modulation of noise source 2 — is given by

$$
\begin{aligned}
\langle |\tilde{r}^{lin}_{n,1}(\xi_p)|^2 \rangle = \; & (R_1 + R)^2 C_1^2 \, \langle |\tilde{w}_1(\xi_p)|^2 \rangle \\
& + \frac{1}{4} R^2 (R_1 + R)^2 \, C_2^2 \varrho_{1,1}^2 \Big[\langle |\tilde{w}_2(\xi_p - \vartheta_{1,1})|^2 \rangle \\
& + \langle |\tilde{w}_2(\xi_p + \vartheta_{1,1})|^2 \rangle \Big]
\end{aligned}
\tag{8.147}
$$

which can be derived by setting $a_2 = a_3 = 0$.

Simulation 1 The autocorrelation $\langle |\tilde{r}_{n,1}(\xi_p)|^2 \rangle_\Gamma$ with $p = 3$ is simulated versus the number of iterations Γ. The data and result of the simulation are shown in Figure 8.12. This simulation is for a linear system, i.e. $a_2 = a_3 = 0$, but where noise source 2 is being modulated by the applied deterministic signal. The simulated result $\langle |\tilde{r}_{n,1}(\xi_3)|^2 \rangle^{40000}_{39901} = 1.7523 \times 10^{-15}$ agrees with the theoretical result $\langle |\tilde{r}_{n,1}(\xi_3)|^2 \rangle = 1.7596 \times 10^{-15}$ with a deviation of -0.42 %.

Simulation 2 The autocorrelation $\langle |\tilde{r}_{n,1}(\xi_p)|^2 \rangle^{40000}_{39901}$ with $p = 3$ is simulated versus the amplitude $\varrho_{1,1}$ of the deterministic input sinusoidal signal. The data and result of the simulation are shown in Figure 8.13. All simulated values $\langle |\tilde{r}_{n,1}(\xi_3)|^2 \rangle^{40000}_{39901}$ agree with the theoretical result $\langle |\tilde{r}_{n,1}(\xi_3)|^2 \rangle$ in Equation (8.144) with a deviation less than ± 0.39 %. As seen from Figure 8.12 the noise level $\langle |\tilde{n}_1(\xi_7)|^2 \rangle$ increases with the amplitudes $\varrho_{1,1} = \varrho_{1,2}$ of the deterministic excitation signals. For very small $\varrho_{1,1}$ only noise source 1 is of importance since it, as opposed to noise source 2, is independent of a modulating signal.

Simulation 3 The autocorrelation $\langle |\tilde{r}_{n,1}(\xi_p)|^2 \rangle^{40000}_{39901}$ with $p = 3$ is simulated versus the standard deviations $\sigma_1 = \sigma_2$. The data and result of the simulation are shown in Figure 8.14. All simulated values $\langle |\tilde{r}_{n,1}(\xi_3)|^2 \rangle^{40000}_{39901}$ in the range of $\sigma_1 = \sigma_2$ from 10^{-9} to 10^{-2} agree with the theoretical result $\langle |\tilde{r}_{n,1}(\xi_3)|^2 \rangle$ in Equation (8.144) with deviations less than ± 0.14 %. The simulation for $\sigma_1 = \sigma_2 = 10^{-1}$ deviates from the theoretical result by 6.0 %. This rather large deviation is due to the fact that the mixing of noise with noise for increasing standard deviations is not considered in the theoretical prediction but is included in the numerical simulation.

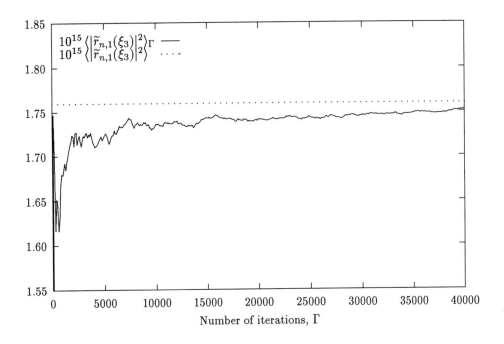

Figure 8.12: Example 3: A numerical experiment to illustrate the properties of noise response $\tilde{r}_{n,1}(\xi_p)$. The full line gives the simulated values $\langle|\tilde{r}_{n,1}(\xi_p)|^2\rangle_\Gamma$ versus number of iterations, and the dotted line is the theoretical result $\langle|\tilde{r}_{n,1}(\xi_p)|^2\rangle$ predicted from Equation (8.144). The simulated $\langle|\tilde{n}_1(\xi_p)|^2\rangle_{39901}^{40000} = 1.7523 \times 10^{-15}$ with a standard deviation of 2.7810×10^{-19}. The theoretical result is $\langle|\tilde{n}_1(\xi_{17})|^2\rangle = 1.7596 \times 10^{-15}$. This corresponds to a deviation of the simulated result from the theoretical result of -0.42 %. The data for the experiment are: $\Lambda = 64$, $\sigma_1 = 5.0 \times 10^{-9}$, $\sigma_2 = 7.0 \times 10^{-8}$, $C_1 = 0.3$, $C_2 = 0.2$, $R_1 = 13.0$, $R = 10.3$, $a_2 = 0$, $a_3 = 0$, $p = 3$, $\tau = 0.5$, $\varrho_{1,1} = 0.2$, $\vartheta_{1,1} = 4.0$, and $\varphi_{1,1} = 0.0$.

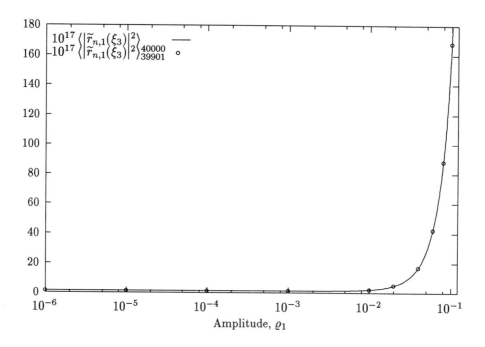

Figure 8.13: Example 3: A numerical experiment to illustrate the properties of noise response $\tilde{r}_{n,1}(\xi_p)$. The full line gives the theoretical values for $\langle|\tilde{r}_{n,1}(\xi_p)|^2\rangle$ versus the amplitude of the input signal, and the circles give the simulated values $\langle|\tilde{r}_{n,1}(\xi_p)|^2\rangle_{39901}^{40000}$ predicted from Equation (8.144). The simulated values agree with the theoretical predictions with a deviation less than ±0.39 % for the amplitudes seen from the figure. The data for the experiment are: $\Lambda = 64$, $\sigma_1 = 5.4 \times 10^{-9}$, $\sigma_2 = 7.3 \times 10^{-8}$, $C_1 = 0.33$, $C_2 = 0.27$, $R_1 = 13.1$, $R = 10.3$, $a_2 = 27.4$, $a_3 = 76.9$, $p = 3$, $\tau = 0.5$, $\vartheta_{1,1} = 4.0$, and $\varphi_{1,1} = 0.0$.

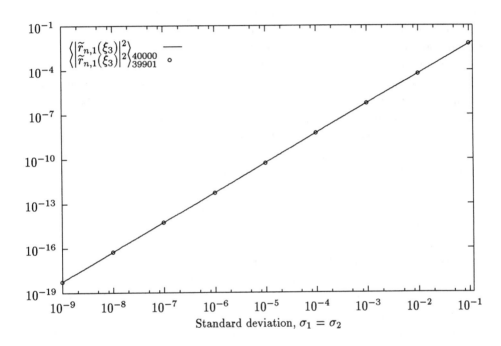

Figure 8.14: Example 3: A numerical experiment to illustrate the properties of noise response $\tilde{r}_{n,1}(\xi_p)$. The full line gives the theoretical values for $\langle|\tilde{r}_{n,1}(\xi_p)|^2\rangle$ versus the standard deviations of the noise sources $\sigma_1 = \sigma_2$, and the circles give the simulated values $\langle|\tilde{r}_{n,1}(\xi_p)|^2\rangle_{39901}^{40000}$ predicted from Equation (8.144). The simulated values agree with the theoretical predictions with deviations less than ±0.17 % for standard deviations up to 10^{-2}. For $\sigma_1 = \sigma_2 = 10^{-1}$ the deviation between simulation and theory is 6.0 %. The data for the experiment are: $\Lambda = 64$, $C_1 = 0.33$, $C_2 = 0.27$, $R_1 = 13.1$, $R = 10.3$, $a_2 = 27.4$, $a_3 = 76.9$, $p = 3$, $\tau = 0.5$, $\vartheta_{1,1} = 4.0$, and $\varphi_{1,1} = 0.0$.

8.4 Conclusion

A method has been presented to analyze noise in non-autonomous non-linear multiport networks and systems where low level noise can be assumed. This means that the systems must be small signal linear which is true for many types of systems.

The system description and mathematical representation of noise have been discussed. The non-linear noisy system under consideration must be transferred to an equivalent non-linear noise free system with external deterministic and noise generators (sources). This is required to apply the Volterra series technique. The deterministic and noise signals are represented mathematically as Fourier series which are computationally very efficient when Volterra series are used. The mathematical representation of noise sources has been treated in detail. A noise source is given as the response from a non-linear noise free multi-port system with a fundamental (unmodulated) noise source and possibly modulating signals as inputs. In this way it is possible to represent unmodulated as well as modulated (dependent) noise sources. The cross-correlation between Fourier series coefficients of any two (possibly) modulated noise source signals can be described by a cross-correlation matrix for the two fundamental noise sources at various frequencies, and vectors which describe the transfers from fundamental noise sources to modulated noise sources. This way to describe (possibly) modulated noise sources is very flexible, and it is easy to describe even very complicated multi-signal modulations.

Expressions for the noise and deterministic signal response at arbitrary response ports have been determined. The cross-correlation between two Fourier series coefficients at arbitrary response ports has been determined from noise cross-correlation matrices for the fundamental noise sources, and vectors describing the transfer from fundamental to modulated noise sources and from modulated noise sources to the noise responses at the response ports. Also, it has been shown how these rather complicated transfers must be specified for the Volterra series technique.

Expressions have been derived for various average noise powers and noise power densities. The quantities are usually of high interest in the analysis of noise. The average noise powers and noise power densities have been determined from noise cross-correlation matrices for the fundamental noise sources, and from the non-linear transfer of the fundamental noise sources to the given response port.

Three examples have been shown to illustrate the use of the presented method. One example shows the representation and the properties of a type of modulated noise source. Two examples show how to determine noise cross-correlations at a response port for two type of circuits. All examples have been chosen to make it possible to find an alternative (analytical) solution to the given examples. The results for the presented method and from the alternative (analytical) solutions are in agreement, which indicates the correctness of the presented method. Generally it is not possible to find alternative analytical solutions, and all three examples have been carefully chosen to make comparisons possible.

9

Multi-port Volterra transfer functions

This final chapter deals with the determination of frequency domain Volterra transfer functions of non-linear multi-port networks containing non-linear multi-port elements (subsystems). A pure frequency domain method is derived which allows
commensurate as well as incommensurate frequencies. The method is based on an
extension of the probing method to allow multi-port networks and commensurate
frequencies. A computer implementation of the method in an algebraic programming language is made which allows determination of Volterra transfer functions in
algebraic form up to eighth order on a low end workstation. Examples are presented and the results are compared with existing literature in the special cases where
comparison is possible.

9.1   Introduction

In this chapter a method and algorithm are developed to determine multi-port
Volterra transfer functions of non-linear multi-port networks which may contain non-
linear multi-port elements (subsystems) — that is, non-linear elements (subsystems)
which are controlled by arbitrary variables. The purpose of this is twofold: (i)
multi-port Volterra transfer functions are a fundamental requirement for the low-
and high-level noise analysis, and (ii) it has been pointed out in the literature that
the use of Volterra series is limited to one-port non-linear elements which precludes,
for example, the analysis of MESFET transistors which ought to contain (at least)
two-dimensional non-linear elements [1,2]. The work of the present chapter is also
of importance in the analysis of the convergence properties of Volterra series since
relatively high order (> 4) Volterra transfer functions may be determined. In the
literature, much material is available on the one-port Volterra series representation,
but little on the general multi-port Volterra series representation.

Previously, Bussgang, Ehrman and Graham [3], Maas [4,5] and Chua and Ng

[6] have   investigated the determination of one-port Volterra transfer functions
using the method of non-linear currents by a combination of time and frequency
domain analysis. These analysis methods have been limited to one-port networks
containing one-port non-linear elements and generators controlled by one variable.
The method derived in the present work is based on an extension of the probing
method to allow arbitrary (also commensurate) frequencies.[1] A recursively based
algorithm for the determination of Volterra transfer functions of non-linear multi-
port networks is derived. A computer implementation of the method in a symbolic
programming language is presented. This makes it possible to determine algebraic
expressions of the multi-port Volterra transfer functions. Finally, three examples
are considered.

The chapter is organized as follows. Section 9.2 gives some preliminaries regarding multi-port Volterra series where the symmetry properties of the transfer
functions are analyzed, and a modification of the probing method to allow commensurate frequencies is presented. In section 9.3 the theoretical formulation for
the determination of multi-port Volterra transfer functions is presented. Section 9.4
discusses the implementation of the method in a symbolic programming language
on a digital computer. Section 9.5 presents some types of time domain multi-port
Volterra kernels and the corresponding frequency domain transfer functions. In section 9.6 three examples are presented to illustrate the method. Finally, section 9.7
presents some concluding remarks.

9.2   Preliminaries

The frequency domain response $v(f)$ at a frequency f from a non-linear multi-port
Volterra system (a system which can be described by a convergent Volterra series)
with K input ports and input signals $\{s_1(f), \ldots, s_K(f)\}$ is given by

---

[1]In Chua and Ng [7] a set of frequencies $\{\Omega_1, \ldots, \Omega_m\}$ is called incommensurate (and a frequency base), if there does not exist a set of rational numbers $\{r_1, \ldots, r_m\}$ (not all zero) such that
$r_1\Omega_1 + \cdots + r_m\Omega_m = 0$. This is indeed a sufficient but actually not a necessary condition for the
conventional probing method to be valid. It can be shown that a sufficient and necessary condition
for the conventional probing method to be valid is that there does not exist a set of integer numbers
$q_1, \ldots, q_m$ where $q_1 + 1, \ldots, q_m + 1 \in \mathcal{Z}_{0+}$ (not all zero) such that $q_1\Omega_1 + \cdots + q_m\Omega_m = 0$. For
example, the Chua and Ng condition says that the probing method can not be used for the set
of frequencies $\{\Omega_1 = 4, \Omega_2 = 7\}$ since $r_1\Omega_1 + r_2\Omega_2 = 0$ for $\{r_1 = -1/4, r_2 = 1/7\}$. According to
the sufficient and necessary condition, the probing method can be used for the set of frequencies
$\{\Omega_1 = 4, \Omega_2 = 7\}$ since $q_1\Omega_1 + q_2\Omega_2 \neq 0$ for all integer numbers q_1, q_2 where $q_1 + 1, q_2 + 1 \in \mathcal{Z}_{0+}$
except for $q_1 = q_2 = 0$. This is also confirmed from Equations (9.9) and (9.10) in section 9.2.

$$v(f) = \sum_{m_1=0}^{\infty} \cdots \sum_{m_K=0}^{\infty} \int_{-\infty}^{\infty} \cdots \int_{-\infty}^{\infty} \cdots \cdots \int_{-\infty}^{\infty} \cdots \int_{-\infty}^{\infty} \mathcal{L}_{1,\infty}(\|m\|)$$
$$\mathcal{H}_{m_1,\dots,m_K}(f_{1,1},\dots,f_{1,m_1};\cdots\cdots;f_{K,1},\dots,f_{K,m_K})$$
$$s_1(f_{1,1})\cdots s_1(f_{1,m_1})\cdots\cdots s_K(f_{K,1})\cdots s_K(f_{K,m_K})$$
$$\delta\Big(f - f_{1,1} - \cdots - f_{1,m_1} - \cdots\cdots - f_{K,1} - \cdots - f_{K,m_K}\Big)$$
$$df_{1,1}\cdots df_{1,m_1}\cdots\cdots df_{K,1}\cdots df_{K,m_K} \tag{9.1}$$

where

$$m = [m_1,\dots,m_K]^T \in \mathcal{Z}_{0+}^{K\times 1} \tag{9.2}$$

$$\mathcal{L}_{\alpha,\beta}(\gamma) = \begin{cases} 1 & \text{for} \quad \gamma \in \{\alpha, \alpha+1, \alpha+2,\dots,\beta-1,\beta\} \\ 0 & \text{otherwise} \end{cases} \tag{9.3}$$

and $\|\cdot\|$ denotes the sum of all elements in the given vector, i.e.

$$\|m\| = m_1 + \cdots + m_K \tag{9.4}$$

In Equation (9.1) the quantity $\mathcal{H}_{m_1,\dots,m_K}(\cdot)$ is a (possibly) unsymmetrical multi-port frequency domain Volterra transfer function of order $\|m\|$ and $\delta(\cdot)$ is the Dirac δ-function [8]. In Equation (9.1) the zeroth order contribution corresponding to $\|m\| = 0$ has been excluded due to the factor $\mathcal{L}_{1,\infty}(\|m\|)$. This contribution, which is usually not of interest, is identical to the response $v(f)$ when no input signals are applied (the contribution may be called a mathematical offset). As pointed out in the previous section, this is no restriction in the noise analysis since the internal sources in the non-linear systems are applied at external ports.

9.2.1 Symmetry properties

For one-port Volterra systems $(K = 1)$ it is usually assumed, without loss of generality, that the one-port Volterra transfer function $\mathcal{H}_{m_1}(f_{1,1},\dots,f_{1,m_1})$ is symmetrical. If the transfer function is not symmetrical it can be replaced by $\frac{1}{m_1!}$ times the sum of all permutations of the unsymmetrical transfer function. In this way a symmetrical one-port transfer function can be obtained. This transfer function can directly be substituted with the given (and possibly unsymmetrical) transfer function in Equation (9.1).

For multi-port Volterra systems this approach for obtaining a symmetrical multi-port Volterra transfer function is not directly possible. As seen from Equation (9.1) the response $v(f)$ remains the same for any arbitrary permutation of the transfer function $\mathcal{H}_{m_1,\dots,m_K}(f_{1,1},\dots,f_{1,m_1};\cdots\cdots;f_{K,1},\dots,f_{K,m_K})$ with respect to variables from the same set $\{f_{k,1}, \ldots, f_{k,m_k}\}$ where $k \in \{1, 2, \ldots, K\}$. This means, for example, that two signals may generally be interchanged to yield the same response $v(f)$ unless they enter the system through two different input ports. Thus a partly symmetrical Volterra transfer function, indicated by $\text{Sym}\{\mathcal{H}_{m_1,\ldots,m_K}(\cdot)\}$, can be obtained as

$$
\begin{aligned}
&\text{Sym}\Big\{\mathcal{H}_{m_1,\ldots,m_K}(f_{1,1}, \ldots, f_{1,m_1}; \cdots\cdots; f_{K,1}, \ldots, f_{K,m_K})\Big\} \\
&= H_{m_1,\ldots,m_K}(f_{1,1}, \ldots, f_{1,m_1}; \cdots\cdots; f_{K,1}, \ldots, f_{K,m_K}) \\
&= \frac{1}{m_1! \cdots m_K!} \sum_{l_1=1}^{m_1!} \cdots \sum_{l_K=1}^{m_K!} \\
&\quad \mathcal{H}_{m_1,\ldots,m_K}\Big(\mathcal{P}_{1,l_1}\{f_{1,1}, \ldots, f_{1,m_1}\}; \cdots\cdots; \mathcal{P}_{K,l_K}\{f_{K,1}, \ldots, f_{K,m_K}\}\Big) \quad (9.5)
\end{aligned}
$$

where $\mathcal{P}_{k,l_k}\{f_{k,1}, \ldots, f_{k,m_k}\}$ indicates permutation number $l_k \in \{1, 2, \ldots, m_k!\}$ of $\{f_{k,1}, \ldots, f_{k,m_k}\}$ for $k \in \{1, 2, \ldots, K\}$. All permutations $\mathcal{P}_{k,1}\{\cdot\}, \ldots, \mathcal{P}_{k,m_k!}\{\cdot\}$ are required to be different for all $k \in \{1, 2, \ldots, K\}$ to assure (partial) symmetry of $H_{m_1,\ldots,m_K}(\cdot)$. The symmetrical transfer function $H_{m_1,\ldots,m_K}(\cdot)$ can directly be substituted for the corresponding (possibly) unsymmetrical transfer function $\mathcal{H}_{m_1,\ldots,m_K}(\cdot)$ in Equation (9.1).

It should be noted that the (possibly) unsymmetrical Volterra transfer function $\mathcal{H}_{m_1,\ldots,m_K}(\cdot)$ is not unique in the sense that there may be several distinct transfer functions which give the same response $v(f)$ in Equation (9.1) for the same input. However, the (partly) symmetrical multi-port Volterra transfer function $H_{m_1,\ldots,m_K}(\cdot)$ is unique since

$$
\begin{aligned}
&H_{m_1,\ldots,m_K}(f_{1,1}, \ldots, f_{1,m_1}; \cdots\cdots; f_{K,1}, \ldots, f_{K,m_K}) \\
&= H_{m_1,\ldots,m_K}\Big(\mathcal{P}_{1,l_1}\{f_{1,1}, \ldots, f_{1,m_1}\}; \cdots\cdots; \mathcal{P}_{K,l_K}\{f_{K,1}, \ldots, f_{K,m_K}\}\Big) \quad (9.6)
\end{aligned}
$$

for all permutations $\mathcal{P}_{k,l_k}\{f_{k,1}, \ldots, f_{k,m_k}\}$ where $k \in \{1, 2, \ldots, K\}$, and $l_k \in \{1, 2, \ldots, m_k!\}$. The use of partly symmetrical multi-port Volterra transfer functions, and not (possibly) unsymmetrical Volterra transfer functions, implies that the amount of computations needed to determine the response $v(f)$ to sinusoidal or complex exponential inputs may be substantially reduced by exploiting the symmetry properties to avoid recalculating contributions which are just permutations of variables from the same set.

9.2.2 Determination of transfer functions

The time domain signal applied at input port $k \in \{1, 2, \ldots, K\}$ is given by a sum of I_k complex exponentials:

$$s_k(t) \;=\; \sum_{i_k=1}^{I_k} \exp\!\left[j\left(2\pi\psi_{k,i_k}t + \phi_{k,i_k}\right)\right] \tag{9.7}$$

Thus $s_k(f) = \mathcal{F}\{s_k(t)\} = \int_{-\infty}^{\infty} s_k(t)\exp[-j2\pi ft]\,dt$ where $\mathcal{F}\{\cdot\}$ denotes the (integral) Fourier transform is given by

$$s_k(f) \;=\; \sum_{i_k=1}^{I_k} \exp[j\,\phi_{k,i_k}]\,\delta(f - \psi_{k,i_k}) \tag{9.8}$$

Insertion of Equation (9.8) into (9.1) leads to

$$
\begin{aligned}
v(f) \;=\; & \sum_{m_1=0}^{\infty}\cdots\sum_{m_K=0}^{\infty}\sum_{i_{1,1}=1}^{I_1}\cdots\sum_{i_{1,m_1}=1}^{I_1}\cdots\cdots\sum_{i_{K,1}=1}^{I_K}\cdots\sum_{i_{K,m_K}=1}^{I_K} \\[4pt]
& \mathcal{L}_{1,\infty}(\|\boldsymbol{m}\|) \\[4pt]
& \mathcal{H}_{m_1,\ldots,m_K}(\psi_{1,i_{1,1}},\ldots,\psi_{1,i_{1,m_1}};\cdots\cdots;\psi_{K,i_{K,1}},\ldots,\psi_{K,i_{K,m_K}}) \\[4pt]
& \exp\!\left[j\sum_{k=1}^{K}\sum_{l=1}^{m_k}\phi_{k,i_{k,l}}\right]\delta\!\left(f - \sum_{k=1}^{K}\sum_{l=1}^{m_k}\psi_{k,i_{k,l}}\right)
\end{aligned}
\tag{9.9}
$$

It should be noted that in Equation (9.9) it is generally not possible to reduce the amount of computations required by permutation of the variables, e.g. $i_{1,1},\ldots,i_{1,m_1}$, since the multi-port Volterra transfer function $\mathcal{H}_{m_1,\ldots,m_K}(\cdot)$ is generally not partly symmetrical in the sense described earlier. Choosing $I_k = m_k$ for all $k \in \{1,2,\ldots,K\}$ a partly symmetrical multi-port Volterra transfer function $H_{m_1,\ldots,m_K}(\cdot)$ can be determined as

$$
\begin{aligned}
& H_{m_1,\ldots,m_K}(\psi_{1,1},\ldots,\psi_{1,m_1};\cdots\cdots;\psi_{K,1},\ldots,\psi_{K,m_K}) \\[4pt]
=\; & \frac{1}{m_1!\cdots m_K!} \\[4pt]
& \times\left\{\text{coefficient of }\;\exp\!\left[j\sum_{k=1}^{K}\sum_{l=1}^{m_k}\phi_{k,l}\right]\delta\!\left(f - \sum_{k=1}^{K}\sum_{l=1}^{m_k}\psi_{k,l}\right)\;\text{ in }\;v(f)\right\}
\end{aligned}
\tag{9.10}
$$

for $\|\boldsymbol{m}\| \in \{1,2,\ldots,\infty\}$ and provided that the set of phases $\{\phi_{1,1},\ldots,\phi_{1,m_1},\ldots$ $\ldots,\phi_{K,1},\ldots,\phi_{K,m_K}\}$ is chosen as a phase base up to order $\|\boldsymbol{m}\|$ as defined below.

Definition 9.1 A set of phases $\{\phi_{1,1},\ldots,\phi_{1,m_1},\ldots\ldots,\phi_{K,1},\ldots,\phi_{K,m_K}\}$ **is defined as a phase base up to order** $\mu \in \mathcal{Z}_+$ **if for positive (zero included) integers**

$$\{q_{1,1},\ldots,q_{1,m_1},\ldots\ldots,q_{K,1},\ldots,q_{K,m_K}\}$$

$$\sum_{k=1}^{K}\sum_{l=1}^{m_k} q_{k,l}\,\phi_{k,l} \;=\; \sum_{k=1}^{K}\sum_{l=1}^{m_k}\phi_{k,l} \tag{9.11}$$

only for $q_{k,l} = 1$ **for all** $k \in \{1, 2, \ldots, K\}$ **and** $l \in \{1, 2, \ldots, m_k\}$ **with the restriction that** $\sum_{k=1}^{K} \sum_{l=1}^{m_k} q_{k,l} \in \{0, 1, 2, \ldots, \mu\}$.

When Equation (9.10) is used to determine the multi-port frequency domain Volterra transfer function it will never be necessary to actually choose specific values of the phase base. Only the properties which are associated with $\{\phi_{1,1}, \ldots, \phi_{1,m_1}, \ldots$ $\ldots, \phi_{K,1}, \ldots, \phi_{K,m_K}\}$ being a phase base will be used in the following. Note that if $\{\phi_{1,1}, \ldots, \phi_{1,m_1}, \ldots\ldots, \phi_{K,1}, \ldots, \phi_{K,m_K}\}$ is a phase base then $\phi_{k,l} \neq 0$ for all $k \in \{1, 2, \ldots, K\}$ and $l \in \{1, 2, \ldots, m_k\}$.

Note that $I_k = m_k$ for all $k \in \{1, 2, \ldots, K\}$ is used in the derivation of Equation (9.10). If $I_k < m_k$ for one or more $k \in \{1, 2, \ldots, K\}$ then at least one frequency in $\{\psi_{k,1}, \ldots, \psi_{k,m_k}\}$ appears twice in the argument to $\mathcal{H}_{m_1, \ldots, m_K}(\cdot)$. In this case Equation (9.10) does not follow since frequencies in the set $\{\psi_{k,1}, \ldots, \psi_{k,m_k}\}$ for the symmetrical $\mathcal{H}_{m_1, \ldots, m_K}(\cdot)$ are not different. This is necessary to obtain a generally correct $\mathcal{H}_{m_1, \ldots, m_K}(\cdot)$ independent of the argument frequencies. If $I_k > m_k$ for $k \in \{1, 2, \ldots, K\}$ then Equation (9.10) can be derived. However, if $I_k > m_k$ for $k \in \{1, 2, \ldots, K\}$ then the determination of $v(f)$ in Equation (9.9) requires more calculations than are actually necessary. Thus, choosing $I_k = m_k$ for all $k \in \{1, 2, \ldots, K\}$ is the proper choice for the analysis which leads to Equation (9.10).

The conventional probing method to determine one-port Volterra transfer functions uses only a sum of δ-functions and no phases in the probing signal [3,6]. This situation is a special case of Equation (9.8) when $\phi_{k,i_k} = 0$ for all $k \in \{1, 2, \ldots, K\}$ and $i_k \in \{1, 2, \ldots, I_k\}$, and precludes the situation of commensurate frequencies as noted by Chua and Ng [6].

9.3   Theory

In Figure 9.1 the overall non-linear network under consideration is shown. The non-linear network is separated into a purely linear network and Q purely non-linear subsystems.[2] Non-linear subsystem number $q \in \{1, 2, \ldots, Q\}$ is controlled by a number of $P(q)$ variables. There are R controlling variables for the overall non-linear network represented by $\{x_1(f), \ldots, x_R(f)\}$. Thus $P(q) \in \{1, 2, \ldots, R\}$ for all $q \in \{1, 2, \ldots, Q\}$. There are Q controlled variables, identical to the number of non-linear subsystems, for the overall non-linear network represented by $\{y_1(f), \ldots, y_Q(f)\}$. Thus there are not necessarily equally many controlling as controlled variables. For example, one controlling variable may control several controlled variables, and one controlled variable may be controlled by several control-

---

[2]In case a non-linear subsystem contains a first order (linear) contribution this *must* be included in the linear network and not in the non-linear subsystem.

ling variables. At the signal input ports independent voltage or current generators $\{s_1(f),\ldots,s_K(f)\}$ are applied. The complementary currents or voltages of the (voltage or current respectively) signal generators $\{s_1(f),\ldots,s_K(f)\}$ are not of interest in this analysis. At the designated output port the response $v(f)$ is either the open circuited voltage or the short circuited current. Thus the complementary current or voltage of the (voltage or current respectively) response $v(f)$ is zero.

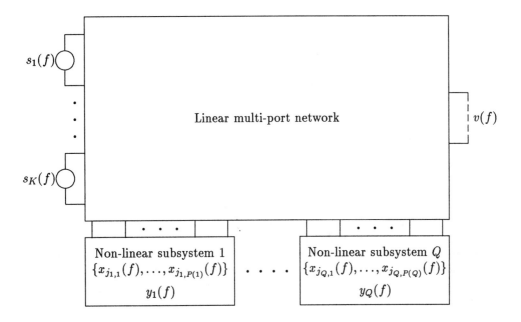

Figure 9.1: Non-linear network separated into a purely linear network and purely non-linear subsystems.

Each non-linear subsystem in the overall non-linear network is described by the general non-linear multi-port type shown in Figure 9.2. The variables $\{x_{j_{q,1}}(f),\ldots,x_{j_{q,P(q)}}(f)\}$ where $j_{q,r} \in \{1,2,\ldots,R\}$ for $q \in \{1,2,\ldots,Q\}$ and $r \in \{1,2,\ldots,P(q)\}$ are the controlling voltage or current variables and $y_q(f)$ is the corresponding controlled voltage or current output variable for non-linear subsystem $q \in \{1,2,\ldots,Q\}$. The subscripts for the controlling variables are collected in the vectors $\{\boldsymbol{j}_1,\ldots,\boldsymbol{j}_Q\}$ where $\boldsymbol{j}_q = [j_{q,1},\ldots,j_{q,P(q)}]^T \in \mathcal{Z}_+^{P(q)\times 1}$ for $q \in \{1,2,\ldots,Q\}$. Each controlling input port is either short or open circuited and hence the port voltage or current respectively is zero. The complementary current or voltage of the controlled (voltage or current respectively) output variable $y_q(f)$ where $q \in \{1,2,\ldots,Q\}$ is not of interest in the analysis. The general type of non-linear multi-port subsystem shown in Figure 9.2 can be used for one-port non-linear elements as well as multi-port non-linear subsystems. One-port non-linear elements are represented from Figure 9.2 by appropriate use of feedback between the output and the input port.

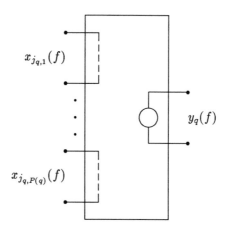

Figure 9.2: The used type of general non-linear multi-port subsystem. At the controlling signal input ports, $\{x_{j_{q,1}}(f), \ldots, x_{j_{q,P(q)}}(f)\}$ are the controlling variables where $j_{q,r} \in \{1, 2, \ldots, R\}$, $r \in \{1, 2, \ldots, P(q)\}$ and $q \in \{1, 2, \ldots, Q\}$. At the output port, $y_q(f)$ is the controlled output variable for non-linear subsystem $q \in \{1, 2, \ldots, Q\}$. $P(q)$ is the number of controlling variables for non-linear subsystem $q \in \{1, 2, \ldots, Q\}$. This multi-port non-linear subsystem can also be used to represent non-linear one-port elements, e.g. non-linear resistors and capacitors, by appropriate feedback between the output and input ports.

The problem to be solved in the following is to determine the multi-port frequency domain Volterra transfer function relating the input signals $\{s_1(f), \ldots, s_K(f)\}$ to the output response $v(f)$.

9.3.1   Relations for the linear system

By appropriate arrangement of the network variables and using the above properties for the non-linear subsystems, the linear system can, after some considerations, be described by the matrix equation

$$\left[\begin{array}{c} y(f) \\ \cdots\cdots \\ v(f) \end{array} \right] = \left[\begin{array}{c} A(f) \\ \cdots\cdots \\ a^T(f) \end{array} \right] x(f) + \left[\begin{array}{c} B(f) \\ \cdots\cdots \\ b^T(f) \end{array} \right] s(f) \qquad (9.12)$$

where

$$
\begin{array}{rcll}
\boldsymbol{y}(f) & = & [y_1(f),\ldots,y_Q(f)]^T & \in \quad \mathcal{C}^{Q\times 1} \\
\boldsymbol{x}(f) & = & [x_1(f),\ldots,x_R(f)]^T & \in \quad \mathcal{C}^{R\times 1} \\
\boldsymbol{s}(f) & = & [s_1(f),\ldots,s_K(f)]^T & \in \quad \mathcal{C}^{K\times 1}
\end{array}
$$

(9.13)
(9.14)
(9.15)

and

$$
\boldsymbol{A}(f) = \begin{bmatrix} A_{1,1}(f) & \cdots & A_{1,R}(f) \\ \vdots & & \vdots \\ A_{Q,1}(f) & \cdots & A_{Q,R}(f) \end{bmatrix} \in \mathcal{C}^{Q\times R} \tag{9.16}
$$

$$
\boldsymbol{B}(f) = \begin{bmatrix} B_{1,1}(f) & \cdots & B_{1,K}(f) \\ \vdots & & \vdots \\ B_{Q,1}(f) & \cdots & B_{Q,K}(f) \end{bmatrix} \in \mathcal{C}^{Q\times K} \tag{9.17}
$$

$$
\boldsymbol{a}(f) = [a_1(f),\ldots,a_R(f)]^T \in \mathcal{C}^{R\times 1} \tag{9.18}
$$

$$
\boldsymbol{b}(f) = [b_1(f),\ldots,b_K(f)]^T \in \mathcal{C}^{K\times 1} \tag{9.19}
$$

As seen from Equations (9.16)–(9.19) the sizes of the system matrices and vectors depend on the number of controlling variables, the number of controlled variables and the number of signal input ports. Thus the sizes of $\boldsymbol{A}(f)$, $\boldsymbol{B}(f)$, $\boldsymbol{a}(f)$ and $\boldsymbol{b}(f)$ do not depend on the number of linear elements in the overall network. However, it is obvious that the linear system matrices and vectors get more complicated the more linear elements there are in the network. That is, the number of multiplications and additions required to determine $\boldsymbol{A}(f)$, $\boldsymbol{B}(f)$, $\boldsymbol{a}(f)$ and $\boldsymbol{b}(f)$ increases with the number of linear network elements. The system matrices and vectors of the linear network $\boldsymbol{A}(f)$, $\boldsymbol{B}(f)$, $\boldsymbol{a}(f)$ and $\boldsymbol{b}(f)$ may be determined using standard techniques [9]. However, when the system vectors and matrices are determined, $\{y_1(f),\ldots,y_Q(f)\}$ should be interpreted as the port voltages or currents and not as generators since $\{y_1(f),\ldots,y_Q(f)\}$ are dependent variables — the variables $\{s_1(f),\ldots,s_K(f)\}$ can still be viewed as generators since these are independent variables.

9.3.2 Non-linear response

An expression for the non-linear response $v(f)$ is determined with the signals at the K signal input ports as given by Equation (9.8) and $I_k = m_k$ for all $k \in \{1,2,\ldots,K\}$. Thus the output response $v(f)$ is given by

$$v(f) = \sum_{m_1=0}^{\infty} \cdots \sum_{m_K=0}^{\infty} \sum_{i_{1,1}=1}^{m_1} \cdots \sum_{i_{1,m_1}=1}^{m_1} \cdots\cdots \sum_{i_{K,1}=1}^{m_K} \cdots \sum_{i_{K,m_K}=1}^{m_K}$$

$$\mathcal{L}_{1,\infty}(\|\boldsymbol{m}\|)$$

$$H_{m_1,\ldots,m_K}(\psi_{1,i_{1,1}},\ldots,\psi_{1,i_{1,m_1}};\cdots\cdots;\psi_{K,i_{K,1}},\ldots,\psi_{K,i_{K,m_K}})$$

$$\exp\left[j\sum_{k=1}^{K}\sum_{l=1}^{m_k}\phi_{k,i_{k,l}}\right]\delta\left(f - \sum_{k=1}^{K}\sum_{l=1}^{m_k}\psi_{k,i_{k,l}}\right) \tag{9.20}$$

From Equation (9.12) it is seen that another expression for $v(f)$ is

$$v(f) = \boldsymbol{a}^T(f)\,\boldsymbol{x}(f) + \boldsymbol{b}^T(f)\,\boldsymbol{s}(f) \tag{9.21}$$

Thus

$$v(f) = \sum_{r=1}^{R} a_r(f)\,x_r(f) + \sum_{k=1}^{K} b_k(f)\,s_k(f) \tag{9.22}$$

The controlling variable $x_r(f)$ where $r \in \{1,2,\ldots,R\}$ in Equation (9.22) is related to the known input signals $\{s_1(f),\ldots,s_K(f)\}$ as a multi-port Volterra series. The controlling variable $x_r(f)$ where $r \in \{1,2,\ldots,R\}$ is given as the response from a non-linear multi-port Volterra system with frequency domain input signals $\{s_1(f),\ldots,s_K(f)\}$ where $s_k(f)$ with $k \in \{1,2,\ldots,K\}$ is given by Equation (9.8) with $I_k = m_k$. Thus

$$x_r(f) = \sum_{o_1=0}^{\infty} \cdots \sum_{o_K=0}^{\infty} \sum_{i_{1,1}=1}^{m_1} \cdots \sum_{i_{1,o_1}=1}^{m_1} \cdots\cdots \sum_{i_{K,1}=1}^{m_K} \cdots \sum_{i_{K,o_K}=1}^{m_K}$$

$$\mathcal{L}_{1,\infty}(o_1 + \cdots + o_K)$$

$$(H_r)_{o_1,\ldots,o_K}(\psi_{1,i_{1,1}},\ldots,\psi_{1,i_{1,o_1}};\cdots\cdots;\psi_{K,i_{K,1}},\ldots,\psi_{K,i_{K,o_K}})$$

$$\exp\left[j\sum_{k=1}^{K}\sum_{l=1}^{o_k}\phi_{k,i_{k,l}}\right]\delta\left(f - \sum_{k=1}^{K}\sum_{l=1}^{o_k}\psi_{k,i_{k,l}}\right) \tag{9.23}$$

where $(H_r)_{o_1,\ldots,o_K}(\cdot)$ is the multi-port frequency domain Volterra transfer function between input signals $\{s_1(f),\ldots,s_K(f)\}$ and the controlling variable $x_r(f)$. From Equation (9.23) it can be seen that

$$\sum_{k=1}^{K}\sum_{l=1}^{o_k}\psi_{k,i_{k,l}} = \sum_{k=1}^{K}\sum_{l=1}^{m_k} q_{k,l}\,\psi_{k,l} \tag{9.24}$$

for $i_{k,l} \in \{1,2,\ldots,m_k\}$ and where

$$q_{k,l} = \text{number of } \{i_{k,1},\ldots,i_{k,o_k}\} \text{ which are equal to } l \tag{9.25}$$

for $k \in \{1, 2, \ldots, K\}$ and $l \in \{1, 2, \ldots, m_k\}$. This means that the $q_{k,l}$ variable where $k \in \{1, 2, \ldots, K\}$ and $l \in \{1, 2, \ldots, m_k\}$ fulfils the following two properties

$$q_{k,l} \in \{0, 1, \ldots, o_k\} \tag{9.26}$$

$$\sum_{l=1}^{m_k} q_{k,l} = o_k \tag{9.27}$$

Thus

$$
\sum_{i_{1,1}=1}^{m_1} \cdots \sum_{i_{1,o_1}=1}^{m_1} \cdots \cdots \sum_{i_{K,1}=1}^{m_K} \cdots \sum_{i_{K,o_K}=1}^{m_K} \sum_{k=1}^{K} \sum_{l=1}^{o_k} \psi_{k, i_{k,l}}
$$
$$
= \underbrace{\sum_{q_{1,1}=0}^{o_1} \cdots \sum_{q_{1,m_1}=0}^{o_1}}_{q_{1,1}+\cdots+q_{1,m_1}=o_1} \cdots \cdots \underbrace{\sum_{q_{K,1}=0}^{o_K} \cdots \sum_{q_{K,m_K}=0}^{o_K}}_{q_{K,1}+\cdots+q_{K,m_K}=o_K} \sum_{k=1}^{K} \sum_{l=1}^{m_k} q_{k,l} \, \psi_{k,l} \tag{9.28}
$$

Then the response $x_r(f)$ where $r \in \{1, 2, \ldots, R\}$ can be written as

$$
x_r(f) = \sum_{o_1=0}^{\infty} \cdots \sum_{o_K=0}^{\infty} \underbrace{\sum_{q_{1,1}=0}^{o_1} \cdots \sum_{q_{1,m_1}=0}^{o_1}}_{q_{1,1}+\cdots+q_{1,m_1}=o_1} \cdots \cdots \underbrace{\sum_{q_{K,1}=0}^{o_K} \cdots \sum_{q_{K,m_K}=0}^{o_K}}_{q_{K,1}+\cdots+q_{K,m_K}=o_K}
$$
$$
\mathcal{L}_{1,\infty}(o_1 + \cdots + o_K)
$$
$$
(x_r)_{o_1+\cdots+o_K}(\psi^T q) \; \exp[j\,\phi^T q] \; \delta(f - \psi^T q) \tag{9.29}
$$

where

$$q = [q_{1,1}, \ldots, q_{1,m_1}, \ldots \ldots, q_{K,1}, \ldots, q_{K,m_K}]^T \in \mathcal{Z}_{0+}^{\|m\| \times 1} \tag{9.30}$$

$$\psi = [\psi_{1,1}, \ldots, \psi_{1,m_1}, \ldots \ldots, \psi_{K,1}, \ldots, \psi_{K,m_K}]^T \in \mathcal{R}^{\|m\| \times 1} \tag{9.31}$$

$$\phi = [\phi_{1,1}, \ldots, \phi_{1,m_1}, \ldots \ldots, \phi_{K,1}, \ldots, \phi_{K,m_K}]^T \in \mathcal{R}^{\|m\| \times 1} \tag{9.32}$$

and $(x_r)_{o_1+\cdots+o_K}(\psi^T q)$ has been introduced as a coefficient instead of $(H_r)_{m_1,\ldots,m_K}(\cdot)$ which is just a coefficient to $\exp[j\,\phi^T q] \; \delta(f - \psi^T q)$. Then, $x_r(f)$ can, after some careful considerations, be rewritten into the following more convenient form:

$$x_r(f) = \sum_{o=1}^{\infty} \sum_{q} \Lambda_o(\|q\|) \; (x_r)_o(\psi^T q) \; \exp[j\,\phi^T q] \; \delta(f - \psi^T q) \tag{9.33}$$

where

$$\Lambda_\alpha(\beta) \quad = \quad \begin{cases} 1 & \text{for} \quad \alpha = \beta \\ 0 & \text{otherwise} \end{cases} \tag{9.34}$$

$$\sum_{\boldsymbol{q}} \quad \equiv \quad \sum_{q_{1,1}=0}^{o} \cdots \sum_{q_{1,m_1}=0}^{o} \cdots\cdots \sum_{q_{K,1}=0}^{o} \cdots \sum_{q_{K,m_K}=0}^{o} \tag{9.35}$$

The variables α and β in Equation (9.34) may generally be scalars or vectors. In Equation (9.33), $(x_r)_o(\boldsymbol{\psi}^T\boldsymbol{q})$ is the (presently unknown) oth order coefficient of $\exp[j\,\boldsymbol{\phi}^T\boldsymbol{q}]\ \delta(f - \boldsymbol{\psi}^T\boldsymbol{q})$ in $x_r(f)$. Substituting Equations (9.33) and (9.8) with $I_k = m_k$ into Equation (9.22) gives $v(f)$ as

$$\begin{aligned} v(f) \quad = \quad & \sum_{r=1}^{R}\sum_{o=1}^{\infty}\sum_{\boldsymbol{q}} \Lambda_o(\|\boldsymbol{q}\|)\ a_r(\boldsymbol{\psi}^T\boldsymbol{q})\ (x_r)_o(\boldsymbol{\psi}^T\boldsymbol{q})\ \exp[j\,\boldsymbol{\phi}^T\boldsymbol{q}]\ \delta(f - \boldsymbol{\psi}^T\boldsymbol{q}) \\ & + \sum_{k=1}^{K}\sum_{l=1}^{m_k} b_k(\psi_{k,l})\ \exp[j\,\phi_{k,l}]\ \delta(f - \psi_{k,l}) \end{aligned} \tag{9.36}$$

Thus two expressions for the output response $v(f)$ are obtained. In Equation (9.9) the response $v(f)$ is given by the multi-port Volterra transfer function which is to be determined (with the substitutions mentioned earlier in this section), and in Equation (9.36) the response $v(f)$ is given by network specific vectors and the non-linear controlling variables.

9.3.3   Transfer functions

From Equations (9.9), (9.10) and (9.36) the expression for an arbitrary first order ($\|\boldsymbol{m}\| = 1$) frequency domain Volterra transfer function can be determined as

$$H_{0,\dots,0,m_k=1,0,\dots,0}(\psi_{k,l}) \quad = \quad \sum_{r=1}^{R} a_r(\psi_{k,l})\ (x_r)_1(\psi_{k,l}) + b_k(\psi_{k,l}) \tag{9.37}$$

for given $k \in \{1, 2, \dots, K\}$ and $l \in \{1, 2, \dots, m_k\}$. Using vector notation the expression for a first order frequency domain Volterra transfer function is given by

$$H_{0,\dots,0,m_k=1,0,\dots,0}(\psi_{k,l}) \quad = \quad \boldsymbol{a}^T(\psi_{k,l})\ \boldsymbol{x}_1(\psi_{k,l}) + b_k(\psi_{k,l}) \tag{9.38}$$

and where

$$\boldsymbol{x}_o(\boldsymbol{\psi}^T\boldsymbol{q}) \quad = \quad [(x_1)_o(\boldsymbol{\psi}^T\boldsymbol{q}), \dots, (x_R)_o(\boldsymbol{\psi}^T\boldsymbol{q})]^T \quad \in \quad C^{R\times 1} \tag{9.39}$$

is the oth order vector of the controlling variables at frequency $\boldsymbol{\psi}^T\boldsymbol{q}$. Thus to determine the first order transfer function from a given input port $k \in \{1, 2, \dots, K\}$

and a given $l \in \{1, 2, \ldots, m_k\}$, the quantities $a(\psi_{k,l})$, $x_1(\psi_{k,l})$ and $b_k(\psi_{k,l})$ must be determined.

The multi-port Volterra transfer functions of second and higher order ($\|m\| \in \{2, 3, \ldots, \infty\}$) can be determined from Equations (9.9), (9.10) and (9.36) as

$$H_{m_1,\ldots,m_K}(\psi_{1,1}, \ldots, \psi_{1,m_1}; \cdots\cdots; \psi_{K,1}, \ldots, \psi_{K,m_K})$$

$$= \frac{1}{m_1! \cdots m_K!} \sum_{r=1}^{R} a_r(\|\psi\|) \ (x_r)_{\|m\|}(\|\psi\|) \tag{9.40}$$

Equation (9.40) can also be written in vector notation as

$$H_{m_1,\ldots,m_K}(\psi_{1,1}, \ldots, \psi_{1,m_1}; \cdots\cdots; \psi_{K,1}, \ldots, \psi_{K,m_K})$$

$$= \frac{1}{m_1! \cdots m_K!} \ a^T(\|\psi\|) \ x_{\|m\|}(\|\psi\|) \tag{9.41}$$

for $\|m\| \in \{2, 3, \ldots, \infty\}$. Thus to determine the multi-port frequency domain Volterra transfer function of order 2 and higher, the system vector $a(\|\psi\|)$ and the $\|m\|$th order coefficients of the controlling variables at frequency $\|\psi\|$ must be determined.

9.3.4 Controlled variables

From Equation (9.12) the controlled variable vector $y(f)$ is given by

$$y(f) = A(f) x(f) + B(f) s(f) \tag{9.42}$$

Thus

$$y_q(f) = \sum_{r=1}^{R} A_{q,r}(f) \ x_r(f) + \sum_{k=1}^{K} B_{q,k}(f) \ s_k(f) \tag{9.43}$$

for a given $q \in \{1, 2, \ldots, Q\}$. Substituting Equations (9.8) with $I_k = m_k$ and (9.33) into Equation (9.43) gives

$$y_q(f) = \sum_{r=1}^{R} \sum_{o=1}^{\infty} \sum_{q} \Lambda_o(\|q\|) \ A_{q,r}(\psi^T q) \ (x_r)_o(\psi^T q) \ \exp[j \ \phi^T q] \ \delta(f - \psi^T q)$$

$$+ \sum_{k=1}^{K} \sum_{l=1}^{m_k} B_{q,k}(\psi_{k,l}) \ \exp[j \ \phi_{k,l}] \ \delta(f - \psi_{k,l}) \tag{9.44}$$

The response $y_q(f)$ from the purely non-linear subsystem $q \in \{1, 2, \ldots, Q\}$ can also be described by a multi-port Volterra series as

$$
y_q(f) = \sum_{n_1=0}^{\infty} \cdots \sum_{n_{P(q)}=0}^{\infty} \int_{-\infty}^{\infty} \cdots \int_{-\infty}^{\infty} \cdots\cdots \int_{-\infty}^{\infty} \cdots \int_{-\infty}^{\infty}
$$

$$
\mathcal{L}_{2,\infty}(\|\boldsymbol{n}\|)
$$

$$
(G_q)_{n_1,\ldots,n_{P(q)}}(\theta_{1,1},\ldots,\theta_{1,n_1};\cdots\cdots;\theta_{P(q),1},\ldots,\theta_{P(q),n_{P(q)}})
$$

$$
\prod_{r=1}^{P(q)}\prod_{p=1}^{n_r} x_{j_{q,r}}(\theta_{r,p})\ \delta\!\left(f - \sum_{r=1}^{P(q)}\sum_{p=1}^{n_r}\theta_{r,p}\right)\ \prod_{r=1}^{P(q)}\prod_{p=1}^{n_r} d\theta_{r,p} \qquad (9.45)
$$

where

$$
\boldsymbol{n} = [n_1,\ldots,n_{P(q)}]^T \ \in\ \mathcal{Z}_{0+}^{P(q)\times 1} \qquad (9.46)
$$

and $(G_q)_{n_1,\ldots,n_{P(q)}}(\cdot)$ is the partly symmetrical multi-port frequency domain Volterra transfer function between inputs $\{x_{j_{q,1}}(f),\ldots,x_{j_{q,P(q)}}(f)\}$ and output $y_q(f)$ for non-linear subsystem $q \in \{1,2,\ldots,Q\}$.[3] In case the original $(G_q)_{n_1,\ldots,n_{P(q)}}(\cdot)$ transfer function is not partly symmetrical, it *must* be made partly symmetrical using for example Equation (9.5). The reason for requiring a partly symmetrical $(G_q)_{n_1,\ldots,n_{P(q)}}(\cdot)$ transfer function is that it significantly reduces the amount of computations that must be carried out to determine the overall $H_{m_1,\ldots,m_K}(\cdot)$ partly symmetrical transfer function. When $(G_q)_{n_1,\ldots,n_{P(q)}}(\cdot)$ is partly symmetrical then $H_{m_1,\ldots,m_K}(\cdot)$ automatically becomes partly symmetrical when the present method is used. Note that $y_q(f)$ where $q \in \{1,2,\ldots,Q\}$ contains only non-linear contributions. Insertion of Equation (9.33) into (9.45) gives

---

[3]In section 9.5 some common types of non-linear time domain relations between inputs $\{x_{j_{q,1}}(t),\ldots,x_{j_{q,P(q)}}(t)\}$ and output $y_q(t)$ are given as well as the corresponding multi-port $(G_q)_{n_1,\ldots,n_{P(q)}}(\cdot)$ transfer functions.

$$y_q(f)$$

$$= \sum_{n_1=0}^{\infty} \cdots \sum_{n_{P(q)}=0}^{\infty} \sum_{o_{1,1}=1}^{\infty} \sum_{\boldsymbol{q}_{1,1}} \cdots \sum_{o_{1,n_1}=1}^{\infty} \sum_{\boldsymbol{q}_{1,n_1}} \cdots$$

$$\cdots \sum_{o_{P(q),1}=1}^{\infty} \sum_{\boldsymbol{q}_{P(q),1}} \cdots \sum_{o_{P(q),n_{P(q)}}=1}^{\infty} \sum_{\boldsymbol{q}_{P(q),n_{P(q)}}}$$

$$\mathcal{L}_{2,\infty}(\|\boldsymbol{n}\|) \prod_{r=1}^{P(q)} \prod_{p=1}^{n_r} \Lambda_{o_{r,p}}(\|\boldsymbol{q}_{r,p}\|)$$

$$(G_q)_{n_1,\ldots,n_{P(q)}} \left(\boldsymbol{\psi}^T \boldsymbol{q}_{1,1}, \ldots, \boldsymbol{\psi}^T \boldsymbol{q}_{1,n_1}; \cdots \right.$$

$$\left. \cdots ; \boldsymbol{\psi}^T \boldsymbol{q}_{P(q),1}, \ldots, \boldsymbol{\psi}^T \boldsymbol{q}_{P(q),n_{P(q)}} \right)$$

$$\prod_{r=1}^{P(q)} \prod_{p=1}^{n_r} (x_{j_{q,r}})_{o_{r,p}} (\boldsymbol{\psi}^T \boldsymbol{q}_{r,p})$$

$$\exp\left[j \, \boldsymbol{\phi}^T \sum_{r=1}^{P(q)} \sum_{p=1}^{n_r} \boldsymbol{q}_{r,p} \right] \delta\left(f - \boldsymbol{\psi}^T \sum_{r=1}^{P(q)} \sum_{p=1}^{n_r} \boldsymbol{q}_{r,p} \right) \qquad (9.47)$$

where

$$\boldsymbol{q}_{r,p} \; = \; \left[q_{r,p,1,1}, \ldots, q_{r,p,1,m_1}, \ldots \ldots, q_{r,p,K,1}, \ldots, q_{r,p,K,m_K} \right]^T \; \in \; \mathcal{Z}_{0+}^{\|\boldsymbol{m}\| \times 1}$$

$$(9.48)$$

and the summation over $\boldsymbol{q}_{r,p}$ is defined symbolically as

$$\sum_{\boldsymbol{q}_{r,p}} \; \equiv \; \sum_{q_{r,p,1,1}=0}^{o_{r,p}} \cdots \sum_{q_{r,p,1,m_1}=0}^{o_{r,p}} \cdots \cdots \sum_{q_{r,p,K,1}=0}^{o_{r,p}} \cdots \sum_{q_{r,p,K,m_K}=0}^{o_{r,p}} \qquad (9.49)$$

where $r \in \{1, 2, \ldots, P(q)\}$ and $p \in \{1, 2, \ldots, n_r\}$.

9.3.5 Controlling variables

As seen from Equations (9.38) and (9.41) it is necessary to determine $\boldsymbol{x}_{\|\boldsymbol{m}\|}(\|\boldsymbol{\psi}\|)$. This is done using the fact that Equations (9.44) and (9.47) must be identical:

$$\sum_{n_1=0}^{\infty} \cdots \sum_{n_{P(q)}=0}^{\infty} \sum_{o_{1,1}=1}^{\infty} \sum_{\boldsymbol{q}_{1,1}} \cdots \sum_{o_{1,n_1}=1}^{\infty} \sum_{\boldsymbol{q}_{1,n_1}} \cdots \qquad (9.50)$$

$$\cdots \sum_{o_{P(q),1}=1}^{\infty} \sum_{\boldsymbol{q}_{P(q),1}} \cdots \sum_{o_{P(q),n_{P(q)}}=1}^{\infty} \sum_{\boldsymbol{q}_{P(q),n_{P(q)}}}$$

$$\mathcal{L}_{2,\infty}(\|\boldsymbol{n}\|) \prod_{r=1}^{P(q)} \prod_{p=1}^{n_r} \Lambda_{o_{r,p}}(\|\boldsymbol{q}_{r,p}\|)$$

$$(G_q)_{n_1,\ldots,n_{P(q)}} \left(\boldsymbol{\psi}^T \boldsymbol{q}_{1,1}, \ldots, \boldsymbol{\psi}^T \boldsymbol{q}_{1,n_1}; \cdots \right.$$

$$\left. \cdots ; \boldsymbol{\psi}^T \boldsymbol{q}_{P(q),1}, \ldots, \boldsymbol{\psi}^T \boldsymbol{q}_{P(q),n_{P(q)}} \right)$$

$$\prod_{r=1}^{P(q)} \prod_{p=1}^{n_r} (x_{j_{q,r}})_{o_{r,p}}(\boldsymbol{\psi}^T \boldsymbol{q}_{r,p})$$

$$\exp\left[j\,\boldsymbol{\phi}^T \sum_{r=1}^{P(q)} \sum_{p=1}^{n_r} \boldsymbol{q}_{r,p}\right] \delta\left(f - \boldsymbol{\psi}^T \sum_{r=1}^{P(q)} \sum_{p=1}^{n_r^{\cdot}} \boldsymbol{q}_{r,p}\right)$$

$$= \sum_{r=1}^{R} \sum_{o=1}^{\infty} \sum_{\boldsymbol{q}} \Lambda_o(\|\boldsymbol{q}\|) \; A_{q,r}(\boldsymbol{\psi}^T \boldsymbol{q}) \; (x_r)_o(\boldsymbol{\psi}^T \boldsymbol{q}) \; \exp[j\,\boldsymbol{\phi}^T \boldsymbol{q}] \; \delta(f - \boldsymbol{\psi}^T \boldsymbol{q})$$

$$+ \sum_{k=1}^{K} \sum_{l=1}^{m_k} B_{q,k}(\psi_{k,l}) \; \exp[j\,\phi_{k,l}] \; \delta(f - \psi_{k,l}) \qquad (9.51)$$

The controlling variables can be determined by using the fact that the coefficient of $\exp[j\,\|\boldsymbol{\phi}\|]\,\delta(f - \|\boldsymbol{\psi}\|)$ on both sides of Equation (9.51) must be identical.

To determine the first order controlling variables $x_1(\psi_{k,l})$ for given $k \in \{1,2,\ldots, K\}$ and $l \in \{1,2,\ldots, m_k\}$ Equation (9.51) is used to yield

$$\sum_{r=1}^{R} A_{q,r}(\psi_{k,l}) \; (x_r)_1(\psi_{k,l}) \quad = \quad -B_{q,k}(\psi_{k,l}) \qquad (9.52)$$

where $q \in \{1,2,\ldots,Q\}$. Now all the first order controlling variables $x_1(\psi_{k,l})$ can be determined as

$$\boldsymbol{x}_1(\psi_{k,l}) \quad = \quad -\boldsymbol{A}^{-1}(\psi_{k,l})\,\boldsymbol{b}_k(\psi_{k,l}) \qquad (9.53)$$

where

$$\boldsymbol{b}_k(\psi_{k,l}) \quad = \quad [B_{1,k}(\psi_{k,l}), \ldots, B_{Q,k}(\psi_{k,l})]^T \quad \in \quad \mathcal{C}^{Q \times 1} \qquad (9.54)$$

Thus the vector $\boldsymbol{b}_k(\psi_{k,l})$ is the kth column vector of the $\boldsymbol{B}(\psi_{k,l})$ matrix. The solution vector in Equation (9.53), $\boldsymbol{x}_1(\psi_{k,l})$, can be determined provided the matrix $\boldsymbol{A}(\psi_{k,l})$ is unique and non-singular. If this is not the case, the original network can be perturbed by some linear elements to obtain a unique and non-singular

$A(\psi_{k,l})$ matrix. (Practically, the matrix $A(\psi_{k,l})$ is not determined since it is actually only $A^{-1}(\psi_{k,l})$ that is used. This matrix may be determined by setting $s(f) = [0, 0, \ldots, 0]^T \in \{0\}^{K \times 1}$ and then using $y(f)$ as excitation signals instead of $x(f)$. This technique is used in Example 2 in Section 9.6. If $Q \neq R$ then $A(f)$ is non-quadratic and then the inverse does not exist. This problem can be solved by introducing fictitious controlling or controlled variables as follows:

- If $Q < R$ then introduce $R - Q$ fictitious controlled sources which depend on one or more of the controlling variables $x_1(f), \ldots, x_R(f)$.

- If $Q > R$ then introduce $Q - R$ fictitious controlling variables which control one or more of the controlled variables $y_1(f), \ldots, y_Q(f)$.

To determine the second and higher order controlling variables $x_{\|m\|}(\|\psi\|)$ where $\|m\| \in \{2, 3, \ldots, \infty\}$ use Equation (9.51) to yield

$$\sum_{n_1=0}^{\infty} \cdots \sum_{n_{P(q)}=0}^{\infty} \sum_{o_{1,1}=1}^{\infty} \sum_{q_{1,1}} \cdots \sum_{o_{1,n_1}=1}^{\infty} \sum_{q_{1,n_1}} \cdots$$

$$\cdots \sum_{o_{P(q),1}=1}^{\infty} \sum_{q_{P(q),1}} \cdots \sum_{o_{P(q),n_{P(q)}}=1}^{\infty} \sum_{q_{P(q),n_{P(q)}}}$$

$$\mathcal{L}_{2,\infty}(\|n\|) \prod_{r=1}^{P(q)} \prod_{p=1}^{n_r} \Lambda_{o_{r,p}}(\|q_{r,p}\|)$$

$$(G_q)_{n_1,\ldots,n_{P(q)}} \left(\psi^T q_{1,1}, \ldots, \psi^T q_{1,n_1}; \cdots \right.$$

$$\left. \cdots; \psi^T q_{P(q),1}, \ldots, \psi^T q_{P(q),n_{P(q)}} \right)$$

$$\prod_{r=1}^{P(q)} \prod_{p=1}^{n_r} (x_{j_{q,r}})_{o_{r,p}}(\psi^T q_{r,p})$$

$$\exp\left[j \phi^T \sum_{r=1}^{P(q)} \sum_{p=1}^{n_r} q_{r,p} \right] \delta\left(f - \psi^T \sum_{r=1}^{P(q)} \sum_{p=1}^{n_r} q_{r,p} \right)$$

$$= \sum_{r=1}^{R} \sum_{o=1}^{\infty} \sum_{q} \Lambda_o(\|q\|) \, A_{q,r}(\psi^T q) \, (x_r)_o(\psi^T q) \, \exp[j \phi^T q] \, \delta(f - \psi^T q)$$

$$(9.55)$$

The coefficient of $\exp[j \|\phi\|] \, \delta(f - \|\psi\|)$ must be the same on both sides of Equation (9.55). From this, the following properties can be derived regarding which terms should be included on the left-hand side of Equation (9.55):

Property 9.1

$$\sum_{r=1}^{P(q)} \sum_{p=1}^{n_r} q_{r,p} \quad = \quad 1 \tag{9.56}$$

where

$$1 \quad = \quad \underbrace{[1,1,\dots\dots,1]}_{\|\boldsymbol{m}\|}^T \quad \in \quad \{1\}^{\|\boldsymbol{m}\| \times 1} \tag{9.57}$$

for any given $q \in \{1, 2, \dots, Q\}$.

Proof 9.1 *Follows directly since* $\{\phi_{1,1}, \dots, \phi_{1,m_1}, \dots\dots, \phi_{K,1}, \dots, \phi_{K,m_K}\}$ *forms a phase base.*   □

Property 9.2

$$q_{r,p,k,l} \quad \in \quad \{0, 1\} \tag{9.58}$$

for all $r \in \{1, 2, \dots, P(q)\}$, $p \in \{1, 2, \dots, n_r\}$, $k \in \{1, 2, \dots, K\}$, $l \in \{1, 2, \dots, m_k\}$.

Proof 9.2 *Since* $\{\phi_{1,1}, \dots, \phi_{1,m_1}, \dots\dots, \phi_{K,1}, \dots, \phi_{K,m_K}\}$ *forms a phase base, then only one of the coefficients of any* $\phi_{k,l}$ *where* $k \in \{1, 2, \dots, K\}$ *and* $l \in \{1, 2, \dots, m_k\}$ *must be equal to 1 and all others must be equal to 0.*   □

Property 9.3

$$\|q_{r,p}\| \quad = \quad o_{r,p} \quad \in \quad \{1, 2, \dots, \infty\} \tag{9.59}$$

for all $r \in \{1, 2, \dots, P(q)\}$ and $p \in \{1, 2, \dots, n_r\}$.

Proof 9.3 *Follows directly from Equation (9.55).*   □

Property 9.4

$$\sum_{r=1}^{P(q)} \sum_{p=1}^{n_r} \|q_{r,p}\| \quad = \quad \|\boldsymbol{o}\| \tag{9.60}$$

$$= \quad \|\boldsymbol{m}\| \tag{9.61}$$

where the *o*-vector is defined in Equation (9.65).

Proof 9.4 *Equation (9.60) follows directly from property 9.3 and Equation (9.61) follows directly from property 9.2.* □

Property 9.5

$$o_{r,p} \in \{1, 2, \ldots, \|m\| - 1\} \tag{9.62}$$

for all $r \in \{1, 2, \ldots, P(q)\}$ and $p \in \{1, 2, \ldots, n_r\}$.

Proof 9.5 *From Equation (9.55) it is given that $o_{r,p} \in \{1, 2, \ldots, \infty\}$ for all $r \in \{1, 2, \ldots, P(q)\}$ and $p \in \{1, 2, \ldots, n_r\}$. From properties 9.3 and 9.4 and as $\|n\| \in \{2, 3, \ldots, \infty\}$ then $\|o\| \in \{2, 3, \ldots, \infty\}$. As $\|o\| = \|m\|$ from Equations (9.60) and (9.61) then $o_{r,p} \in \{1, 2, \ldots, \|m\| - 1\}$ for all $r \in \{1, 2, \ldots, P(q)\}$ and $p \in \{1, 2, \ldots, n_r\}$.* □

Property 9.6

$$\|n\| \in \{2, 3, \ldots, \|m\|\} \tag{9.63}$$

for any $q \in \{1, 2, \ldots, Q\}$.

Proof 9.6 *From Equation (9.55) it is directly seen that $\|n\| \in \{2, 3, \ldots, \infty\}$ for any $q \in \{1, 2, \ldots, Q\}$. Similarly, from properties 9.3 and 9.4 it follows that $\|n\| \in \{2, 3, \ldots, \|m\|\}$ for any $q \in \{1, 2, \ldots, Q\}$.* □

Using these properties, the second and higher order controlling variables $x_{\|m\|}(\|\psi\|)$ can be determined as

$$\sum_{n_1=0}^{N_1} \cdots \sum_{n_{P(q)}=0}^{N_{P(q)}} \sum_{o_{1,1}=1}^{O_{1,1}} \cdots \sum_{o_{1,n_1}=1}^{O_{1,n_1}} \cdots \cdots \sum_{o_{P(q),1}=1}^{O_{P(q),1}} \cdots \sum_{o_{P(q),n_{P(q)}}=1}^{O_{P(q),n_{P(q)}}}$$

$$\sum_{\boldsymbol{q}_{1,1}} \cdots \sum_{\boldsymbol{q}_{1,n_1}} \cdots \cdots \sum_{\boldsymbol{q}_{P(q),1}} \cdots \sum_{\boldsymbol{q}_{P(q),n_{P(q)}}}$$

$$\mathcal{L}_{2,\infty}(\|\boldsymbol{n}\|) \; \Lambda_1\Bigg(\sum_{r=1}^{P(q)}\sum_{p=1}^{n_r} \boldsymbol{q}_{r,p}\Bigg) \; \prod_{r=1}^{P(q)}\prod_{p=1}^{n_r} \Lambda_{o_{r,p}}(\|\boldsymbol{q}_{r,p}\|)$$

$$(G_q)_{n_1,\ldots,n_{P(q)}}\Big(\boldsymbol{\psi}^T\boldsymbol{q}_{1,1},\ldots,\boldsymbol{\psi}^T\boldsymbol{q}_{1,n_1};\cdots\cdots;\boldsymbol{\psi}^T\boldsymbol{q}_{P(q),1},\ldots,\boldsymbol{\psi}^T\boldsymbol{q}_{P(q),n_{P(q)}}\Big)$$

$$\prod_{r=1}^{P(q)}\prod_{p=1}^{n_r}(x_{j_{q,r}})_{o_{r,p}}(\boldsymbol{\psi}^T\boldsymbol{q}_{r,p})$$

$$= \sum_{r=1}^{R} A_{q,r}(\|\boldsymbol{\psi}\|) \; (x_r)_{\|\boldsymbol{m}\|}(\|\boldsymbol{\psi}\|) \tag{9.64}$$

where $\mathbf{1}$ and $\Lambda_1(\cdot)$ are defined in Equations (9.57) and (9.34) respectively, and

$$\boldsymbol{o} = [o_{1,1},\ldots,o_{1,n_1},\ldots\ldots,o_{P(q),1},\ldots,o_{P(q),n_{P(q)}}]^T \in \mathbb{Z}_+^{\|\boldsymbol{n}\|\times 1} \tag{9.65}$$

$$N_r = \|\boldsymbol{m}\| - (n_1 + \cdots + n_{r-1}) \tag{9.66}$$

$O_{r,p} =$
$$\begin{cases} \|\boldsymbol{m}\| - (n_r + \cdots + n_{P(q)}) + 1 - (o_{1,1} + \cdots + o_{r-1,n_{r-1}}) & \text{if } p = 1 \\ \|\boldsymbol{m}\| - (n_r + \cdots + n_{P(q)}) + p - (o_{1,1} + \cdots + o_{r,p-1}) & \text{if } p \in \{2,3,\ldots,\infty\} \end{cases} \tag{9.67}$$

for $r \in \{1,2,\ldots,P(q)\}$ and $p \in \{1,2,\ldots,n_r\}$. Equation (9.66) is derived from property 9.6, and Equation (9.67) is derived from properties 9.4 and 9.5.

Now $x_{\|\boldsymbol{m}\|}(\|\boldsymbol{\psi}\|)$ is determined by using all the sets of equations for $q \in \{1,2,\ldots,Q\}$ in Equation (9.64) from which

$$\boldsymbol{x}_{\|\boldsymbol{m}\|}(\|\boldsymbol{\psi}\|) = \boldsymbol{A}^{-1}(\|\boldsymbol{\psi}\|) \; \boldsymbol{u}_{\|\boldsymbol{m}\|}(\|\boldsymbol{\psi}\|) \tag{9.68}$$

where

$$\boldsymbol{u}_{\|\boldsymbol{m}\|}(\|\boldsymbol{\psi}\|) = [(u_1)_{\|\boldsymbol{m}\|}(\|\boldsymbol{\psi}\|),\ldots,(u_Q)_{\|\boldsymbol{m}\|}(\|\boldsymbol{\psi}\|)]^T \in \mathbb{C}^{Q\times 1} \tag{9.69}$$

has elements

$$(u_q)_{\|\boldsymbol{m}\|}(\|\boldsymbol{\psi}\|)$$

$$= \sum_{n_1=0}^{N_1} \cdots \sum_{n_{P(q)}=0}^{N_{P(q)}} \sum_{o_{1,1}=1}^{O_{1,1}} \cdots \sum_{o_{1,n_1}=1}^{O_{1,n_1}} \cdots\cdots \sum_{o_{P(q),1}=1}^{O_{P(q),1}} \cdots \sum_{o_{P(q),n_{P(q)}}=1}^{O_{P(q),n_{P(q)}}}$$

$$\sum_{\boldsymbol{q}_{1,1}} \cdots \sum_{\boldsymbol{q}_{1,n_1}} \cdots\cdots \sum_{\boldsymbol{q}_{P(q),1}} \cdots \sum_{\boldsymbol{q}_{P(q),n_{P(q)}}} \prod_{r=1}^{P(q)} n_r! \prod_{r=1}^{P(q)} \mathcal{Q}(\boldsymbol{q}_{r,1}, \ldots, \boldsymbol{q}_{r,n_r})$$

$$\mathcal{L}_{2,\infty}(\|\boldsymbol{n}\|)\, \Lambda_1\Big(\sum_{r=1}^{P(q)}\sum_{p=1}^{n_r} \boldsymbol{q}_{r,p}\Big) \prod_{r=1}^{P(q)}\prod_{p=1}^{n_r} \Lambda_{o_{r,p}}(\|\boldsymbol{q}_{r,p}\|)$$

$$(G_q)_{n_1,\ldots,n_{P(q)}}\Big(\boldsymbol{\psi}^T\boldsymbol{q}_{1,1}, \ldots, \boldsymbol{\psi}^T\boldsymbol{q}_{1,n_1}; \cdots\cdots; \boldsymbol{\psi}^T\boldsymbol{q}_{P(q),1}, \ldots, \boldsymbol{\psi}^T\boldsymbol{q}_{P(q),n_{P(q)}}\Big)$$

$$\prod_{r=1}^{P(q)}\prod_{p=1}^{n_r} (x_{j_{q,r}})_{o_{r,p}}(\boldsymbol{\psi}^T\boldsymbol{q}_{r,p}) \tag{9.70}$$

for $q \in \{1,2,\ldots,Q\}$ where

$$\mathcal{Q}(\boldsymbol{q}_{r,1},\ldots,\boldsymbol{q}_{r,n_r}) = \begin{cases} 1 & \text{for } \boldsymbol{c}^T\boldsymbol{q}_{r,1} < \cdots < \boldsymbol{c}^T\boldsymbol{q}_{r,n_r} \\ 0 & \text{otherwise} \end{cases} \tag{9.71}$$

for $r \in \{1,2,\ldots,P(q)\}$ and

$$\boldsymbol{c} = \left[2^0, 2^1, \ldots, 2^{\|\boldsymbol{m}\|-1}\right]^T \in \mathcal{Z}_+^{\|\boldsymbol{m}\|\times 1} \tag{9.72}$$

Equation (9.70) has used the fact that the multi-port frequency domain Volterra transfer function $(G_q)_{n_1,\ldots,n_{P(q)}}(\cdot)$ is partly symmetrical for all $q \in \{1,2,\ldots,Q\}$. The condition in Equation (9.71) is used to avoid contributions which are simply permutations of $\{\boldsymbol{q}_{r,1},\ldots,\boldsymbol{q}_{r,n_r}\}$ where $r \in \{1,2,\ldots,P(q)\}$. As seen from Equations (9.68) and (9.70) there must be summation over many q-variables. From Equation (9.70) it is seen that there are $\|\boldsymbol{m}\| \times \|\boldsymbol{n}\|$ q-variables. According to the properties, only $\|\boldsymbol{m}\|$ of these q-variables are different from 0 (those q-variables that are not 0 are 1). Thus it would be more convenient to use a pointer $w_{k,l}$ to indicate which q-variable in the set $\{q_{1,1,k,l},\ldots,q_{1,n_1,k,l},\cdots\cdots,q_{P(q),1,k,l},\ldots,q_{P(q),n_{P(q)},k,l}\}$ is equal to 1. In this case only $\|\boldsymbol{m}\|$ w-pointers are needed instead of $\|\boldsymbol{m}\| \times \|\boldsymbol{n}\|$ q-variables.

Now the requirements given for the q-variables in Equation (9.70) — and in turn in properties 9.2, 9.1, 9.3 and 9.4 — must be transferred to the w-pointers. As seen from the properties

$$w_{k,l} \in \{1,2,\ldots,\|\boldsymbol{n}\|\} \tag{9.73}$$

for $k \in \{1,2,\ldots,K\}$ and $l \in \{1,2,\ldots,m_k\}$.

Next the correct $\boldsymbol{q}_{r,p}$ vectors where $r \in \{1,2,\ldots,P(q)\}$ and $p \in \{1,2,\ldots,n_r\}$ must be determined. To do this define vectors

$$z_{r,p}(\boldsymbol{w}) = \Big[z_{r,p,1,1}(\boldsymbol{w}),\ldots,z_{r,p,1,m_1}(\boldsymbol{w}),\ldots$$
$$\ldots,z_{r,p,K,1}(\boldsymbol{w}),\ldots,z_{r,p,K,m_K}(\boldsymbol{w})\Big]^T \in \mathcal{Z}_+^{\|\boldsymbol{m}\|\times 1} \tag{9.74}$$

for $r \in \{1,2,\ldots,P(q)\}$ and $p \in \{1,2,\ldots,n_r\}$, where

$$\boldsymbol{w} = [w_{1,1},\ldots,w_{1,m_1},\ldots\ldots,w_{K,1},\ldots,w_{K,m_K}]^T \in \mathcal{Z}_+^{\|\boldsymbol{m}\|\times 1} \tag{9.75}$$

$$z_{r,p,k,l}(\boldsymbol{w}) = \begin{cases} 1 & \text{for} \quad w_{k,l} = n_1 + \cdots + n_{r-1} + p \\ 0 & \text{otherwise} \end{cases} \tag{9.76}$$

for $k \in \{1,2,\ldots,K\}$ and $l \in \{1,2,\ldots,m_k\}$. Thus

$$\boldsymbol{q}_{r,p} = z_{r,p}(\boldsymbol{w}) \tag{9.77}$$

Using the above, a modified version of Equation (9.70) is obtained as

$$(u_q)_{\|\boldsymbol{m}\|}(\|\boldsymbol{\psi}\|)$$

$$= \sum_{n_1=0}^{N_1}\cdots\sum_{n_{P(q)}=0}^{N_{P(q)}}\sum_{o_{1,1}=1}^{O_{1,1}}\cdots\sum_{o_{1,n_1}=1}^{O_{1,n_1}}\cdots\cdots\sum_{o_{P(q),1}=1}^{O_{P(q),1}}\cdots\sum_{o_{P(q),n_{P(q)}}=1}^{O_{P(q),n_{P(q)}}}$$

$$\sum_{w_{1,1}=1}^{\|\boldsymbol{n}\|}\cdots\sum_{w_{1,m_1}=1}^{\|\boldsymbol{n}\|}\cdots\cdots\sum_{w_{K,1}=1}^{\|\boldsymbol{n}\|}\cdots\sum_{w_{K,m_K}=1}^{\|\boldsymbol{n}\|}$$

$$\prod_{r=1}^{P(q)} n_r! \prod_{r=1}^{P(q)} \mathcal{Q}(z_{1,1}(\boldsymbol{w}),\ldots,z_{r,n_r}(\boldsymbol{w}))$$

$$\mathcal{L}_{2,\infty}(\|\boldsymbol{n}\|) \prod_{r=1}^{P(q)}\prod_{p=1}^{n_r} \Lambda_{o_{r,p}}(\|z_{r,p}(\boldsymbol{w})\|)$$

$$(G_q)_{n_1,\ldots,n_{P(q)}}\Big(\boldsymbol{\psi}^T z_{1,1}(\boldsymbol{w}),\ldots,\boldsymbol{\psi}^T z_{1,n_1}(\boldsymbol{w});\cdots$$
$$\cdots;\boldsymbol{\psi}^T z_{P(q),1}(\boldsymbol{w}),\ldots,\boldsymbol{\psi}^T z_{P(q),n_{P(q)}}(\boldsymbol{w})\Big)$$

$$\prod_{r=1}^{P(q)}\prod_{p=1}^{n_r}(x_{j_{q,r}})_{o_{r,p}}(\boldsymbol{\psi}^T z_{r,p}(\boldsymbol{w})) \tag{9.78}$$

In Equation (9.78) it is not necessary to include $\Lambda_1(\cdot)$ from Equation (9.70) because the use of w-pointers ensures that $\Lambda_1(\cdot) = 1$.

9.3.6 Algorithm

To determine the non-linear multi-port frequency domain Volterra transfer function $H_{m_1,...,m_K}(\cdot)$ the following algorithm can be used.

1. Specify the number of controlling variables for all Q non-linear subsystems, $\{P(1),\ldots,P(Q)\}$. Specify $j_q = [j_{q,1},\ldots,j_{q,P(q)}]^T$ for all $q \in \{1,2,\ldots,Q\}$ where j_q contains all subscript numbers for the relevant controlling variables $\boldsymbol{x}(f) = [x_1(f),\ldots,x_R(f)]^T$. For example, if $j_q = [1,2,4]^T$ for non-linear subsystem $q \in \{1,2,\ldots,Q\}$ then the corresponding controlling variables are $\{x_1(f), x_2(f), x_4(f)\}$.

2. Determine expressions for the system matrices and vectors of the linear system: $\boldsymbol{A}^{-1}(f)$, $\boldsymbol{B}(f)$, $\boldsymbol{a}(f)$ and $\boldsymbol{b}(f)$.

3. Specify the multi-port frequency domain Volterra transfer function $(G_q)_{n_1,\ldots,n_{P(q)}}(\cdot)$ for all the non-linear subsystems $q \in \{1,2,\ldots,Q\}$.

4. Determine the controlling vector $\boldsymbol{x}_{\|\boldsymbol{m}\|}(\|\boldsymbol{\psi}\|)$. This is done by a recursive method where higher order variables are determined from lower order variables as follows:

   - Determine $x_1(\psi_{k,l})$ for all $k \in \{1,2,\ldots,K\}$ and $l \in \{1,2,\ldots,m_k\}$ using Equation (9.53).

   - Determine $x_2(\boldsymbol{\psi}^T\boldsymbol{e})$ for all $\boldsymbol{e}$ where $\|\boldsymbol{e}\| = 2$ using Equations (9.68) and (9.78). The vector $\boldsymbol{e}$ is given by

$$\boldsymbol{e} = [e_{1,1},\ldots,e_{1,m_1},\ldots\ldots,e_{K,1},\ldots,e_{K,m_K}]^T \in \{0,1\}^{\|\boldsymbol{m}\|\times 1} \tag{9.79}$$

   where the element $e_{k,l} \in \{0,1\}$ for all $k \in \{1,2,\ldots,K\}$ and $l \in \{1,2,\ldots,m_k\}$.

   - Determine $x_3(\boldsymbol{\psi}^T\boldsymbol{e})$ for all $\boldsymbol{e}$ where $\|\boldsymbol{e}\| = 3$ using Equations (9.68) and (9.78).

   - Determine $x_4(\boldsymbol{\psi}^T\boldsymbol{e})$ and so on.

   - • •

   - Determine $x_{\|\boldsymbol{m}\|}(\boldsymbol{\psi}^T\boldsymbol{e})$ for all $\boldsymbol{e}$ where $\|\boldsymbol{e}\| = \|\boldsymbol{m}\|$ using Equations (9.68) and (9.78). The only $\boldsymbol{e}$ which fulfils this is $\boldsymbol{e} = 1$. Thus the desired controlling variables $\boldsymbol{x}_{\|\boldsymbol{m}\|}(\|\boldsymbol{\psi}\|)$ are determined.

5. Determine $H_{m_1,...,m_K}(\psi_{1,1},\ldots,\psi_{1,m_1};\cdots\cdots;\psi_{K,1},\ldots,\psi_{K,m_K})$ using Equation (9.38) if $\|\boldsymbol{m}\| = 1$ and Equation (9.41) if $\|\boldsymbol{m}\| \in \{2,3,\ldots,\infty\}$.

9.4   Computer implementation

This section briefly discusses the computer implementation of the theory in Section 9.3 to determine non-linear multi-port frequency domain Volterra transfer functions. The implementation is made in the *MAPLE V Release 3* symbolic programming language which, for example, makes it possible to determine algebraic expressions for the multi-port Volterra transfer functions. The source code for the program is included in appendix E.

9.4.1   Some introductory considerations

It can be derived from the method described in section 9.3 that the number of contributions of $x_o(\boldsymbol{\psi}^T \boldsymbol{e})$ for a given $o = \|\boldsymbol{e}\| \in \{1, 2, \ldots, \|\boldsymbol{m}\|\}$ is given by

$$T_o(\|\boldsymbol{m}\|) \quad = \quad \frac{\|\boldsymbol{m}\|!}{o! \, (\|\boldsymbol{m}\| - o)!} \tag{9.80}$$

Thus the total number of evaluations required to determine $x_{\|\boldsymbol{m}\|}(\|\boldsymbol{\psi}\|)$ is given by

$$T(\|\boldsymbol{m}\|) \quad = \quad \sum_{o=1}^{\|\boldsymbol{m}\|} T_o(\|\boldsymbol{m}\|) \tag{9.81}$$

$$= \quad 2^{\|\boldsymbol{m}\|} - 1 \tag{9.82}$$

In Table 9.1, $T_o(\|\boldsymbol{m}\|)$ and $T(\|\boldsymbol{m}\|)$ are listed for some values of o and $\|\boldsymbol{m}\|$. When the multi-port frequency domain Volterra transfer function of a given order $\|\boldsymbol{m}\|$ is to be determined, all lower order Volterra transfer functions at any permutation of $\{m_1, \ldots, m_K\}$ are determined at a very low computational cost. This is because all the required controlling x vectors have already been calculated. Thus $H_{o_1, \ldots, o_K}(\cdot)$ where $o_1 + \cdots + o_K \in \{1, 2, \ldots, \|\boldsymbol{m}\| - 1\}$ is easily determined using Equation (9.38) if $o_1 + \cdots + o_K = 1$ and Equation (9.41) if $o_1 + \cdots + o_K \in \{2, 3, \ldots, \infty\}$.

It can also be shown that the number of $H_{m_1, \ldots, m_K}(\cdot)$ with different frequency arguments that must be determined to evaluate the response $v(f)$ is given by

$$\mathcal{D}(m_1, \ldots, m_K; I_1, \ldots, I_K) \quad = \quad \prod_{k=1}^{K} \frac{1}{m_k!} I_k(I_k + 1) \cdots (I_k + m_k - 1) \tag{9.83}$$

In Table 9.2 examples are given of the number of different transfer functions ('different' in the sense that the frequency arguments are different) that must be determined versus the number of input ports and applied (incommensurate) frequencies.

When $K > 1$ there have to be determined several $H_{m_1, \ldots, m_K}(\cdot)$ transfer functions of the same order $\|\boldsymbol{m}\|$. The number of $H_{m_1, \ldots, m_K}(\cdot)$ of order $\|\boldsymbol{m}\|$ that must be determined for given K is shown in Table 9.3.

Order	$T_o(\|m\|)$							$T(\|m\|)$
$\|m\|$	$o=1$	$o=2$	$o=3$	$o=4$	$o=5$	$o=6$	$o=7$	
1	1							1
2	2	1						3
3	3	3	1					7
4	4	6	4	1				15
5	5	10	10	5	1			31
6	6	15	20	15	6	1		63
7	7	21	35	35	21	7	1	127

Table 9.1: Number of controlling vectors which must be determined, $T_o(\|m\|)$, for a given order $\|m\|$ versus the orders of the individual controlling variables o. If the frequencies form a frequency base then the number of controlling variables listed is the same as the number of different frequencies at which the controlling variables must be determined. The rows indicate that a parallel computer could be efficient in computing the Volterra transfer functions as the calculation of different contributions of the same order o can run in parallel.

9.4.2 Precalculated tables

To make the program run relatively fast, tables are precalculated to determine valid n, o and w vectors from Equation (9.78) where

$$\mathcal{L}_{2,\infty}(\|n\|) \prod_{r=1}^{P(q)} \prod_{p=1}^{n_r} \Lambda_{o_{r,p}}(\|z_{r,p}(w)\|) \prod_{r=1}^{P(q)} \mathcal{Q}(z_{r,1}(w),\ldots,z_{r,n_r}(w)) = 1 \quad (9.84)$$

In the computer implementation the valid o- and w-values are combined in one pointer to give the location of each which controlling x-variable involved in the

K	$I_1 = \cdots = I_K$	$\#H_1$	$\#H_2$	$\#H_3$	$\#H_4$	$\#H_5$	$\#H_6$
1	2	2	3	4	5	6	7
1	4	4	10	20	35	56	84
1	6	6	21	56	126	252	462
2	2	4	10	20	35	56	84
2	4	8	36	120	330	792	1716
2	6	12	78	364	1365	4368	12376

Table 9.2: $\#H_o$ is the number of different $H_{m_1,\ldots,m_K}(\cdot)$ where $m_1 + \cdots + m_K = o$ that must be evaluated at different frequencies to determine $v(f)$ when the symmetry properties of the Volterra transfer functions are utilized.

Order	Number of $H_{m_1,\ldots,m_K}(\cdot)$ of order $\|m\|$					
$\|m\|$	$K = 1$	$K = 2$	$K = 3$	$K = 4$	$K = 5$	$K = 6$
1	1	2	3	4	5	6
2	1	3	6	10	15	21
3	1	4	10	20	35	56
4	1	5	15	35	70	126
5	1	6	21	56	126	252
6	1	7	28	84	210	462

Table 9.3: Number of $H_{m_1,\ldots,m_K}(\cdot)$ that must be determined for a given order m and number of input ports K.

calculation of $(u_q)_{\|m\|}(\|\psi\|)$ where $q \in \{1, 2, \ldots, Q\}$. These table entries are of the form

$$\{\{n_1, \ldots, n_{P(q)}\}, \ \{n_1! \cdots n_{P(q)}!\}, \ \{\|n\| \text{ pointers to } x\text{-variables}\}\} \qquad (9.85)$$

for a given $P(q)$ and $\|m\|$. In (9.85) the coefficient is derived from the fact that there are $n_1! \cdots n_{P(q)}!$ identical contributions to $(u_q)_{\|m\|}(\|\psi\|)$ due to permutations of the partly symmetrical $(G_q)_{n_1,\ldots,n_{P(q)}}(\cdot)$ transfer function as seen from Equation (9.78). The number of entries in this type of table versus the number of controlling variables $P(q)$ for the given non-linear subsystem and the order $\|m\|$ are shown in Table 9.4.

Order	Number of table entries		
$\|m\|$	$P(q) = 1$	$P(q) = 2$	$P(q) = 3$
2	1	4	9
3	4	20	54
4	14	92	306
5	51	452	1863
6	202	2428	12348
7	876	14212	88560

Table 9.4: Number of table entries for the determination of $u_{\|m\|}(\|\psi\|)$ versus the number of controlling variables $P(q)$ and order $\|m\|$.

Also tables are precalculated for the locations of the $2^{\|m\|} - 1$ evaluations required to determine $x_{\|m\|}(\|\psi\|)$ as seen from Equation (9.82). To illustrate why this type of table is necessary consider the following example.

For a given order, o, compare $\boldsymbol{x}_o(\Omega_1)$ and $\boldsymbol{x}_o(\Omega_2)$ where $\Omega_1 \neq \Omega_2$. As has been described previously, the oth order contribution is calculated from contributions of order $1, 2, \ldots, o-1$. Thus the frequencies at which $\boldsymbol{x}$ must be known for lower order contributions to determine $\boldsymbol{x}_o(\Omega_1)$ are not the same as those used to determine $\boldsymbol{x}_o(\Omega_2)$. Thus it is necessary to know where to find the lower order contributions dependent on the given frequency (actually a pointer to the given frequency).

During calculation reference from the former table to the latter is made.

9.4.3 Program

The program named **volfun** consists of three table-generating procedures, an initialization procedure which defines arrays and reads tables into memory and a procedure to calculate the multi-port frequency domain Volterra transfer function. Once the tables are calculated only the latter two are used. The current version of the program **volfun** allows determination of up to tenth order transfer functions and non-linear subsystems with up to two input ports.

The program can be used in three different modes:

- 'algf' – A full algebraic evaluation where the multi-port frequency domain Volterra transfer function is given by a single expression as a function of the elements in the linear network and Volterra transfer functions of the non-linear subsystems in the overall network. This is usually used for visual inspection of relatively low order Volterra transfer functions.

- 'algr' – A recursive algebraic evaluation where a given oth order contribution is given by lower order contributions of the controlling $\boldsymbol{x}$-vectors. The result from this mode can be directly translated into *FORTRAN 77* code using an internal *Maple V* procedure. This mode is used for relatively high orders and where a high speed *FORTRAN 77* version is to be determined.

- 'num' – A numerical evaluation of the Volterra transfer function. As this version does not use hardware floating-point arithmetic this mode is usually used for numerical evaluation of relatively low order Volterra transfer functions. For higher order numerical evaluation the 'algr' mode is used to produce *FORTRAN 77* code.

The user must specify: (i) the number K of signal input ports, (ii) the number Q of controlled variables, (iii) the number R of controlling variables, (iv) index lists $\boldsymbol{j}_q$ for all $q \in \{1, 2, \ldots, Q\}$ to specify which variables control each of the non-linear subsystems, (v) procedures to calculate $\boldsymbol{A}^{-1}(f)$, $\boldsymbol{B}(f)$, $\boldsymbol{a}(f)$ and $\boldsymbol{b}(f)$, (vi) procedures to calculate the multi-port frequency domain Volterra transfer functions $(G_q)_{n_1,\ldots,n_{P(q)}}(\cdot)$ for all $q \in \{1, 2, \ldots, Q\}$ of the non-linear subsystems, and (vii) mode of operation which may be either of the types 'algf', 'algr' or 'num'.

9.5   Some types of non-linear subsystems

This section gives some examples of types of non-linear subsystems. The time domain relation between inputs $\{x_{j_{q,1}}(t), \ldots, x_{j_{q,P(q)}}(t)\}$ where $j_{q,r} \in \{1, 2, \ldots, R\}$ for $r \in \{1, 2, \ldots, P(q)\}$ and output $y_q(t)$ for subsystem $q \in \{1, 2, \ldots, Q\}$ is given, and the corresponding multi-port Volterra transfer function from Equation (9.45), $(G_q)_{n_1, \ldots, n_{P(q)}}(\theta_{1,1}, \ldots, \theta_{1,n_1}; \cdots \cdots; \theta_{P(q),1}, \ldots, \theta_{P(q),n_{P(q)}})$, is presented without derivation. In the following $(\mathcal{K}_q)_{n_1, \ldots, n_{P(q)}}$ is a real constant related to the transfer of signals for the multi-port non-linear system in Figure 9.2.

9.5.1   Type 1

A common type of non-linearity is given by the time domain input-output relation

$$y_q(t) \quad = \quad \sum_{n_1=0}^{\infty} \cdots \sum_{n_{P(q)}=0}^{\infty} (\mathcal{K}_q)_{n_1, \ldots, n_{P(q)}} \prod_{r=1}^{P(q)} x_{j_{q,r}}^{n_r}(t) \qquad (9.86)$$

It can be shown that the corresponding multi-port frequency domain Volterra transfer function is given by

$$(G_q)_{n_1, \ldots, n_{P(q)}}(\theta_{1,1}, \ldots, \theta_{1,n_1}; \cdots \cdots; \theta_{P(q),1}, \ldots, \theta_{P(q),n_{P(q)}}) \quad = \quad (\mathcal{K}_q)_{n_1, \ldots, n_{P(q)}} \qquad (9.87)$$

This type of non-linearity (note from Equation (9.86) that it does not have memory) can be used to represent e.g. a non-linear resistance, conductance, transresistance, transconductance, and current and voltage gain transfer.

9.5.2   Type 2

Consider the following type of non-linearity which contains memory:

$$y_q(t) \quad = \quad \sum_{n_1=0}^{\infty} \cdots \sum_{n_{P(q)}=0}^{\infty} (\mathcal{K}_q)_{n_1, \ldots, n_{P(q)}} \frac{d}{dt}\left[\prod_{r=1}^{P(q)} x_{j_{q,r}}^{n_r}(t)\right] \qquad (9.88)$$

It can be shown that the corresponding multi-port frequency domain Volterra transfer function is given by

$$(G_q)_{n_1, \ldots, n_{P(q)}}(\theta_{1,1}, \ldots, \theta_{1,n_1}; \cdots \cdots; \theta_{P(q),1}, \ldots, \theta_{P(q),n_{P(q)}})$$
$$= \quad j2\pi (\mathcal{K}_q)_{n_1, \ldots, n_{P(q)}} \sum_{r=1}^{P(q)} \sum_{p=1}^{n_r} \theta_{r,p} \qquad (9.89)$$

This type of non-linearity which contains memory can be used to represent e.g. a non-linear current–voltage capacitance.

9.5.3 Type 3

Consider the following type of non-linearity which contains memory:

$$y_q(t) \;=\; \sum_{n_1=0}^{\infty} \cdots \sum_{n_{P(q)}=0}^{\infty} (\mathcal{K}_q)_{n_1,\ldots,n_{P(q)}} \int_{-\infty}^{t} \prod_{r=1}^{P(q)} x_{j_{q,r}}^{n_r}(\tau)\, d\tau \qquad (9.90)$$

It can be shown that the corresponding multi-port frequency domain Volterra transfer function is given by

$$
(G_q)_{n_1,\ldots,n_{P(q)}}(\theta_{1,1},\ldots,\theta_{1,n_1}; \cdots\cdots; \theta_{P(q),1},\ldots,\theta_{P(q),n_{P(q)}})
$$
$$
= \frac{(\mathcal{K}_q)_{n_1,\ldots,n_{P(q)}}}{j2\pi \sum_{r=1}^{P(q)} \sum_{p=1}^{n_r} \theta_{r,p}} \qquad (9.91)
$$

This type of non-linearity which contains memory can be used to represent e.g. a non-linear voltage–current capacitance.

9.5.4 Type 4

Consider the following type of non-linearity which contains memory:

$$y_q(t) \;=\; \sum_{n_1=0}^{\infty} \cdots \sum_{n_{P(q)}=0}^{\infty} (\mathcal{K}_q)_{n_1,\ldots,n_{P(q)}} \prod_{r=1}^{P(q)} \left[\frac{d}{dt} x_{j_{q,r}}(t) \right]^{n_r} \qquad (9.92)$$

It can be shown that the corresponding multi-port frequency domain Volterra transfer function is given by

$$
(G_q)_{n_1,\ldots,n_{P(q)}}(\theta_{1,1},\ldots,\theta_{1,n_1}; \cdots\cdots; \theta_{P(q),1},\ldots,\theta_{P(q),n_{P(q)}})
$$
$$
= (j2\pi)^{n_1+\cdots+n_{P(q)}} (\mathcal{K}_q)_{n_1,\ldots,n_{P(q)}} \prod_{r=1}^{P(q)} \prod_{p=1}^{n_r} \theta_{r,p} \qquad (9.93)
$$

This type of non-linearity which contains memory can be used to represent e.g. a non-linear current–voltage inductance.

9.5.5 Type 5

Consider the following type of non-linearity which contains memory

$$y_q(t) \quad = \quad \sum_{n_1=0}^{\infty} \cdots \sum_{n_{P(q)}=0}^{\infty} (\mathcal{K}_q)_{n_1,\dots,n_{P(q)}} \prod_{r=1}^{P(q)} \left[\int_{-\infty}^{t} x_{j_{q,r}}(\tau)\, d\tau \right]^{n_r} \qquad (9.94)$$

It can be shown that the corresponding multi-port frequency domain Volterra transfer function is given by

$$(G_q)_{n_1,\dots,n_{P(q)}} \left(\theta_{1,1}, \dots, \theta_{1,n_1}; \cdots \cdots; \theta_{P(q),1}, \dots, \theta_{P(q),n_{P(q)}} \right)$$

$$= \quad \frac{(\mathcal{K}_q)_{n_1,\dots,n_{P(q)}}}{(j2\pi)^{n_1+\cdots+n_{P(q)}} \prod_{r=1}^{P(q)} \prod_{p=1}^{n_r} \theta_{r,p}} \qquad (9.95)$$

This type of non-linearity which contains memory can be used to represent e.g. a non-linear voltage–current inductance.

9.6 Examples

The theory presented in the previous sections will be used to derive the non-linear Volterra transfer functions in four examples. *Maple V* source code for a program to determine algebraic expressions for transfer functions and for the example are included i appendix E.

9.6.1 Example 1

Here the non-linear Volterra transfer function between input s_1 and response v_o in Figure 9.3(a) will be determined. The time domain relation for the non-linear conductance is given by

$$i_o(t) \quad = \quad i_o(v_o) \quad = \quad g_1\, v_o(t) + g_2\, v_o^2(t) \qquad (9.96)$$

The modified non-linear system is shown in Figure 9.3(b) where the linear and non-linear parts of the overall network have been separated. Using the algorithm in section 9.3 the relation for the linear system is given by

$$y_1(f) \quad = \quad A_{1,1}(f)\, x_1(f) + B_{1,1}(f)\, s_1(f) \qquad (9.97)$$

where $x_1(f)$ is the controlling voltage across and $y_1(f)$ is the controlled current through the non-linear conductance. Thus it can be shown from Figure 9.3(b) that

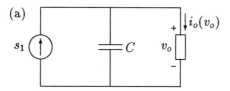

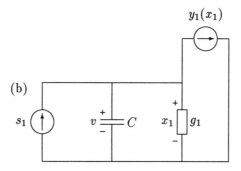

Figure 9.3: Example 1: Non-linear network with one signal input port and one non-linear element. (a) Original network; (b) modified network where the non-linear subsystem is separated from the linear system.

$$A_{1,1}(f) = -(g_1 + j2\pi fC) \tag{9.98}$$
$$B_{1,1}(f) = 1 \tag{9.99}$$

Furthermore

$$v(f) = a_1(f)x_1(f) + b_1(f)s_1(f) \tag{9.100}$$

Thus as $v(f) = x_1(f)$ then

$$a_1(f) = 1 \tag{9.101}$$
$$b_1(f) = 0 \tag{9.102}$$

As seen from Equation (9.96) and Type 1 in section 9.5 the non-linear transfer function relating $x_1(f)$ and $y_1(f)$, $(G_1)_{n_1}(\theta_{1,1}, \ldots, \theta_{1,n_1})$, is given by

$$(G_1)_{n_1}(\theta_{1,1}, \ldots, \theta_{1,n_1}) = \begin{cases} g_2 & \text{for } n_1 = 2 \\ 0 & \text{otherwise} \end{cases} \tag{9.103}$$

Using Equations (9.38) and (9.53) the first order transfer function is given by

$$H_1(\psi_{1,1}) \quad = \quad (x_1)_1(\psi_{1,1}) \quad = \quad \frac{1}{g_1 + j2\pi\psi_{1,1}C} \tag{9.104}$$

The second order non-linear frequency domain Volterra transfer function is determined using Equations (9.41), (9.68) and (9.78) as

$$H_2(\psi_{1,1}, \psi_{1,2}) \quad = \quad \frac{1}{2}(x_1)_2(\psi_{1,1} + \psi_{1,2}) \tag{9.105}$$

where $(x_1)_2(\psi_{1,1} + \psi_{1,2})$ can be derived from

$$
\begin{aligned}
(x_1)_2(\psi_{1,1} + \psi_{1,2}) \quad = \quad & -2\,(G_1)_2(\psi_{1,1}, \psi_{1,2})\,H_1(\psi_{1,1}) \\
& \times H_1(\psi_{1,2})\,H_1(\psi_{1,1} + \psi_{1,2})
\end{aligned}
\tag{9.106}
$$

Using Equation (9.103) gives

$$H_2(\psi_{1,1}, \psi_{1,2}) \quad = \quad -g_2\,H_1(\psi_{1,1})\,H_1(\psi_{1,2})\,H_1(\psi_{1,1} + \psi_{1,2}) \tag{9.107}$$

The third order transfer function can be determined in a similar way as

$$H_3(\psi_{1,1}, \psi_{1,2}, \psi_{1,3}) \quad = \quad \frac{1}{6}(x_1)_3(\psi_{1,1} + \psi_{1,2} + \psi_{1,3}) \tag{9.108}$$

where $(x_1)_3(\psi_{1,1} + \psi_{1,2} + \psi_{1,3})$ can be derived from

$$
\begin{aligned}
&(x_1)_3(\psi_{1,1} + \psi_{1,2} + \psi_{1,3}) \\
& = \quad -2\,\Big\{ (G_1)_2(\psi_{1,1}, \psi_{1,2} + \psi_{1,3})\,(x_1)_1(\psi_{1,1})\,(x_1)_2(\psi_{1,2} + \psi_{1,3}) \\
& \qquad + (G_1)_2(\psi_{1,2}, \psi_{1,1} + \psi_{1,3})\,(x_1)_1(\psi_{1,2})\,(x_1)_2(\psi_{1,1} + \psi_{1,3}) \\
& \qquad + (G_1)_2(\psi_{1,3}, \psi_{1,1} + \psi_{1,2})\,(x_1)_1(\psi_{1,3})\,(x_1)_2(\psi_{1,1} + \psi_{1,2}) \Big\} \\
& \qquad \times H_1(\psi_{1,1} + \psi_{1,2} + \psi_{1,3}) \\
& \qquad - 6\,(G_1)_3(\psi_{1,1}, \psi_{1,2}, \psi_{1,3})\,(x_1)_1(\psi_{1,1})\,(x_1)_1(\psi_{1,2})\,(x_1)_1(\psi_{1,3}) \\
& \qquad \times H_1(\psi_{1,1} + \psi_{1,2} + \psi_{1,3})
\end{aligned}
\tag{9.109}
$$

Thus using Equation (9.103) the third order frequency domain Volterra transfer function is given as

$$
\begin{aligned}
H_3(\psi_{1,1}, \psi_{1,2}, \psi_{1,3}) \quad = \quad & \frac{2}{3}\,g_2^2\,H_1(\psi_{1,1})\,H_1(\psi_{1,2})\,H_1(\psi_{1,3}) \\
& \times H_1(\psi_{1,1} + \psi_{1,2} + \psi_{1,3}) \\
& \times \Big\{ H_1(\psi_{1,1} + \psi_{1,2}) + H_1(\psi_{1,1} + \psi_{1,3}) \\
& \qquad + H_1(\psi_{1,2} + \psi_{i,3}) \Big\}
\end{aligned}
\tag{9.110}
$$

These results for the transfer functions are the same as those obtained by Bussgang, Ehrman and Graham [3] using the traditional probing method.[4]

A somewhat similar example to the one in Figure 9.3 has been given by Maas [10, pp. 179–186] which includes a Type 5 non-linearity given in section 9.5. Using the method in this chapter, results are obtained which are in agreement with Maas.

9.6.2 Example 2

Here the non-linear Volterra transfer function between input s_1 and response $v_c(i_c)+v_z$ in Figure 9.4(a) will be determined. The overall network contains two single-port non-linear elements where the time domain relations are given as

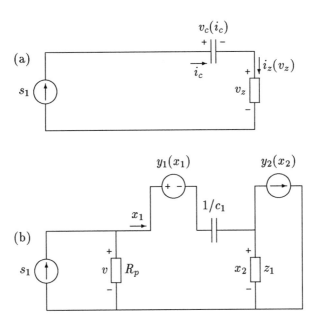

Figure 9.4: Example 2: Non-linear network with one signal input port and two non-linear elements. (a) Original network; (b) modified network where the non-linear subsystems are separated from the linear system (note that the original network is perturbed by the resistance R_p to avoid non-unique system matrices and vectors).

$$v_c(t) \quad = \quad v_c(i_c) \quad = \quad \sum_{n_1=1}^{\infty} c_{n_1} \left[\int_{-\infty}^{t} i_c(\tau) \, d\tau \right]^{n_1} \tag{9.111}$$

---

[4]It should be noted that there are two errors in [3] regarding this example. In Equation (3.19) the factor $-\frac{1}{3}$ should be $-\frac{2}{3}$, and in Equation (3.20) the factor $-\frac{1}{3}$ should be $+\frac{2}{3}$.

where c_{n_1} is a real constant for all n_1 and

$$i_z(t) \quad = \quad i_z(v_z) \quad = \quad \sum_{n_1=1}^{\infty} \left\{ l_{n_1} \left[\int_{-\infty}^{t} v_z(\tau)\, d\tau \right]^{n_1} + g_{n_1} v_z^{n_1}(t) \right\} \qquad (9.112)$$

where l_{n_1} and g_{n_1} are real constants for all n_1. A linear resistor, R_p, is introduced in the modified network in Figure 9.4(b). This is necessary because the system matrices and vectors otherwise will be non-unique (infinity elements in the matrices and vectors). Each Volterra transfer function from Figure 9.4(b) is denoted with a prime. Thus the wanted Volterra transfer function $H_{m_1}(\psi_{1,1}, \ldots, \psi_{1,m_1})$ is determined as

$$H_{m_1}(\psi_{1,1}, \ldots, \psi_{1,m_1}) \quad = \quad \lim_{R_p \to \infty} H'_{m_1}(\psi_{1,1}, \ldots, \psi_{1,m_1}) \qquad (9.113)$$

From Figure 9.4(b) the system matrices and vectors can be determined as

$$\boldsymbol{A}(f) \quad = \quad \frac{1}{j2\pi f} \begin{bmatrix} -(j2\pi f R_p + c_1) & -j2\pi f \\ j2\pi f & -(j2\pi f g_1 + l_1) \end{bmatrix} \qquad (9.114)$$

$$\boldsymbol{B}(f) \quad = \quad \begin{bmatrix} R_p \\ 0 \end{bmatrix} \qquad (9.115)$$

and

$$\boldsymbol{a}(f) \quad = \quad [-R_p,\ 0]^T \qquad (9.116)$$
$$\boldsymbol{b}(f) \quad = \quad [R_p]^T \qquad (9.117)$$

The frequency domain Volterra transfer functions for the two non-linear elements are determined from Types 1 and 5 in section 9.5 as

$$(G_1)_{n_1}(\theta_{1,1}, \ldots, \theta_{1,n_1}) \quad = \quad \frac{c_{n_1}}{(j2\pi)^{n_1} \theta_{1,1} \cdots \theta_{1,n_1}} \qquad (9.118)$$

and

$$(G_2)_{n_1}(\theta_{1,1}, \ldots, \theta_{1,n_1}) \quad = \quad \frac{l_{n_1}}{(j2\pi)^{n_1} \theta_{1,1} \cdots \theta_{1,n_1}} + g_{n_1} \qquad (9.119)$$

Using Equation (9.38) the first order controlling variables are determined as

$$\boldsymbol{x}_1(\psi_{1,1}) \quad = \quad \frac{-4\pi^2 \psi_{1,1}^2 R_p}{(j2\pi\psi_{1,1} g_1 + l_1)(j2\pi\psi_{1,1} R_p + c_1) - 4\pi^2\psi_{1,1}^2}$$
$$\times \left[g_1 + \frac{l_1}{j2\pi\psi_{1,1}},\ 1 \right]^T \qquad (9.120)$$

Then by use of Equation (9.53) the first order Volterra transfer function is determined as

$$H_1(\psi_{1,1}) \quad = \quad \lim_{R_p \to \infty} \left([-R_p, \ 0] \ x_1(\psi_{1,1}) + R_p \right) \tag{9.121}$$

$$= \quad \frac{c_1}{j2\pi\psi_{1,1}} + \frac{j2\pi\psi_{1,1}}{j2\pi\psi_{1,1}g_1 + l_1} \tag{9.122}$$

Then, using Equations (9.68) and (9.78) the second order controlling variables are determined as

$$x_2(\psi_{1,1} + \psi_{1,2}) \quad = \quad \left[\begin{array}{l} 2\,(G_1)_2(\psi_{1,1}, \psi_{1,2})\ (x_1)_1(\psi_{1,1})\ (x_1)_1(\psi_{1,2}) \\ 2\,(G_2)_2(\psi_{1,1}, \psi_{1,2})\ (x_2)_1(\psi_{1,1})\ (x_2)_1(\psi_{1,2}) \end{array} \right] \tag{9.123}$$

Thus by use of Equations (9.41) and (9.120) the second order frequency domain Volterra transfer function is determined as

$$H_2(\psi_{1,1}, \psi_{1,2}) \quad = \quad \lim_{R_p \to \infty} \left(\frac{1}{2} [-R_p, \ 0]\ x_2(\psi_{1,1} + \psi_{1,2}) \right) \tag{9.124}$$

$$= \quad \frac{-c_2}{4\pi^2\psi_{1,1}\psi_{1,2}} + \left(\frac{l_2}{4\pi^2\psi_{1,1}\psi_{1,2}} - g_2 \right)$$

$$\times \frac{j2\pi(\psi_{1,1} + \psi_{1,2})}{j2\pi(\psi_{1,1} + \psi_{1,2})g_1 + l_1}$$

$$\times \frac{j2\pi\psi_{1,1}}{j2\pi\psi_{1,1}g_1 + l_1} \ \frac{j2\pi\psi_{1,2}}{j2\pi\psi_{1,2}g_1 + l_1} \tag{9.125}$$

Higher order Volterra transfer functions can be determined in the same way. The results obtained are in agreement with Chua and Ng [6] who have determined the Volterra transfer functions up to second order.

9.6.3   Example 3

Here the non-linear Volterra transfer functions between inputs $\{s_1, s_2, s_3\}$ and response v_{ds} in Figure 9.5(a) will be determined. The overall network contains two one-port non-linear elements and one non-linear transconductance with the following time domain relations:

$$i_g(t) \quad = \quad i_g(v_g) \quad = \quad \sum_{n_1=1}^{3} c_{g,n_1} \frac{d}{dt} v_g^{n_1}(t) \tag{9.126}$$

$$i_d(t) \quad = \quad i_d(v_g) \quad = \quad \sum_{n_1=1}^{3} g_{m,n_1}\, v_g^{n_1}(t) \tag{9.127}$$

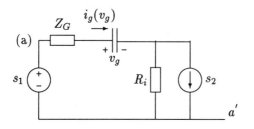

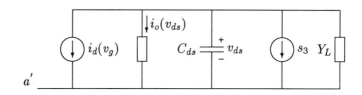

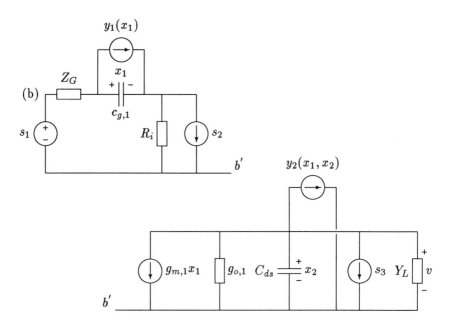

Figure 9.5: Example 3: Non-linear network with three non-linear elements and three signal input ports. (a) Original network; (b) modified network where the non-linear subsystems are separated from the linear system.

and

$$i_o(t) \;=\; i_o(v_{ds}) \;=\; \sum_{n_1=1}^{3} g_{o,n_1}\, v_{ds}^{n_1}(t) \tag{9.128}$$

where c_{g,n_1}, g_{m,n_1} and g_{o,n_1} are real constants for $n_1 \in \{1,2,3\}$. The modified overall network is shown in Figure 9.5(b) where the non-linear conductance and the non-linear transconductance are represented at the same controlled generator, $y_2(x_1,x_2)$. Thus the overall network has two controlling variables and two controlled variables. It should be noted that the controlling variable x_1 controls both the non-linear capacitor and the non-linear transconductance. From Figure 9.5(b) the system matrices and vectors can be determined as

$$\boldsymbol{A}(f) \;=\; \begin{bmatrix} -Y_i(f) & 0 \\ -g_{m,1} & -Y_o(f) \end{bmatrix} \tag{9.129}$$

$$\boldsymbol{B}(f) \;=\; \begin{bmatrix} Y_s(f) & R_i\, Y_s(f) & 0 \\ 0 & 0 & -1 \end{bmatrix} \tag{9.130}$$

where

$$Y_s(f) \;=\; \frac{1}{R_i + Z_G(f)} \tag{9.131}$$

$$Y_i(f) \;=\; Y_s(f) + j2\pi f c_{g,1} \tag{9.132}$$

$$Y_o(f) \;=\; g_{o,1} + Y_L(f) + j2\pi f\, C_{ds} \tag{9.133}$$

and

$$\boldsymbol{a}(f) \;=\; [0,\ 1]^T \tag{9.134}$$

$$\boldsymbol{b}(f) \;=\; [0,\ 0,\ 0]^T \tag{9.135}$$

From Equations (9.126), (9.127) and (9.128) and by use of Types 1 and 2 in section 9.5 the Volterra transfer functions for the two non-linear subsystems are derived as

$$(G_1)_{n_1}(\theta_{1,1},\ldots,\theta_{1,n_1}) \;=\; \begin{cases} j2\pi\, c_{g,n_1}(\theta_{1,1} + \cdots + \theta_{1,n_1}) & \text{for} \quad n_1 \in \{2,3\} \\ 0 & \text{otherwise} \end{cases} \tag{9.136}$$

and

$$(G_2)_{n_1,n_2}(\theta_{1,1},\ldots,\theta_{1,n_1},\theta_{2,1},\ldots,\theta_{2,n_2}) =$$
$$\begin{cases} g_{m,n_1} & \text{for } n_1 \in \{2,3\} \quad \wedge \quad n_2 = 0 \\ g_{o,n_2} & \text{for } n_1 = 0 \quad \wedge \quad n_2 \in \{2,3\} \\ 0 & \text{otherwise} \end{cases} \tag{9.137}$$

In this example there are $K = 3$ signal input ports. This means that there are three first order, six second order and ten third order Volterra transfer functions as seen from Table 9.3. In the following only a few of these are shown as the equations are quite lengthy. Using the algorithm in section 9.3 the following Volterra transfer functions are obtained. The first order frequency domain Volterra transfer function from port 1 is given by

$$H_{1,0,0}(\psi_{1,1}) = \frac{-g_{m,1}\, Y_s(\psi_{1,1})}{Y_i(\psi_{1,1})\, Y_o(\psi_{1,1})} \tag{9.138}$$

The second order frequency domain Volterra transfer function from port 1 is given by

$$H_{2,0,0}(\psi_{1,1},\psi_{1,2}) =$$
$$\frac{Y_s(\psi_{1,1})\, Y_s(\psi_{1,2})}{Y_i(\psi_{1,1}+\psi_{1,2})\, Y_o(\psi_{1,1}+\psi_{1,2})\, Y_i(\psi_{1,1})\, Y_i(\psi_{1,2})\, Y_o(\psi_{1,1})\, Y_o(\psi_{1,2})}$$
$$\times \Big\{ j2\pi(\psi_{1,1}+\psi_{1,2})\, c_{g,2}\, g_{m,1}\, Y_o(\psi_{1,1})\, Y_o(\psi_{1,2})$$
$$- g_{o,2}\, g_{m,1}^2\, Y_i(\psi_{1,1}+\psi_{1,2})$$
$$+ g_{m,2}\, Y_i(\psi_{1,1}+\psi_{1,2})\, Y_o(\psi_{1,1})\, Y_o(\psi_{1,2}) \Big\} \tag{9.139}$$

and the second order multi-port frequency domain Volterra transfer function from input ports 1 and 3 is given by

$$H_{1,0,1}(\psi_{1,1},\psi_{3,1}) = \frac{-2\, g_{m,1}\, g_{o,2}\, Y_s(\psi_{1,1})}{Y_o(\psi_{1,1}+\psi_{3,1})\, Y_i(\psi_{1,1})\, Y_o(\psi_{1,1})\, Y_o(\psi_{3,1})} \tag{9.140}$$

The results for the Volterra transfer functions from port 1 from this example are in agreement with Minasian [11] for orders up to three which is the highest order considered by him. It should be noted that Minasian has represented the non-linear capacitor and transconductance as two separate one-port non-linear generators and not by one two-port generator as in this work. Using one two-port non-linear element instead of two one-port non-linear elements, the number of non-linear elements in the modified network is reduced by 1 which reduces the sizes of the system matrices $A(f)$ and $B(f)$.

9.6.4   Example 4

Here the non-linear Volterra transfer function between input s_1 and response v_r in Figure 9.6(a) will be determined. In Figure 9.6, $R_1 = 4\ \Omega$, $R_2 = 2\ \Omega$, and $C = 1/2$ F.

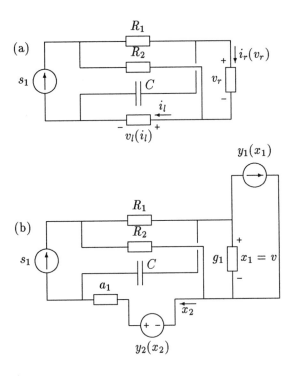

Figure 9.6: Example 4: Non-linear bridge network with one signal input port and two non-linear elements. (a) Original network; (b) modified network where the non-linear subsystems are separated from the linear network.

The overall network contains two single-port non-linear elements where the time domain relations are given as

$$i_r(t) \quad = \quad i_r(v_r) \quad = \quad \sum_{n_1=1}^{2} g_{n_1} v_r^{n_1}(t) \qquad (9.141)$$

$$v_l(t) \quad = \quad v_l(i_l) \quad = \quad \frac{d}{dt} \sum_{n_1=1}^{2} a_{n_1} i_l^{n_1}(t) \qquad (9.142)$$

where $a_1 = 1$ H, $a_2 = 1/3$ HA, $g_1 = 1/2$ S and $g_2 = 1/5$ SV. From Figure 9.6(b) the system matrices and vectors can be determined as

$$A^{-1}(s) = \frac{1}{D(s)} \begin{bmatrix} -3(s^2 + 2) & -3s \\ 3s & -2s \end{bmatrix} \tag{9.143}$$

$$B(s) = \begin{bmatrix} -2/3 \\ 2/s \end{bmatrix} \tag{9.144}$$

where $D(s) = 2s^2 + 3s + 4$ and $s = j2\pi f$ has been used as frequency variable, and

$$a(s) = [1, 0]^T \tag{9.145}$$
$$b(s) = [0]^T \tag{9.146}$$

The Volterra transfer functions for the two non-linear elements are determined from Equations (9.141)–(9.142) as

$$(G_1)_{n_1}(s_{1,1}, \ldots, s_{1,n_1}) = \begin{cases} 1/5 & \text{for} \quad n_1 = 2 \\ 0 & \text{otherwise} \end{cases} \tag{9.147}$$

and

$$(G_2)_{n_1}(s_{1,1}, \ldots, s_{1,n_1}) = \begin{cases} (s_{1,1} + s_{1,2})/3 & \text{for} \quad n_1 = 2 \\ 0 & \text{otherwise} \end{cases} \tag{9.148}$$

The results obtained are in agreement with Chua and Ng [6, Section 6.2] who have determined the Volterra transfer functions up to second order.

9.7   Conclusion

A method has been presented to determine the frequency domain Volterra transfer functions of non-linear multi-port networks containing non-linear multi-port subsystems. The method is based on a generalization of the probing method to allow arbitrary frequencies. The Volterra transfer functions are described as functions of network specific vectors of the linear part of the overall network and as functions of variables which control the non-linear multi-port subsystems. A method has been derived to determine these controlling variables in recursive form.

The method has been implemented in a symbolic programming language (*Maple V Release 3*), which makes it possible to determine algebraic expressions for the Volterra transfer functions. This implementation can directly translate the resulting Volterra transfer functions in recursive algebraic form into *FORTRAN 77* code.

Four examples have been presented. The results obtained are in agreement with existing literature in the cases where comparison has been possible.

9.8   References

[1] Steer, M. B., Chang, C.-R. & Rhyne, G. W.: "Computer-aided analysis of nonlinear microwave circuits using frequency-domain nonlinear analysis techniques: the state of the art", *Int. J. Microwave and Millimeter-wave Computer-aided Engineering*, vol. 1, no. 2, pp. 181–200, 1991.

[2] Gilmore, R. J. & Steer, M. B.: "Nonlinear circuit analysis using the method of harmonic balance — a review of the art. II. Advanced concepts", *Int. J. Microwave and Millimeter-wave Computer-aided Engineering*, vol. 1, no. 2, pp. 159–180, 1991.

[3] Bussgang, J. J., Ehrman, L. & Graham, J. W.: "Analysis of nonlinear systems with multiple inputs", *Proc. IEEE*, vol. 62, no. 8, pp. 1088–1119, 1974.

[4] Maas, S. A.: "A general-purpose computer program for the Volterra-series analysis of nonlinear microwave circuits", *IEEE Microwave Theory and Techniques Symposium Digest*, pp. 311–314. 1988.

[5] Maas, S. A.: "C/NL linear and nonlinear microwave circuit analysis and optimization", Artech House, 1990.

[6] Chua, L. O. & Ng, C.-Y.: "Frequency domain analysis of nonlinear systems: formulation of transfer functions", *IEE J. Electronic Circuits and Systems*, vol. 3, no. 6, pp. 257–269, 1979.

[7] Chua, L. O. & Ng, C.-Y.: "Frequency domain analysis of nonlinear systems: general theory", *IEE J. Electronic Circuits and Systems*, vol. 3, no. 4, pp. 165–185, 1979.

[8] Lighthill, M. J.: "Introduction to Fourier analysis and generalized functions", Cambridge University Press, 1958.

[9] Chua, L. O. & Lin, P.-L.: "Computer-aided analysis of electronic circuits — algorithms and computational techniques", Prentice Hall, New Jersey, 1975.

[10] Maas, S. A.: "Nonlinear microwave circuits", Artech House, 1988.

[11] Minasian, R. A.: "Intermodulation distortion analysis of MESFET amplifiers using the Volterra series representation", *IEEE Trans. Microwave Theory and Techniques*, vol. 28, no. 1, pp. 1–8, 1980.

A

Mathematical concepts

This appendix discusses the mathematical representation of noise. First there is a short introduction to stochastic processes which is the basis for the mathematical treatment of noise. Next the representation of noise in the frequency domain using a Fourier series expansion is discussed in detail. Finally, there is an investigation of signal energies and powers of random variables.

A.1 Stochastic processes

Noise in electrical networks and systems can be considered as random fluctuations in time of, for example, voltage, current or charge. Statistical methods are used to describe the influence of such random noise signals in a qualitative way. These are, for example, to analyze the mean value (which is usually assumed identical to zero), variance, autocorrelation and spectral density of the noise signals.

Using stochastic processes an ensemble of realizations $\{_1\lambda(t_0), _2\lambda(t_0), \ldots, _N\lambda(t_0)\}$ with similar statistical properties are observed at the same instant of time t_0. The ensemble average of the random noise variable $\lambda(t_0)$ is given as

$$\langle \lambda(t_0) \rangle = \lim_{N \to \infty} \frac{1}{N} \sum_{n=1}^{N} {}_n\lambda(t_0) \tag{A.1}$$

$$= \int_{-\infty}^{\infty} \lambda_{t_0} \, \mathcal{P}(\lambda_{t_0}) \, d\lambda_{t_0} \tag{A.2}$$

where $\langle \cdots \rangle$ denotes the ensemble average over realizations with similar statistical properties, and $\mathcal{P}(\lambda_{t_0})$ is the probability density function for the random variable $\lambda(t_0)$. Another important quantity is the autocorrelation function defined as

$$R(t_1, t_2) = \langle \lambda(t_1)\,\lambda^*(t_2) \rangle \tag{A.3}$$

$$= \lim_{N \to \infty} \frac{1}{N} \sum_{n=1}^{N} {}_n\lambda(t_1)\, {}_n\lambda^*(t_2) \tag{A.4}$$

$$= \int_{-\infty}^{\infty} \int_{-\infty}^{\infty} \lambda_{t_1}\, \lambda_{t_2}^*\, \mathcal{P}(\lambda_{t_1}, \lambda_{t_2}^*)\, d\lambda_{t_1}\, d\lambda_{t_2}^* \tag{A.5}$$

where $\mathcal{P}(\lambda_{t_1}, \lambda_{t_2})$ is the joint probability density function for the random variables $\lambda(t_1)$ and $\lambda(t_2)$.

Sometimes, the time average of a random noise signal is used which is a somewhat different concept from the ensemble average. The time average of the noise signal ${}_n\lambda(t)$ of a given realization $n \in \{1, 2, \ldots, N\}$ in the ensemble is defined as

$$\overline{{}_n\lambda(t)} = \lim_{\tau \to \infty} \frac{1}{2\tau} \int_{-\tau}^{\tau} {}_n\lambda(t)\, dt \tag{A.6}$$

where the bar denotes the time average. An auto-correlation function is defined as

$$\overline{{}_n\lambda(t)\, {}_n\lambda(t + \tau')} = \lim_{\tau \to \infty} \frac{1}{2\tau} \int_{-\tau}^{\tau} {}_n\lambda(t)\, {}_n\lambda(t + \tau')\, dt \tag{A.7}$$

As the random noise fluctuations are usually observed versus time, it may seem closest to the physical reality to use time averages instead of ensemble averages. However, ensemble averages are very useful from a theoretical point of view. This is because the (joint) probability density functions for the random variables in many cases can be deduced from theoretical considerations. Thereby it is possible to evaluate the statistical properties of the random variables without the need for time averaging. Another important factor is that the ensemble average usually equals the time average with regard to electrical networks. In case $\langle \lambda(t_0) \rangle = \overline{{}_n\lambda(t)}$ the noise process is called ergodic, which means that the ensemble average is equal to the time average for any realization $n \in \{1, 2, \ldots, N\}$.

The fact that a set of random variables are uncorrelated or independent is often used in the analysis of noise. A set of variables $\{\lambda_1(t), \lambda_2(t), \ldots, \lambda_M(t)\}$ are said to be uncorrelated if

$$\langle \lambda_1(t)\, \lambda_2(t) \cdots \lambda_M(t) \rangle = \langle \lambda_1(t) \rangle\, \langle \lambda_2(t) \rangle \cdots \langle \lambda_M(t) \rangle \tag{A.8}$$

and independent if

$$\mathcal{P}(\lambda_1(t), \lambda_2(t), \ldots, \lambda_M(t)) = \mathcal{P}(\lambda_1(t))\, \mathcal{P}(\lambda_2(t)) \cdots \mathcal{P}(\lambda_M(t)) \tag{A.9}$$

where $\mathcal{P}(\cdots)$ is the (joint) probability density function. This means that if a set of random variables are independent then they are also uncorrelated but not necessarily vice versa.

A.2 Fourier series representation

Generally the noise signals considered are extending over all time, $-\infty < t < \infty$, and have infinite energies. Thus for a real valued random noise signal $\lambda(t)$ where $-\infty < t < \infty$ it is given that

$$\lim_{\tau \to \infty} \int_{-\tau}^{\tau} |\lambda(t)|^2 \, dt \;=\; \infty \tag{A.10}$$

This means that the noise signal $\lambda(t)$ is not square integrable and thus it does not generally have a Fourier transform.[1] Assuming that the noise signal $\lambda(t)$ has finite energy in the finite time interval $-\tau < t < \tau$ where $\tau > 0$ then

$$\int_{-\tau}^{\tau} |\lambda(t)|^2 \, dt \;<\; \infty \tag{A.11}$$

In this case the signal $\lambda(t)$ where $-\tau < t < \tau$ can be represented as a Fourier series given by

$$\lambda(t) \;=\; \sum_{p=-\infty}^{\infty} \tilde{\Lambda}(p\xi) \, \exp[j2\pi p\xi t] \tag{A.12}$$

where

$$\tilde{\Lambda}(p\xi) \;=\; \xi \int_{-\tau}^{\tau} \lambda(t) \, \exp[-j2\pi p\xi t] \, dt \tag{A.13}$$

$$\xi \;=\; \frac{1}{\tau} \tag{A.14}$$

In Equation (A.12) the quantity $\tilde{\Lambda}(p\xi)$ is a complex valued random variable in the random process describing the statistical properties of the random noise signal $\lambda(t)$.

In [1] two suggestions for the frequency domain representation of random noise signals are given.

The first suggestion is to represent $\lambda(t)$ in a time interval $-\tau < t < \tau$ and to assume $\lambda(t) = 0$ for $|t| \geq \tau$. The frequency domain representation in this case is the (integral) Fourier transform of $\lambda(t)$ given as $\Lambda(f) = \int_{-\tau}^{\tau} \lambda(t) \, \exp[-j2\pi ft] \, dt$.

The second suggestion is to assume that the random noise signal $\lambda(t)$ is periodic with period 2τ such that $\lambda(t) = \lambda(t + 2n\tau)$ where n is an integer. From this the frequency domain representation is given as a Fourier series similar to Equations (A.12), (A.13) and (A.14). However, the assumption that $\lambda(t)$ is periodic has the unfortunate consequence that the autocorrelation function is also periodic such that $R(t_1, t_2) = R(t_1, t_2 + 2n\tau)$ where n is an integer. This also means that the coefficients

---

[1]It should be noted that it is a sufficient but actually not a necessary condition for a signal to have a Fourier transform that it is square integrable. For example the signal $\cos(2\pi ft)$ where $-\infty < t < \infty$ has a Fourier transform though it is not square integrable.

of the Fourier series are orthogonal.[2] Since $R(t_1, t_2)$ is generally not periodic for the type of signals considered in the present work this suggestion is not useful. If the systems under consideration are linear (single response) then it is not a problem that $R(t_1, t_2)$ is periodic because there is no need for any evaluation between Fourier series coefficients at different frequencies. However, as some of the systems considered in the present book are non-linear (multi-response) the periodicity of $\lambda(t)$ can not be assumed since there may very well be a correlation between two Fourier series coefficients at different frequencies.

Since it is most useful to have some kind of Fourier series representation in the present work, none of the suggestions made in [1] are useful.

The Fourier transform of the random noise signal $\lambda(t)$ in Equation (A.12) is given by

$$\Lambda(f) \;=\; \sum_{p=-\infty}^{\infty} \tilde{\Lambda}(p\xi)\, \delta(f - p\xi) \tag{A.15}$$

where $\delta(\cdots)$ is the Dirac δ-function.[3] It is seen from Equation (A.15) that the frequency resolution in the spectrum of $\Lambda(f)$ is given by ξ. This frequency resolution can be made arbitrarily small by choosing τ sufficiently large. To prove that Equation (A.12) is fulfilled in the time interval $-\tau < t < \tau$ it suffices to show that

$$\left\langle \left| \lambda(t) - \sum_{p=-\infty}^{\infty} \tilde{\Lambda}(p\xi)\, \exp[j2\pi p\xi t] \right|^2 \right\rangle \;=\; 0 \tag{A.16}$$

The autocorrelation function for $\lambda(t)$ is expressed as a Fourier series as

$$R(t_1, t_2) \;=\; \langle \lambda(t_1)\, \lambda^*(t_2) \rangle \tag{A.17}$$

$$\;=\; \sum_{n=-\infty}^{\infty} \tilde{R}(t_1, n\xi)\, \exp[j2\pi n\xi t_2] \tag{A.18}$$

versus time $-\tau < t_2 < \tau$ for a given $-\tau < t_1 < \tau$, and

$$\tilde{R}(t_1, n\xi) \;=\; \xi \int_{-\infty}^{\infty} R(t_1, t_2)\, \exp[-j2\pi n\xi t_2]\, dt_2 \tag{A.19}$$

---

[2] Two Fourier series coefficients $\tilde{\Lambda}_1(p_1\xi)$ and $\tilde{\Lambda}_2(p_2\xi)$ are said to be orthogonal if $\langle \tilde{\Lambda}_1(p_1\xi)\, \tilde{\Lambda}_2^*(p_2\xi) \rangle = 0$ for all integers p_1 and p_2 except when $p_1 = p_2$.
[3] The Dirac δ-function is a generalized function defined by

$$\int_{-\infty}^{\infty} g(a)\, \delta(f - a)\, da \;=\; g(f)$$

This as a consequence implies that

$$\int_{-\infty}^{\infty} \delta(a)\, da \;=\; 1 \qquad \wedge \qquad \delta(c\,a) \;=\; \frac{1}{c}\,\delta(a) \qquad \wedge \qquad g(a)\, \delta(f - a) \;=\; g(f)\, \delta(f - a)$$

An extensive analysis of generalized functions has been made by Lighthill [2].

Using the fact that $|a|^2 = a\,a^*$ in Equation (A.16) gives four terms. The first term can be determined as

$$\langle |\lambda(t)|^2 \rangle \quad = \quad R(t,t) \qquad\qquad (A.20)$$

The second and third terms express the correlation between the time domain signal and a Fourier series coefficient as

$$\left\langle \lambda(t) \sum_{p=-\infty}^{\infty} \tilde{\Lambda}^*(p\xi)\,\exp[-j2\pi p\xi t] \right\rangle$$

$$= \quad \sum_{p=-\infty}^{\infty} \left\langle \lambda(t)\,\xi \int_{-\tau}^{\tau} \lambda^*(t_1)\,\exp[j2\pi p\xi t_1]\,dt_1 \right\rangle \exp[-j2\pi p\xi t]$$

$$= \quad \sum_{p=-\infty}^{\infty} \tilde{R}^*(t,p\xi)\,\exp[-j2\pi p\xi t] \quad = \quad R^*(t,t) \qquad\qquad (A.21)$$

and

$$\left\langle \lambda^*(t) \sum_{p=-\infty}^{\infty} \tilde{\Lambda}(p\xi)\,\exp[j2\pi p\xi t] \right\rangle \quad = \quad R(t,t) \qquad\qquad (A.22)$$

The fourth and last term expresses the correlation between two Fourier series coefficients as

$$\sum_{p_1=-\infty}^{\infty} \sum_{p_2=-\infty}^{\infty} \langle \tilde{\Lambda}(p_1\xi)\,\tilde{\Lambda}^*(p_2\xi) \rangle\,\exp[j2\pi p_1\xi t]\,\exp[-j2\pi p_2\xi t]$$

$$= \quad \sum_{p_1=-\infty}^{\infty} \xi \int_{-\tau}^{\tau} \sum_{p_2=-\infty}^{\infty} \langle \lambda(t_1)\,\tilde{\Lambda}^*(p_2\xi) \rangle\,\exp[-j2\pi p_2\xi t]\,\exp[-j2\pi p_1\xi t_1]\,dt_1$$

$$\times \exp[j2\pi p_1\xi t]$$

$$= \quad \sum_{p_1=-\infty}^{\infty} \tilde{R}(t,p_1\xi)\,\exp[j2\pi p_1\xi t] \quad = \quad R(t,t) \qquad\qquad (A.23)$$

Thus insertion of Equations (A.20) – (A.23) into Equation (A.16) using $R(t,t) = R^*(t,t)$ since $\lambda(t)$ is a real valued signal completes the proof.

It can be shown [3] that if the autocorrelation function $R(t_1,t_2)$ is periodic with period τ then

$$R(t_1,t_2) \quad = \quad R(t_1,t_2+\tau) \qquad\qquad (A.24)$$

Thus the coefficients of the Fourier series expansion are orthogonal which means that

$$\langle \tilde{\Lambda}(p_1\xi)\,\tilde{\Lambda}^*(p_2\xi) \rangle \quad = \quad \begin{cases} \langle |\tilde{\Lambda}(p_1\xi)|^2 \rangle & \text{for} \quad p_1 = p_2 \\ 0 & \text{otherwise} \end{cases} \qquad (A.25)$$

However, in general this property is not valid since the autocorrelation function $R(t_1, t_2)$ is generally not periodic.

Example A.1 Consider a white noise random variable $\lambda(t)$ extending over all time $-\infty < t < \infty$ with the autocorrelation function

$$R(t_1, t_2) = c\,\delta(t_2 - t_1) \qquad (A.26)$$

where c is a positive real constant. The correlation between the frequency domain representation of $\lambda(t)$ at two arbitrary frequencies f_1 and f_2 can be determined as

$$
\begin{aligned}
\langle \Lambda(f_1)\,\Lambda^*(f_2) \rangle &= \left\langle \int_{-\infty}^{\infty} \lambda(t_1)\,\exp[-j2\pi f_1 t_1]\,dt_1 \int_{-\infty}^{\infty} \lambda^*(t_2)\,\exp[j2\pi f_2 t_2]\,dt_2 \right\rangle \\
&= \int_{-\infty}^{\infty}\int_{-\infty}^{\infty} \delta(t_2 - t_1)\,\exp\left[-j2\pi(f_1 t_1 - f_2 t_2)\right]\,dt_1\,dt_2 \\
&= \delta(f_1 - f_2) \qquad (A.27)
\end{aligned}
$$

This means that the correlation of white noise at two different frequencies, $f_1 \neq f_2$, is zero and that there is correlation not equal to zero only when the two frequencies are identical, $f_1 = f_2$.

Example A.2 Consider a white noise random variable $\lambda(t)$ in a finite time interval $-\tau < t < \tau$ where the correlation between two Fourier coefficients of arbitrary frequencies f_1 and f_2 is to be determined. The autocorrelation function for the time domain random variable $\lambda(t)$ is

$$
R(t_1, t_2) = \begin{cases} c\,\delta(t_2 - t_1) & \text{for} \quad -\tau < t < \tau \\ 0 & \text{otherwise} \end{cases} \qquad (A.28)
$$

The correlation between the two Fourier series coefficients is given by

$$
\begin{aligned}
\langle \tilde{\Lambda}(f_1)\,\tilde{\Lambda}^*(f_2) \rangle &= \frac{1}{4\,\tau^2}\int_{-\tau}^{\tau}\int_{-\tau}^{\tau} \langle \lambda(t_1)\,\lambda^*(t_2) \rangle \\
&\qquad \times \exp[-j2\pi f_1 t_1]\,\exp[j2\pi f_2 t_2]\,dt_1\,dt_2 \\
&= \frac{c}{4\,\tau^2}\int_{-\tau}^{\tau}\exp[-j2\pi(f_1 - f_2)t_2]\,dt_2 \\
&= \frac{c}{2\,\tau}\,\frac{\sin[\pi(f_1 - f_2)\tau]}{\pi(f_1 - f_2)\tau} \qquad (A.29)
\end{aligned}
$$

The $\sin(a)/a$ function in Equation (A.29) has the following properties as τ approaches infinity:

$$
\lim_{\tau \to \infty} \frac{\sin[\pi(f_1 - f_2)\tau]}{\pi(f_1 - f_2)\tau} = \begin{cases} 1 & \text{for} \quad f_1 = f_2 \\ 0 & \text{otherwise} \end{cases} \qquad (A.30)
$$

As the factor c in Equation (A.29) generally increases with τ, the ensemble average of the two Fourier series coefficients is zero for $f_1 \neq f_2$ and some non-zero quantity for $f_1 = f_2$.

A.3 Signal energy and average power

The total energy $\mathcal{E}_{tot}$ of a random noise signal $\lambda(t)$ averaged over the ensemble of realizations is given by

$$\mathcal{E}_{tot} = \left\langle \int_{-\infty}^{\infty} |\lambda(t)|^2 \, dt \right\rangle \tag{A.31}$$

$$= \left\langle \int_{-\infty}^{\infty} \int_{-\infty}^{\infty} \Lambda(f_1) \, \exp[j2\pi f_1 t] \, df_1 \right. $$
$$\left. \times \int_{-\infty}^{\infty} \Lambda^*(f_2) \, \exp[-j2\pi f_2 t] \, df_2 \, dt \right\rangle \tag{A.32}$$

$$= \int_{-\infty}^{\infty} \langle |\Lambda(f)|^2 \rangle \, df \tag{A.33}$$

Equation (A.33) is based on the assumption that $\lambda(t)$ is square integrable and that $\int_{-\infty}^{\infty} \exp[j2\pi(f_1 - f_2)t] \, dt = \delta(f_1 - f_2)$. In Equation (A.31) the quantity $\mathcal{E}_{tot}$ can be interpreted as the total energy in joules if the right-hand side of Equation (A.31) is divided by 1 Ω where $\lambda(t)$ is a voltage or multiplied by 1 S where $\lambda(t)$ is a current. The case for $\Lambda(f)$ in Equation (A.33) is similar. Note from Equation (A.31) that if $\lambda(t)$ is not square integrable then $\mathcal{E}_{tot} = \infty$. Usually the total energy of a random noise signal is of no interest as the signals are only observed in some finite time interval.

The average power in a random noise signal $\lambda(t)$ observed in the time interval $-\tau < t < \tau$ is given by

$$P = \lim_{\tau \to \infty} \frac{1}{2\tau} \int_{-\tau}^{\tau} \langle |\lambda(t)|^2 \rangle \, dt \tag{A.34}$$

$$= \lim_{\tau \to \infty} \frac{1}{2\tau} \int_{-\infty}^{\infty} \langle |\Lambda(f)|^2 \rangle \, df \tag{A.35}$$

If $\lambda(t)$ is represented as a Fourier series in the time interval $-\tau < t < \tau$ then the average power is given by

$$P = \lim_{\tau \to \infty} \frac{1}{2\tau} \int_{-\tau}^{\tau} \left\langle \sum_{p_1=-\infty}^{\infty} \tilde{\Lambda}(p_1\xi) \, \exp[j2\pi p_1\xi t] \right. $$
$$\left. \times \sum_{p_2=-\infty}^{\infty} \tilde{\Lambda}^*(p_2\xi) \, \exp[-j2\pi p_2\xi t] \right\rangle \, dt \tag{A.36}$$

$$= \sum_{p_1=-\infty}^{\infty} \sum_{p_2=-\infty}^{\infty} \langle \tilde{\Lambda}(p_1\xi) \, \tilde{\Lambda}^*(p_2\xi) \rangle $$
$$\times \lim_{\tau \to \infty} \frac{1}{2\tau} \int_{-\tau}^{\tau} \exp[-j2\pi(p_2 - p_1)\xi t] \, dt \tag{A.37}$$

where

$$\lim_{\tau \to \infty} \frac{1}{2\tau} \int_{-\tau}^{\tau} \exp[-j2\pi(p_2 - p_1)\xi t]\, dt \;\; = \;\; \begin{cases} 1 & \text{for} \;\; p_2 = p_1 \\ 0 & \text{otherwise} \end{cases} \tag{A.38}$$

Insertion of Equation (A.38) into (A.37) gives

$$P \;\; = \;\; \sum_{p=-\infty}^{\infty} \langle |\tilde{\Lambda}(p\xi)|^2 \rangle \tag{A.39}$$

Thus the total average power is determined by summing the ensemble average of the magnitude squared Fourier series coefficients at all the relevant frequencies. From Equation (A.39) an average power density is defined as

$$P'(f) \;\; = \;\; \langle |\tilde{\Lambda}(f)|^2 \rangle \tag{A.40}$$

where it is assumed that τ is chosen so large that there exists an integer p such that $p\xi = f$. It should be noted that f in Equation (A.40) runs over both positive and negative frequencies.

A.4   References

[1] Haus, H. A. (Chairman of IRE Subcommittee 7.9 on Noise) et al.: "Representation of noise in linear twoports", *Proc. IRE*, vol. 48, pp. 69–74, 1960.

[2] Lighthill, M. J.: "Introduction to Fourier analysis and generalized functions", Cambridge University Press, London, 1958.

[3] Papoulis, A.: "Probability, random variables, and stochastic processes", McGraw–Hill, USA, 1965.

B

Expressions for reflection coefficients and exchangeable powers

In the first section of this appendix the reflection coefficients looking into an n-port and also looking into the terminating immittances are derived. In the second section the exchangeable power gain from an arbitrary port is derived.

B.1   Derivation of reflection coefficients

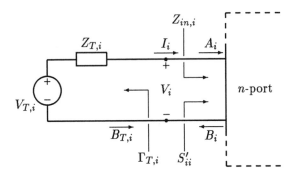

Figure B.1: Reflection coefficients at port i.

The reflection coefficients connected to port i are shown in Figure B.1. The reflection coefficient S'_{ii} is given by

$$S'_{ii} \quad = \quad \frac{B_i}{A_i} \tag{B.1}$$

The power waves A_i and B_i are expressed by the port voltage and current which leads to

$$S'_{ii} = \frac{V_i - Z_i^* I_i}{V_i + Z_i I_i} \tag{B.2}$$

where Z_i is the reference impedance at port i. The input impedance at port i is $Z_{in,i}$ and then

$$S'_{ii} = \frac{Z_{in,i} I_i - Z_i^* I_i}{Z_{in,i} I_i + Z_i I_i}$$

$$= \frac{Z_{in,i} - Z_i^*}{Z_{in,i} + Z_i} \tag{B.3}$$

The reflection coefficient $\Gamma_{T,i}$ is derived from the equation

$$A_i = \Gamma_{T,i} B_i + B_{T,i} \tag{B.4}$$

This means that

$$\Gamma_{T,i} = \left.\frac{A_i}{B_i}\right|_{B_{T,i}=0} \tag{B.5}$$

is correct, as $B_{T,i} = 0$ when $V_{T,i} = 0$.

$$\Gamma_{T,i} = \left.\frac{V_i + Z_i I_i}{V_i - Z_i^* I_i}\right|_{V_{T,i}=0} = \frac{-Z_{T,i} I_i + Z_i I_i}{-Z_{T,i} I_i - Z_i^* I_i}$$

$$= \frac{Z_{T,i} - Z_i}{Z_{T,i} + Z_i^*} \tag{B.6}$$

Comparing Equations (B.3) and (B.6) it is seen that S'_{ii} has the reference impedance conjugated in the numerator and $\Gamma_{T,i}$ in the denominator. Thus care should be shown when using complex reference impedance.

Conjugate match is obtained when

$$Z_{T,i} = Z_{in,i} \quad \text{or} \quad \Gamma_{T,i} = S'^*_{ii}$$

B.2 Incident power wave expressed by exchangeable power

As the incident power wave $B_{T,i}$ in Figure B.1 is an independent node, $B_{T,i}$ can be derived from

$$A_i = \Gamma_{T,i} B_i + B_{T,i} \tag{B.7}$$

and from Equation (B.7)

$$B_{T,i} = A_i\big|_{B_i=0} \tag{B.8}$$

The condition $B_i = 0$ is fulfilled when port i does not reflect power. This means that $S'_{ii} = 0$ or that $Z_{in,i} = Z_i^*$ as shown in Figure B.2.

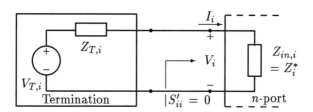

Figure B.2: Reflection free match at port i.

The incident power wave at port i, A_i, can now be written [1]

$$A_i = B_{T,i} = \frac{V_i + Z_i I_i}{2\sqrt{|\mathrm{Re}[Z_i]|}} \tag{B.9}$$

From Figure B.2 V_i and I_i are expressed by $V_{T,i}$ and then inserted in Equation (B.9):

$$
\begin{aligned}
B_{T,i} &= \frac{\frac{Z_i^*}{Z_{T,i}+Z_i^*}V_{T,i} + \frac{Z_i}{Z_{T,i}+Z_i^*}V_{T,i}}{2\sqrt{|\mathrm{Re}[Z_i]|}} \\[2mm]
&= \frac{\mathrm{Re}[Z_i]\,V_{T,i}}{\sqrt{|\mathrm{Re}[Z_i]|}\,V_{T,i}\,(Z_{T,i}+Z_i^*)}
\end{aligned}
$$

Now $\langle|B_{T,i}|^2\rangle$ is derived:

$$
\begin{aligned}
\langle|B_{T,i}|^2\rangle &= \langle B_{T,i}B_{T,i}^*\rangle \\[2mm]
&= \frac{\mathrm{Re}^2[Z_i]\,\langle|V_{T,i}|^2\rangle}{|\mathrm{Re}[Z_i]|\,|Z_{T,i}+Z_i^*|^2} \\[2mm]
&= \frac{\mathrm{Re}[Z_i]}{|\mathrm{Re}[Z_i]|}\cdot\frac{\langle|V_{T,i}|^2\rangle}{4\,\mathrm{Re}[Z_{T,i}]}\cdot\frac{4\,\mathrm{Re}[Z_{T,i}]\,\mathrm{Re}[Z_i]}{|Z_{T,i}+Z_i^*|^2} \tag{B.10}
\end{aligned}
$$

The term $\mathrm{Re}[Z_i]/|\mathrm{Re}[Z_i]|$ ensures that $\langle|B_{T,i}|^2\rangle$ never becomes negative, and the symbol p_i is introduced as

$$
p_i = \frac{\mathrm{Re}[Z_i]}{|\mathrm{Re}[Z_i]|} = \begin{cases} 1 & \text{for } \mathrm{Re}[Z_i] > 0 \\ -1 & \text{for } \mathrm{Re}[Z_i] < 0 \end{cases} \tag{B.11}
$$

The second term is the termination's exchangeable power:

$$P_{e\,S_i} = \frac{\langle |V_{T,i}|^2 \rangle}{4\,\mathrm{Re}[Z_{T,i}]} \tag{B.12}$$

The third term is related to the reflection coefficient of the termination by

$$
\begin{aligned}
\frac{4\,\mathrm{Re}[Z_{T,i}]\,\mathrm{Re}[Z_i]}{|Z_{T,i} + Z_i^*|^2}
&= \frac{(Z_{T,i} + Z_{T,i}^*)(Z_i + Z_i^*)}{|Z_{T,i} + Z_i^*|^2} \\[2mm]
&= \frac{Z_{T,i}Z_i + Z_{T,i}Z_i^* + Z_{T,i}^*Z_i + Z_{T,i}^*Z_i^*}{|Z_{T,i} + Z_i^*|^2} \\[2mm]
&= \frac{(Z_{T,i} + Z_i^*)(Z_{T,i}^* + Z_i) - (Z_{T,i} - Z_i)(Z_{T,i}^* - Z_i^*)}{|Z_{T,i} + Z_i^*|^2} \\[2mm]
&= 1 - \frac{|Z_{T,i} - Z_i|^2}{|Z_{T,i} + Z_i^*|^2} \\[2mm]
&= 1 - |\Gamma_{T,i}|^2 \tag{B.13}
\end{aligned}
$$

Thus

$$\langle |B_{T,i}|^2 \rangle = p_i \cdot P_{e\,S_i} \cdot (1 - |\Gamma_{T,i}|^2) \tag{B.14}$$

As the left side of Equation (B.14) is always positive the signs of the right side add up to be positive as well. This is shown in Table B.1.

$\mathrm{Re}[Z_{T,i}]$	$\mathrm{Re}[Z_i]$	$1 - |\Gamma_{T,i}|^2$	$P_{e\,S_i}$	p_i	$\langle |B_{T,i}|^2 \rangle$
> 0	> 0	> 0	> 0	> 0	> 0
> 0	< 0	< 0	> 0	< 0	> 0
< 0	> 0	< 0	< 0	> 0	> 0
< 0	< 0	> 0	< 0	< 0	> 0

Table B.1: Signs of $\langle |B_{T,i}|^2 \rangle$ and of its three terms as a function of the signs of $\mathrm{Re}[Z_{T,i}]$ and $\mathrm{Re}[Z_i]$.

From Table B.1 it is seen how p_i makes Equation (B.14) positive. p_i is only necessary when $\mathrm{Re}[Z_i] < 0$.

B.3 Reference

[1] Kurokawa, K.: "Power waves and the scattering matrix", *IEEE Trans. on Microwave Theory and Techniques*, vol. MTT-13, pp. 194 – 202, March 1965.

C

Extended noise factor as a hyperboloid of two sheets

The noise factor, F, expressed as a function of the source admittance and the Y noise parameters (Equation (4.11)) is

$$F = 1 + \frac{1}{G_S}\left(G_n + R_n |Y_S + Y_\gamma|^2\right) \tag{C.1}$$

Let $x = G_S$, $y = B_S$ and $z = F$, then Equation (C.1) can be written as

$$x^2 + y^2 - \frac{xz}{R_n} + \left(2G_\gamma + \frac{1}{R_n}\right)x + 2B_\gamma y + G_\gamma^2 + B_\gamma^2 + \frac{G_n}{R_n} = 0 \tag{C.2}$$

In order to reduce Equation (C.2) to canonical form the roots of

$$\begin{vmatrix} 1-\lambda & 0 & \frac{-1}{2R_n} \\ 0 & 1-\lambda & 0 \\ \frac{-1}{2R_n} & 0 & -\lambda \end{vmatrix} = 0 \tag{C.3}$$

are found: $\lambda_1 = 1$, $\lambda_2 = \frac{R_n - \sqrt{R_n^2 + 1}}{2R_n}$ and $\lambda_3 = \frac{R_n + \sqrt{R_n^2 + 1}}{2R_n}$. Thus a translation to the new coordinates $x_1 = x$, $y_1 = y + B_\gamma$ and $z_1 = z - 1 - 2R_nG_\gamma$ moves the quadric surface to its centre.

A rotation given by

$$\left\{ \begin{array}{c} x_1 \\ y_1 \\ z_1 \end{array} \right\} = \left\{ \begin{array}{ccc} 0 & A & B \\ 1 & 0 & 0 \\ 0 & B & A \end{array} \right\} \left\{ \begin{array}{c} x_2 \\ y_2 \\ z_2 \end{array} \right\} \tag{C.4}$$

where

$$A = \frac{1}{\sqrt{2(R_n^2 + 1 + R_n\sqrt{R_n^2 + 1})}}$$

$$B = \sqrt{\frac{R_n^2 + 1 + R_n\sqrt{R_n^2 + 1}}{2(R_n^2 + 1)}}$$

determines the new coordinates (x_2, y_2, z_2).

Equation (C.1) is now rewritten as

$$\frac{y_2^2}{a^2} - \frac{x_2^2}{b^2} - \frac{z_2^2}{c^2} = 1 \qquad\qquad (C.5)$$

where

$$a^2 = \frac{2(R_n G_\gamma^2 + G_n)}{\sqrt{R_n^2 + 1} - R_n}$$

$$b^2 = \frac{R_n}{R_n G_\gamma^2 + G_n}$$

$$c^2 = \frac{2(R_n G_\gamma^2 + G_n)}{\sqrt{R_n^2 + 1} + R_n}$$

and Equation (C.5) is recognized as a hyperboloid of two sheets with centre in $(G_S, B_S, F) = (0, -B_\gamma, 1 + 2\, R_n\, G_\gamma)$.

D

Some useful FORTRAN subroutines

```
      SUBROUTINE DELTA (A,B,C,D,E)
CC    COMPUTES A 2x2 DETERMINANT IN DOUBLE PRECISION
      COMPLEX A,B,C,D,E
      DOUBLE PRECISION AR,AI,BR,BI,CR,CI,DR,DI
      AR=REAL(A)
      AI=AIMAG(A)
      BR=REAL(B)
      BI=AIMAG(B)
      CR=REAL(C)
      CI=AIMAG(C)
      DR=REAL(D)
      DI=AIMAG(D)
      ER=AR*DR-AI*DI-BR*CR+BI*CI
      EI=AR*DI+AI*DR-BR*CI-BI*CR
      E=CMPLX(ER,EI)
      RETURN
      END
C
CC    ***   ***   ***   ***   ***   ***   ***   ***   ***   ***   ***
C
      SUBROUTINE PARALF (Y11,Y12,Y21,Y22,YRN,YGN,YYG,YA,YB,YC,AY11,AY12,
     1AY21,AY22,AYRN,AYGN,AYYG)
CC    COMPUTES SIGNAL AND NOISE PARAMETERS WITH PARALLEL FEEDBACK
      COMPLEX Y11,Y12,Y21,Y22,YYG,YA,YB,YC,AY11,AY12,AY21,AY22,AYYG,AHY
      AY11=Y11+YA+YB
      AY12=Y12-YB
      AY21=Y21-YB
      AY22=Y22+YB+YC
      GA=REAL(YA)
      GB=REAL(YB)
      GC=REAL(YC)
      ADY=(CABS(YB-Y21))**2
      AEY=GB+GC+YRN*((CABS(Y21))**2)
```

271

```
      AHY=GB*(Y11+YA+Y21)+GC*AY11+YRN*(Y11-YYG)*YB*(CONJG(Y21))+YRN*(YYG
     1+YA+YB)*((CABS(Y21))**2)
      ALY=ADY*(GA+YGN)+GB*((CABS(Y11+YA+Y21))**2)+GC*((CABS(AY11))**2)
     1+YRN*((CABS((Y11-YYG)*YB+(YA+YB+YYG)*Y21))**2)
      AYRN=AEY/ADY
      AYGN=ALY/ADY-((CABS(AHY))**2)/(ADY*AEY)
      AYYG=AHY/AEY
      RETURN
      END
C
CC    ***   ***   ***   ***   ***   ***   ***   ***   ***   ***   ***
C
      SUBROUTINE SERIEF (Z11,Z12,Z21,Z22,ZGN,ZRN,ZZG,ZA,ZB,ZC,AZ11,AZ12,
     1AZ21,AZ22,AZGN,AZRN,AZZG)
CC    COMPUTES SIGNAL AND NOISE PARAMETERS WITH SERIES FEEDBACK
      COMPLEX Z11,Z12,Z21,Z22,ZZG,ZA,ZB,ZC,AZ11,AZ12,AZ21,AZ22,AZZG,AHZ,
     1CZ21,C1,C2,C3
      AZ11=Z11+ZA+ZB
      AZ12=Z12+ZB
      AZ21=Z21+ZB
      AZ22=Z22+ZB+ZC
      CZ21=CONJG(Z21)
      B1=(CABS(Z21))**2
      B2=(CABS(Z11+ZA-Z21))**2
      B3=(CABS(AZ11))**2
      C1=(ZZG-Z11)*ZB
      C2=(ZA+ZB+ZZG)*Z21
      C3=-C1*CZ21
      B4=(CABS(C1+C2))**2
      RA=REAL(ZA)
      RB=REAL(ZB)
      RC=REAL(ZC)
      ADZ=(CABS(ZB+Z21))**2
      AEZ=RB+RC+B1*ZGN
      AHZ=(Z11+ZA-Z21)*RB+AZ11*RC-C3*ZGN+(ZA+ZB+ZZG)*B1*ZGN
      ALZ=ADZ*(RA+ZRN)+B2*RB+B3*RC+B4*ZGN
      AZGN=AEZ/ADZ
      AZRN=ALZ/ADZ-((CABS(AHZ))**2)/(ADZ*AEZ)
      AZZG=AHZ/AEZ
      RETURN
      END
C
CC    ***   ***   ***   ***   ***   ***   ***   ***   ***   ***   ***
C
      SUBROUTINE YZTRAN (AY11,AY12,AY21,AY22,AYRN,AYGN,AYYG,AZ11,AZ12,
     1AZ21,AZ22,AZGN,AZRN,AZZG)
CC    TRANSFORMS SIGNAL AND NOISE PARAMETERS FROM Y FORM TO Z FORM
```

```
       COMPLEX AY11,AY12,AY21,AY22,AYYG,AZ11,AZ12,AZ21,AZ22,AZZG,ADY,C
       CALL DELTA (AY11,AY12,AY21,AY22,ADY)
       AZ11=AY22/ADY
       AZ12=-AY12/ADY
       AZ21=-AY21/ADY
       AZ22=AY11/ADY
       A=(CABS(AYYG))**2
       B=A+AYGN/AYRN
       AZGN=B*AYRN
       AZRN=AYGN/B
       C=CONJG(AYYG)
       AZZG=C/B
       RETURN
       END
C
CC     ***   ***   ***   ***   ***   ***   ***   ***   ***   ***   ***
C
       SUBROUTINE ZYTRAN (AZ11,AZ12,AZ21,AZ22,AZGN,AZRN,AZZG,AY11,AY12,
      1AY21,AY22,AYRN,AYGN,AYYG)
CC     TRANSFORMS SIGNAL AND NOISE PARAMETERS FROM Z FORM TO Y FORM
       COMPLEX AZ11,AZ12,AZ21,AZ22,AZZG,AY11,AY12,AY21,AY22,AYYG,ADZ,C
       CALL DELTA (AZ11,AZ12,AZ21,AZ22,ADZ)
       AY11=AZ22/ADZ
       AY12=-AZ12/ADZ
       AY21=-AZ21/ADZ
       AY22=AZ11/ADZ
       A=(CABS(AZZG))**2
       B=A+AZRN/AZGN
       AYRN=B*AZGN
       AYGN=AZRN/B
       C=CONJG(AZZG)
       AYYG=C/B
       RETURN
       END
C
CC     ***   ***   ***   ***   ***   ***   ***   ***   ***   ***   ***
C
       SUBROUTINE YSTRAN (Y11,Y12,Y21,Y22,RN,GN,YG,S11,S12,S21,S22,QN,FO,
      1RO)
CC     TRANSFORMS SIGNAL AND NOISE PARAMETERS FROM Y FORM TO S FORM WITH
CC     F, Q AND GAMMA. REFERENCE FOR S PARAMETERS IS 50 OHM.
       COMPLEX Y11,Y12,Y21,Y22,YG,S11,S12,S21,S22,RO,YA,YB,YC,YN,C
       YR=1./50.
       YA=YR+Y11
       YB=YR+Y22
       YC=Y12*Y21
       YN=YA*YB-YC
```

```
      S11=((YR-Y11)*YB+YC)/YN
      S22=((YR-Y22)*YA+YC)/YN
      S12=-2.*YR*Y12/YN
      S21=-2.*YR*Y21/YN
      GG=REAL(YG)
      BG=AIMAG(YG)
      A=SQRT(RN*GN+((RN*GG)**2))
      F0=1.+2.*(RN*GG+A)
      C=CMPLX(YR+A/RN,BG)
      QN=RN*(CABS(C))**2/YR
      A=YR**2-(CABS(YG))**2-GN/RN
      B=(CABS(C))**2
      C=CMPLX(A,2.*YR*BG)
      R0=C/B
      RETURN
      END
C
CC    ***   ***   ***   ***   ***   ***   ***   ***   ***   ***   ***
C
      SUBROUTINE SYTRAN (S11,S12,S21,S22,QN,F0,R0,Y11,Y12,Y21,Y22,RN,GN,
     1YG)
CC    TRANSFORMS SIGNAL AND NOISE PARAMETERS FROM S FORM WITH F, Q AND
CC    GAMMA TO Y FORM. REFERENCE FOR S PARAMETERS IS 50 OHM.
      COMPLEX S11,S12,S21,S22,R0,Y11,Y12,Y21,Y22,YG,YA,YB,YC,YN
      YR=1./50.
      YA=1.+S11
      YB=1.+S22
      YC=S12*S21
      YN=(YA*YB-YC)/YR
      Y11=((1.-S11)*YB+YC)/YN
      Y22=((1.-S22)*YA+YC)/YN
      Y12=-2.*S12/YN
      Y21=-2.*S21/YN
      A=F0-1.
      B=QN*(1.-(CABS(R0))**2)
      C=QN*(CABS(R0+1.))**2
      GN=YR*A*(B-A)/C
      RN=C/(4.*YR)
      YG=(CMPLX(2.*A-B,2.*QN*(AIMAG(R0))))*YR/C
      RETURN
      END
```

E

Determination of Volterra transfer functions using *Maple*

The present appendix contains program listings for a *Maple V Release 3* program to determine algebraic expressions for Volterra transfer functions. Also the *Maple V* source code for the examples in chapter 9 are included. The two functions listgen and vtf must be in the same file called volfun. Before the functions can be used the user must read the volfun file into a *Maple V* session, and then run the command: volfun[listgen](4):. This generates some tables that are used by the volfun program.

E.1 Program listing for listgen source code

```
##########################################################################
# listgen - listgen procedure for generating o-, m-, and n-tables
##########################################################################
# listgen(MAXORDER)
#
# MAXORDER: Maximum order for the table generation
##########################################################################
volfun[listgen] := proc(MAXORDER)
local i, i1, i2, i3, i4, i5, i6, i7, i8, i9, i10, i21L, j, k, mL, Lambda,
      lm, M, ML, Number, n, n1, n2, no, np, o, ord, order, ordL,
      pt, pt1, pt2, R, Sm, si3,  si4, si5, si6, si7, si8, si9, sm, sn, so3,
      so4, so5, so6, so7, so8, so9, T2, T3, T4, T5, T6, T7, T8, T9, T10, V,
      v1, v2:
global m, ll, ul, iV:

### TEST FOR VALID INPUT DATA #############################################
if (nargs = 1) then
  if not type(MAXORDER, posint) or (MAXORDER < 2) or (MAXORDER > 10) then
```

275

```
      ERROR('Invalid argument')
  fi
else
  ERROR('Invalid number of arguments')
fi:
lprint('working ...'):



### INITIALIZATION #############################################
# Read in combinat package
with(combinat, choose, numbcomb, numbperm, permute):



### Olistgen ##################################################
# Number of elements in i21L list, Sm
Sm := 2^MAXORDER-1:

# Determine i21L and ordL lists
i21L := array(1..Sm):
ordL := array(0..Sm, [(0)=0]):
for m to Sm do ordL[m] := 1 od:

for M to MAXORDER do
  for m from 2^(M-1) to 2^M-1 do
    lm := m - 2^(M-1):
    mL := [1]:
    for i from M-1 by -1 to 1 do
      if (lm >= 2^(i-1)) then
        lm := lm - 2^(i-1):
        mL := [1, op(mL)]:
        ordL[m] := ordL[m] + 1
      else
        mL := [0, op(mL)]
      fi
    od:
    i21L[m] := mL
  od
od:



### Mlistgen ##################################################
# Local procedure ll to determine lower limits for i variables
ll := proc(ino, n, iprev)
if (ino > n) then 0 elif (ino < n) then iprev+1 else 1 fi
end:

# Local procedure ul to determine upper limits for i variables
```

```
ul := proc(ino, n, Sm, inumb, si)
if (ino > n) then 0 else (Sm-inumb-si)/ino fi
end:

# Calculate all M1m.sm lists
for sm from 2 to MAXORDER do
  T.sm := array(1..sm, [(1)=0])
od:
iV := array(1..10):

for sm from 2 to MAXORDER do
  Sm := 2^sm-1:
  no := 0:
  for n from 2 to sm do
    for iV[10] from ll(10, n, 0) to ul(10, n, Sm, 45, 0) do
      for iV[9] from ll(9, n, iV[10]) to ul(9, n, Sm, 36, iV[10]) do
        si9 := iV[10] + iV[9]:
        so9 := ordL[iV[10]] + ordL[iV[9]]:
        if (so9 > sm) then next fi:
        for iV[8] from ll(8, n, iV[9]) to ul(8, n, Sm, 28, si9) do
          si8 := si9 + iV[8]:
          so8 := so9 + ordL[iV[8]]:
          if (so8 > sm) then next fi:
          for iV[7] from ll(7, n, iV[8]) to ul(7, n, Sm, 21, si8) do
            si7 := si8 + iV[7]:
            so7 := so8 + ordL[iV[7]]:
            if (so7 > sm) then next fi:
            for iV[6] from ll(6, n, iV[7]) to ul(6, n, Sm, 15, si7) do
              si6 := si7 + iV[6]:
              so6 := so7 + ordL[iV[6]]:
              if (so6 > sm) then next fi:
              for iV[5] from ll(5, n, iV[6]) to ul(5, n, Sm, 10, si6) do
                si5 := si6 + iV[5]:
                so5 := so6+ordL[iV[5]]:
                if (so5 > sm) then next fi:
                for iV[4] from ll(4, n, iV[5]) to ul(4, n, Sm, 6, si5) do
                  si4 := si5 + iV[4]:
                  so4 := so5 + ordL[iV[4]]:
                  if (so4 > sm) then next fi:
                  for iV[3] from ll(3, n, iV[4]) to ul(3, n, Sm, 3, si4) do
                    si3 := si4 + iV[3]:
                    so3 := so4 + ordL[iV[3]]:
                    if (so3 > sm) then next fi:
                    for iV[2] from ll(2, n, iV[3]) to ul(2, n, Sm, 1, si3)
                    do
                        iV[1] := Sm - si3 - iV[2]:
                        if (so3+ordL[iV[2]]+ordL[iV[1]] = sm) then
```

```
                              no := no + 1:
                              M1m.sm[no] := [[n], [n], [n!], [seq(iV[n-k],
                                                           k=0..n-1)]]
                           fi
                        od
                     od
                  od
               od
            od
         od
      od
   od:
   T.sm[n] := no
 od:
 M1m.sm[0] := no:
od:


# Calculate all M2m.sm lists
for sm from 2 to MAXORDER do
  pt := 0:
  for n1 from 0 to sm do
    for n2 from 0 to sm-n1 do
      sn := n1+n2:
      if (sn >= 2) then
        for no from T.sm[sn-1]+1 to T.sm[sn] do
          V := permute(op(4, M1m.sm[no])):
          for np to numbperm(sn, sn) do
            v1 := [op(1..n1, op(np,V))]:
            v2 := [op(n1+1..sn, op(np,V))]:
            if (sort(v1) = v1) and (sort(v2) = v2) then
              pt := pt+1:
              M2m.sm[pt] := [[sn], [n1,n2], [n1!*n2!], [op(op(np, V))]]
            fi
          od
        od
      fi
    od
  od:
  M2m.sm[0] := pt:
od:


### Nlistgen ######################################################
# Determine Nlistgen lists
for order to MAXORDER do
  no := 0:
```

```
Lambda := [seq(2^(o-1), o=1..order)]:

# case 1
for pt1 to order do
  no := no+1:
  N.order[no] := [[1, op(pt1, Lambda)], [op(pt1, Lambda)]]
od:

# case 2
V := choose(Lambda,2):
for pt1 to numbcomb(order, 2) do
  Number := convert(op(pt1, V), '+'):
  no := no+1:
  N.order[no] := [[2, Number], [op(op(pt1, V))]]
od:

# case>=3
for ord from 3 to order do
  V := choose(Lambda, ord):
  for pt1 to numbcomb(order, ord) do
    v1 := op(pt1, V):
    v2 := choose(v1, 1):
    Number := convert(v1, '+'):
    for pt2 from 2 to ord-1 do
      v2 := [op(v2), op(choose(v1, pt2))]
    od:
    R := [convert(op(1, v2), '+')]:
    for pt2 from 2 to nops(v2) do
      R := [op(R), convert(op(pt2, v2), '+')]
    od:
    R := sort(R):
    no := no+1:
    N.order[no] := [[ord, Number], [op(R)]]
  od
  od
od:


### SAVE RESULTS IN FILE ##########################################
# Save O, M, and N lists in file 'lists.m'
lprint('Saving o-, m-, and n-lists in file: lists.m'):
save i2lL, M1m.(2..MAXORDER), M2m.(2..MAXORDER),
    N.(1..MAXORDER), 'lists.m':
lprint('List generation completed')
end:   # listgen
```

E.2　Program listing for `vtf` source code

```
#####################################################################
# vtf - determination of a given Volterra transfer function
#####################################################################
# vtf(mL, psiL, MAXORDER, sysdefL, caltype)
#
# mL:       List of orders specified as [m1,...,mK]
# psiL:     List of psi frequencies specified as
#           [psi{1,1},...,psi{1,m1},......,psi{K,1},...psi{K,mK}]
# MAXORDER: The maximum order for which the system of analysis is used
# sysdefL:  A system definition list specified as [K, Q, R, jL]
# caltype:  Calculation type specified as either one of: (1) 'algf' for
#           full algebraic evaluation, (2) 'algr' for recursive
#           algebraic evaluation, and (3) 'num' for numerical evaluation
#####################################################################
volfun[vtf] := proc(mL, psiL, MAXORDER, sysdefL, caltype)
local cfak, con, ImdFrqL, i, k, MmL, MptL, NL, NoCgVar, NptL, no, ord,
      order, portL, p, psi, psiGL, q, r, resp, result, Sm, sm, uresult,
      usubresult:
global K, Q, R, jL, AI, B, a, b, u, bk:


### TEST FOR VALID INPUT DATA ######################################
if (nargs = 5) then
  if not type([args],
  [list(nonnegint), list, posint, [posint, posint, posint, list], name])
  then
    ERROR('Invalid arguments')
  fi:
  if (convert(mL, '+') <> nops(psiL)) or (nops(mL) <> sysdefL[1]) then
    ERROR('Unbalanced number of elements in argument lists')
  fi:
  if (nops(sysdefL) <> 4) or (nops(sysdefL[4]) <> sysdefL[2]) then
    ERROR('Invalid sysdefL')
  fi
else
  ERROR('Invalid number of arguments')
fi:


### INITIALIZE #####################################################
if not type(M1m.MAXORDER[0], posint) then
  read 'lists.m':
  if not type(M1m.MAXORDER[0], posint) and (MAXORDER <> 1) then
    ERROR('Execute vsa[listgen](MAXORDER) with MAXORDER = ', MAXORDER)
  fi:
```

```
# Load the necessary procedures from the linalg package
with(linalg, dotprod, multiply, scalarmul, subvector):

# Unassign global variables
unassign('K', 'Q', 'R', 'jL'):

lprint('Initialization completed')
fi:

# Define global variables
if (K <> sysdefL[1]) or (Q <> sysdefL[2]) or (R <> sysdefL[3])
or (jL <> sysdefL[4]) then
  K := sysdefL[1]:
  Q := sysdefL[2]:
  R := sysdefL[3]:
  jL := sysdefL[4]:

# Define vectors and matrices
  AI := array(1..R, 1..Q, sparse):
  B := array(1..Q, 1..K, sparse):
  a := array(1..R, sparse):
  b := array(1..K, sparse):
  u := array(1..Q, sparse):
  bk := array(1..Q, sparse):
  for resp to 2^MAXORDER-1 do
    x.resp := array(1..R):
    X.resp := array(1..R)
  od:

lprint('Global variables updated'):
fi:


### CALCULATE ############################################################
# Calculate the sum sm=m1+...+mK
sm := convert(mL, '+'):

# Calculate the total number of contributions, Sm
Sm := 2^sm - 1:

# Determine the port number list, portL
portL := [seq(seq(k, i=1..op(k, mL)), k=1..K)]:

# Calculate the intermodulation frequency list
ImdFrqL := [seq(dotprod(i2lL[k], [op(1..nops(i2lL[k]), psiL)]), k=1..Sm)]:
```

```
# Calculate all first order x vectors
for no to sm do
  # Determine the input port number, k
  k := op(no, portL):

  # Determine where to put the response, x.resp
  resp := op(2, op(1, N.sm[no])):

  # Determine the frequency at which the x.resp vector is to be determined
  psi := op(resp, ImdFrqL):

  # Update the global inverted A matrix, AI
  AIcal(psi):

  # Update the B matrix and set bk equal to minus the k'th column of B
  Bcal(psi):
  bk := subvector(B, 1..Q, k):
  bk := scalarmul(bk, -1):

  # Determine the x.resp (or X.resp) vector
  if (caltype = 'num') then
    x.resp := map(evalc, multiply(AI, bk))
  elif (caltype = 'algf') then
    x.resp := multiply(AI, bk)
  else
    X.resp := multiply(AI, bk)
  fi
od:
# Now all first order x.resp vectors are determined

# Calculate all higher order x.resp vectors
for no from sm+1 to Sm do
  # Determine the order of the given response
  order := op(1, op(1, N.sm[no])):

  # Determine where to put the response, x.resp
  resp := op(2, op(1, N.sm[no])):

  # Define N list,NL
  NL := op(2, N.sm[no]):

  # Determine the frequency at which the x.resp vector is to be determined
  psi := op(resp, ImdFrqL):

  # Update the global inverted A matrix, AI
  AIcal(psi):
```

```
# Determine the u vector
for q to Q do
  # Determine number of controlling variables for nonlinear element q,
  # NoCgVar
  NoCgVar := nops(op(q, jL)):

  # Set element q in the u vector initially equal to 0, uresult
  uresult := 0:

  for con to M.NoCgVar.m.order[0] do
    # Define Mm list, MmL
    MmL := M.NoCgVar.m.order[con]:

    # Determine order of contribution number con, ord
    ord := op(op(1, MmL)):

    # Determine MptL, NptL and psiGL lists
    MptL := op(4, MmL):
    NptL := [seq(op(i, NL), i=op(MptL))]:
    psiGL := [seq(op(i, ImdFrqL), i=op(NptL))]:

    # Initialize subresult for uresult, usubresult
    usubresult := G.q(op(2, MmL), psiGL):

    if usubresult <> 0 then
      # Determine x multiplication factors
      for i to ord do
        # Determine response number, p
        p := op(i, NptL):

        # Determine the number of the controlling variable, r
        for r while (i > convert([op(1..r, op(2, MmL))], '+')) do od:

        # Update usubresult
        usubresult := usubresult*x.p[op(r, op(q, jL))]
      od:
      uresult := uresult + op(op(3, MmL))*usubresult
    fi
  od:
  u[q] := uresult
od:

# Determine the x.resp (or X.resp) vector
if (caltype = 'num') then
  x.resp := map(evalc, multiply(AI, u))
elif (caltype = 'algf') then
  x.resp := multiply(AI, u)
```

```
      else
        X.resp := multiply(AI, u)
      fi
od:
# Now all higher order x.resp vectors are determined

# Determine the Volterra transfer function, first order
if sm=1 then
  # Determine the frequency psi1+...+psiK
  psi := op(Sm, ImdFrqL):

  # Determine k
  k := op(portL):

  # Update the a and b vectors
  acal(psi):
  bcal(psi):

  # Calculate the first order Volterra transfer function
  if (caltype = 'num') or (caltype = 'algf') then
    result := dotprod(a, x1, 'orthogonal') + b[k]
  else
    result := [seq(x1[no]=X1[no], no=1..R),
               H=dotprod(a, x1, 'orthogonal')+b[k]]
  fi
fi:


# Determine the Volterra transfer function, second and higher order
if (sm > 1) then
  # Determine the frequency psi1+...+psiK
  psi := op(Sm, ImdFrqL):

  # Calculate the multi factorial coefficient, m1! * ... * mK!
  cfak := 1:
  for i in mL do cfak := cfak*i! od:

  # Update the a vector
  acal(psi):

  # Calculate the second and higher order Volterra transfer function
  if (caltype = 'num') or (caltype = 'algf') then
    result := dotprod(a, x.Sm, 'orthogonal')/cfak
  else
    result := [seq(seq(x.resp[no]=X.resp[no], no=1..R), resp=1..Sm),
               H=dotprod(a, x.Sm, 'orthogonal')/cfak]
  fi
fi:
```

```
result
end:   # vtf
```

E.3   Program listing for example 1

```
#############################################################################
# example 1 - Test of vtf procedure
#############################################################################
# Ref.; J. J. Bussgang, L. Ehrman and J. W. Graham: "Analysis of Nonlinear
# Systems with Multiple Inputs", Proc. IEEE, Vol. 62, No. 8, August 1974,
# pp. 1088-1119.
# (! Note: Error in (3.19), -1/3 should be -2/3, and in (3.20), -1/3 should
# be +2/3.)
#############################################################################
read 'volfun':


#############################################################################
# AIcal - calculate the A inverse matrix
#############################################################################
AIcal := proc(f)
global AI:
AI[1,1] := -H(f)   # H(f)=1/(g1 + I*2*Pi*f*C)
end:   # AIcal


#############################################################################
# Bcal - calculate the B matrix
#############################################################################
Bcal := proc(f)
global B:
B[1,1] := 1
end:   # Bcal


#############################################################################
# acal - calculate the a vector
#############################################################################
acal := proc(f)
global a:
a[1] := 1
end:   # acal


#############################################################################
```

```
# bcal - calculate the b vector
#######################################################################
bcal := proc(f)
global b:
b[1] := 0
end:   # bcal



#######################################################################
# G1 - nonlinear Volterra transfer function for nonlinear element 1
#######################################################################
G1 := proc(ordL, psiL)
local order, result:
order := convert(ordL, '+'):
result := 0:
if order = 2 then result := g2 fi:
result
end:   # G1



#######################################################################
# Main
#######################################################################
H1 := volfun[vtf]([1], [f1], 3, [1,1,1,[[1]]], 'algf');
H2 := volfun[vtf]([2], [f1,f2], 3, [1,1,1,[[1]]], 'algf');
H3 := volfun[vtf]([3], [f1,f2,f3], 3, [1,1,1,[[1]]], 'algf');
```

E.4 Program listing for example 2

```
#######################################################################
# example 2 - Test of vtf procedure
#######################################################################
# Ref.; L. O. Chua and C.-Y. Ng: "Frequency-domain analysis of nonlinear
# systems: formulation of transfer functions", IEE Journal on Electronic
# Circuits and Systems, November 1979, Vol.3, No. 6, pp. 257-269.
#######################################################################
read 'volfun':



#######################################################################
# AIcal - calculate the A inverse matrix
#######################################################################
AIcal := proc(s)
```

```
local D:
global AI:
D := s^2*(Rp*g1 + 1) + s*(Rp*l1 + c1*g1) + c1*l1:
AI[1,1] := -s*(s*g1 + l1)/D:
AI[1,2] := s^2/D:
AI[2,1] := -s^2/D:
AI[2,2] := -s*(s*Rp + c1)/D
end:    # AIcal


####################################################################
# Bcal - calculate the B matrix
####################################################################
Bcal := proc(s)
global B:
B[1,1] := Rp:
B[2,1] := 0
end:    # Bcal


####################################################################
# acal - calculate the a vector
####################################################################
acal := proc(s)
global a:
a[1] := -Rp:
a[2] := 0
end:    # acal


####################################################################
# bcal - calculate the b vector
####################################################################
bcal := proc(s)
global b:
b[1] := Rp
end:    # bcal


####################################################################
# G1 - nonlinear Volterra transfer function for nonlinear element 1
####################################################################
G1 := proc(ordL, psiL)
local order, result:
order := op(1, ordL):
result := 0:
if order = 2 then result := c2/(op(1, psiL)*op(2, psiL)) fi:
```

```
if order = 3 then result := c3/(op(1, psiL)*op(2, psiL)*op(3, psiL)) fi:
result
end:    # G1




###################################################################
# G2 - nonlinear Volterra transfer function for nonlinear element 2
###################################################################
G2 := proc(ordL, psiL)
local order, result:
order := op(1, ordL):
result := 0:
if order = 2 then result := 12/(op(1, psiL)*op(2, psiL))+g2 fi:
if order = 3 then result := 13/(op(1, psiL)*op(2, psiL)*op(3, psiL))+g3 fi:
result
end:    # G2




###################################################################
# Main
###################################################################
H1prime := volfun[vtf]([1], [s1], 3, [1,2,2,[[1],[2]]], 'algf'):
H2prime := volfun[vtf]([2], [s1,s2], 3, [1,2,2,[[1],[2]]], 'algf'):
H3prime := volfun[vtf]([3], [s1,s2,s3], 3, [1,2,2,[[1],[2]]], 'algf'):

H1 := limit(H1prime, Rp=infinity);
H2 := limit(H2prime, Rp=infinity);
H3 := limit(H3prime, Rp=infinity);
```

E.5   Program listing for example 3

```
###################################################################
# example 3 - Test of vtf procedure
###################################################################
# Ref.; T. Larsen, "Determination of Multi-Port Volterra Transfer
# Functions", Int. j. cir. theor. appl., 1992.
###################################################################
read 'volfun':




###################################################################
# AIcal - calculate the A inverse matrix
###################################################################
```

```
AIcal := proc(f)
global AI:
AI[1,1] := -1/Yi(f):
AI[1,2] := 0:
AI[2,1] := gm1/(Yi(f)*Yo(f)):
AI[2,2] := -1/Yo(f)
end:    # AIcal


##############################################################
# Bcal - calculate the B matrix
##############################################################
Bcal := proc(f)
global B:
B[1,1] := Ys(f):
B[1,2] := Ri*Ys(f):
B[1,3] := 0:
B[2,1] := 0:
B[2,2] := 0:
B[2,3] := -1
end:    # Bcal


##############################################################
# acal - calculate the a vector
##############################################################
acal := proc(f)
global a:
a[1] := 0:
a[2] := 1
end:    # acal


##############################################################
# bcal - calculate the b vector
##############################################################
bcal := proc(f)
global b:
b[1] := 0:
b[2] := 0:
b[3] := 0
end:    # bcal


##############################################################
# G1 - nonlinear Volterra transfer function for nonlinear element 1
##############################################################
```

```
G1 := proc(ordL, psiL)
local order, result:
order := op(ordL):
result := 0:
if order = 2 then result := I*2*Pi*Cg2*(op(1, psiL)+op(2, psiL)) fi:
if order = 3 then
  result := I*2*Pi*Cg3*(op(1, psiL)+op(2, psiL)+op(3, psiL))
fi:
result
end:   # G1



###############################################################
# G2 - nonlinear Volterra transfer function for nonlinear element 2
###############################################################
G2 := proc(ordL, psiL)
local order1, order2, result:
order1 := op(1, ordL):
order2 := op(2, ordL):
result := 0:
if order1 = 0 and order2 = 2 then result := go2 fi:
if order1 = 0 and order2 = 3 then result := go3 fi:
if order1 = 2 and order2 = 0 then result := gm2 fi:
if order1 = 3 and order2 = 0 then result := gm3 fi:
result
end:   # G2



###############################################################
# Main
###############################################################
H100 := volfun[vtf]([1,0,0], [f11], 3, [3,2,2,[[1],[1,2]]], 'algf');
H200 := volfun[vtf]([2,0,0], [f11,f12], 3, [3,2,2,[[1],[1,2]]], 'algf');
H101 := volfun[vtf]([1,0,1], [f11,f31], 3, [3,2,2,[[1],[1,2]]], 'algf');
```

E.6   Program listing for example 4

```
###############################################################
# example 4 - Test of vtf procedure
###############################################################
# Ref.; L. O. Chua and C.-Y. Ng: "Frequency-domain analysis of nonlinear
# systems: formulation of transfer functions", IEE Journal on Electronic
# Circuits and Systems, November 1979, Vol.3, No. 6, pp. 257-269.
```

```
##################################################################
read 'volfun':


##################################################################
# AIcal - calculate the A inverse matrix
##################################################################
AIcal := proc(s)
local D:
global AI:
D := 2*s^2 + 3*s + 4:
AI[1,1] := -3*(s^2 + 2)/D:
AI[1,2] := -3*s/D:
AI[2,1] := 3*s/D:
AI[2,2] := -2*s/D
end:    # AIcal


##################################################################
# Bcal - calculate the B matrix
##################################################################
Bcal := proc(s)
global B:
B[1,1] := -2/3:
B[2,1] := 2/s
end:    # Bcal


##################################################################
# acal - calculate the a vector
##################################################################
acal := proc(s)
global a:
a[1] := 1:
a[2] := 0
end:    # acal


##################################################################
# bcal - calculate the b vector
##################################################################
bcal := proc(s)
global b:
b[1] := 0
end:    # bcal
```

```
############################################################################
# G1 - nonlinear Volterra transfer function for nonlinear element 1
############################################################################
G1 := proc(ordL, psiL)
local order, result:
order := op(1, ordL):
result := 0:
if order = 2 then result := 1/5 fi:
result
end:    # G1



############################################################################
# G2 - nonlinear Volterra transfer function for nonlinear element 2
############################################################################
G2 := proc(ordL, psiL)
local order, result:
order := op(1, ordL):
result := 0:
if order = 2 then result := (op(1, psiL) + op(2, psiL))/3 fi:
result
end:    # G2



############################################################################
# Main
############################################################################
H1 := volfun[vtf]([1], [s1], 3, [1,2,2,[[1],[2]]], 'algf');
H2 := volfun[vtf]([2], [s1,s2], 3, [1,2,2,[[1],[2]]], 'algf');
H3 := volfun[vtf]([3], [s1,s2,s3], 3, [1,2,2,[[1],[2]]], 'algf');
```

Index